Advances in Pattern Recognition

For other titles published in this series, go to
www.springer.com/series/4205

Liang Wang · Guoying Zhao · Li Cheng ·
Matti Pietikäinen
Editors

Machine Learning for Vision-Based Motion Analysis

Theory and Techniques

Springer

Editors
Dr. Liang Wang
Department of Computer Science
University of Bath
Bath, BA2 7AY
UK
lw356@cs.bath.ac.uk

Dr. Li Cheng
Bioinformatics Institute
A*STAR
30 Biopolis Street, #07-01 Matrix
Singapore, 138671
Singapore
chengli@bii.a-star.edu.sg

Dr. Guoying Zhao
Department of Electrical and
Information Engineering
University of Oulu
Oulu
Finland
gyzhao@ee.oulu.fi

Dr. Matti Pietikäinen
Department of Electrical Engineering,
Machine Vision & Media Processing Unit
University of Oulu
Oulu
Finland
mkp@ee.oulu.fi

Series Editor
Professor Sameer Singh, PhD
Research School of Informatics
Loughborough University
Loughborough
UK

ISSN 1617-7916
ISBN 978-0-85729-056-4 e-ISBN 978-0-85729-057-1
DOI 10.1007/978-0-85729-057-1
Springer London Dordrecht Heidelberg New York

British Library Cataloguing in Publication Data
A catalogue record for this book is available from the British Library

Cover design: VTEX, Vilnius

Printed on acid-free paper

Springer is part of Springer Science+Business Media (www.springer.com)

Preface

Vision-based motion analysis aims to detect, track and identify objects, and more generally, to understand their behaviors, from image sequences. With the ubiquitous presence of video data and its increasing importance in a wide range of real-world applications such as visual surveillance, human-machine interfaces and sport event interpretation, it is becoming increasingly important to automatically analyze and understand object motions from large amount of video footage.

Not surprisingly, this exciting research area has received growing interest in recent years. Although there has been much progress in the past decades, many challenging problems remain unsolved, e.g., robust object detection and tracking, unconstrained object activity recognition, etc. Recently, statistical machine learning algorithms, such as manifold learning, probabilistic graphical models and kernel machines, have been successfully applied in this area for object tracking, activity modeling and recognition. It is fully believed that novel statistical learning technologies have strong potential to further contribute to the development of robust yet flexible vision systems. The process of improving the performance of vision systems has also brought new challenges to the field of machine learning, e.g., learning from partial or limited annotations, online and incremental learning, and learning with very large datasets. Solving the problems involved in object motion analysis will naturally lead to the development of new machine learning algorithms. In return, new machine learning algorithms are able to address more realistic problems in object motion analysis and understanding.

This edited book highlights the development of robust and effective vision-based motion understanding algorithms and systems from a machine learning perspective. Major contributions of this book are as follows: (1) it provides new researchers with a comprehensive review of the recent development in this field, and presents a variety of study cases where the state-of-the-art learning algorithms have been devised to address specific tasks in human motion understanding; (2) it gives the readers a clear picture of the most active research forefronts and discussions of challenges and future directions, which different levels of researchers might find to be useful for guiding their future research; (3) it draws great strength from the research communities of object motion understanding and machine learning and demonstrates the benefits from the interaction and collaboration of both fields.

The targeted audiences of this edited book are mainly researchers, engineers as well as graduate students in the areas of computer vision, pattern recognition and machine learning. The book is also intend to be accessible to a broader audience including practicing professionals working with specific vision applications such as surveillance, sport event analysis, healthcare, video conferencing, and motion video indexing and retrieval.

The origin of this book stems from the great success of the first and second International Workshop on Machine Learning for Vision-Based Motion Analysis (MLVMA'08 and MLVMA'09), held respectively in conjunction with the European Conference on Computer Vision 2008 (ECCV'08) and the IEEE International Conference on Computer Vision 2009 (ICCV'09). These workshops gathered experts from different fields working on machine learning, computer vision, pattern recognition, and related areas.

The book comprises both theoretical advances and practical applications. The organization of the book reflects the combination of analytical and practical topics addressed throughout the book. We have divided the book chapters into four parts as follows; each addresses a specific theme.

Part I: Manifold Learning and Clustering/Segmentation

Chapter "Practical Algorithms of Spectral Clustering: Toward Large-Scale Vision-Based Motion Analysis", presents some practical algorithms of spectral clustering for large-scale data sets. Spectral clustering is a kernel-based method of grouping data on separate nonlinear manifolds. Reducing its computational expense without critical loss of accuracy contributes to its practical use. The presented algorithms exploit random projection and subsampling techniques for reducing dimensionality and the cost for evaluating pairwise similarities of data. The resulting computation time is quasilinear with respect to the data cardinality, and it can be independent of data dimensionality in some appearance-based applications. The efficiency of the algorithms is extensively demonstrated in appearance-based image/video segmentation.

Chapter "Riemannian Manifold Clustering and Dimensionality Reduction for Vision-Based Analysis" focuses on the topic of segmentation, one fundamental aspect of vision-based motion analysis. The goal of segmentation is to group the data into clusters based upon image properties such as intensity, color, texture or motion. Most existing segmentation algorithms proceed by associating a feature vector to each pixel in the image or video and then segmenting the data by clustering these feature vectors. This process can be phrased as a manifold learning and clustering problem, where the objective is to learn a low-dimensional representation of the underlying data structure and to segment the data points into different groups. Over the past few years, various techniques have been developed for learning a low-dimensional representation of a nonlinear manifold embedded in a high-dimensional space. Unfortunately, most of these techniques are limited to the analysis of a single connected nonlinear manifold and suffer from degeneracies when applied to linear manifolds. To address this problem, algorithms for performing simultaneous nonlinear dimensionality reduction and clustering of data sampled from multiple linear and nonlinear manifolds have been recently proposed. In this chapter, a summary of these newly developed algorithms are given and their applications to vision-based motion analysis are demonstrated.

Chapter "Manifold Learning for Multi-dimensional Auto-regressive Dynamical Models", presents a general differential-geometric framework for learning distance functions for dynamical models. Given a training set of models, the optimal metric is selected among a family of pullback metrics induced by the Fisher information tensor through a parameterized automorphism. The problem of classifying motions, encoded as dynamical models of a certain class, can then be posed on the learnt manifold. In particular, the class of multidimensional autoregressive models of order 2 is considered. Experimental results concerning identity recognition are shown that prove how such optimal pullback Fisher metrics greatly improve classification performances.

Part II: Tracking

When analyzing motion observations extracted from image sequences one notes that the histogram of the velocity magnitude at each pixel shows a large probability mass at zero velocity, while the rest of the motion values may be appropriately modeled with a continuous distribution. This suggests the introduction of mixed-state random variables that have probability mass concentrated in discrete states, while they have a probability density over a continuous range of values. In the first part of chapter "Mixed-State Markov Models in Image Motion Analysis", a comprehensive description of the theory behind mixed-state statistical models, in particular the development of mixed-state Markov models that permits to take into account spatial and temporal interaction, is given. The presentation generalizes the case of simultaneous modeling of continuous values and any type of discrete symbolic states. For the second part, the application of mixed-state models to motion texture analysis is presented. Motion textures correspond to the instantaneous apparent motion maps extracted from dynamic textures. They depict mixed-state motion values with a discrete state at zero and a Gaussian distribution for the rest. Mixed-state Markov random fields and mixed-state Markov chains are defined and applied to motion texture recognition and tracking.

Chapter "Learning to Track Objects in Surveillance Image Streams at Very Low Frame Rate", studies on the problem of object tracking. Some camera surveillance systems are designed to be autonomous—both from the energy and memory point of view. Autonomy allows operation in environments where wiring cameras for power and data transmission is neither feasible nor desirable. In these contexts, for cameras to work unattended over long periods requires choosing an adequately low frame rate to match the speed of the process to be supervised while minimizing energy and memory consumption. The result of surveillance is then a large stream of images acquired sparsely over time with limited visual continuity from one frame to the other. Reviewing these images to detect events of interest requires techniques that do not assume traceability of objects by visual similarity. If the process to be surveyed shows recurrent patterns of events over time, as it is often the case in industrial settings, other possibilities open up. Since images are time-stamped, this suggests techniques which use temporal data to help detecting relevant events. This contribution presents an image review tool that combines in cascade a scene change detector (SCD) with a temporal filter. The temporal filter learns to recognize relevant SCD events by their time distribution on the image stream. The learning phase is supported by image annotations provided by end-users during past reviews. The concept is tested on a benchmark of real surveillance images stemming from a nuclear safeguards context. Experimental results show that the combined SCD-temporal filter significantly reduces the workload necessary to detect safeguards-relevant events in large image streams.

In chapter "Discriminative Multiple Target Tracking", a metric learning framework is introduced to learn a single discriminative appearance model for robust visual tracking of multiple targets. The single appearance model effectively captures the discriminative visual information among the different visual targets as well as the background. The appearance modeling and the tracking of the multiple targets are all cast in a discriminative metric learning framework. An implicit exclusive principle is naturally reinforced in the proposed framework, which renders the tracker to be robust to cross occlusions among the multiple targets. The efficacy of the proposed multi-target tracker is demonstrated on benchmark visual tracking sequences and real-world video sequences as well.

Guidewire tracking in fluoroscopy is important to image guided interventions. In chapter "Applications of Wire Tracking in Image Guided Interventions", a semantic guidewire model, is introduced, based on which a probabilistic method is presented to integrate measurements of three guidewire parts, i.e., a catheter tip, a guidewire body and a guidewire tip, in a Bayesian framework to track a whole guidewire. This tracking framework is robust to measurement noises at individual guidewire parts. Learning based measurement models are used to track the guidewire. The learning-based measurement models are trained from a database of guidewires, to detect guidewire parts in low-quality images. The method further incorporates online measurement models, which are based on guidewire appearances, as a complementary to learning based measurements to improve the tracking robustness. A hierarchical and multi-resolution scheme is developed to track a deforming guidewire. By decomposing the guidewire motion into two major components, the hierarchical tracking starts from a rigid alignment, followed by a refined nonrigid tracking. At each stage, a multi-resolution searching strategy is applied by using variable bandwidths in a kernel-based measurement smoothing method, to effectively and efficiently track the deforming guidewire. The guidewire tracking framework is validated on a test set containing 47 sequences that are captured in real-life interventional scenario. Quantitative evaluation results show that the mean tracking error on guidewires is less than 2 pixels, i.e., 0.4 mm.

Part III: Motion Analysis and Behavior Modeling

Chapter "An Integrated Approach to Visual Attention Modeling for Saliency Detection in Videos", presents a framework to learn and predict regions of interest in videos, based on human eye movements. The eye gaze information of several users are recorded as they are watching videos that belong to a similar application domain. This information is used to train a classifier to learn low-level video features from regions that attracted the visual attention of users. Such a classifier is combined with vision-based approaches to provide an integrated framework to detect salient regions in videos. To date, saliency prediction has been viewed from two different perspectives, namely visual attention modeling and spatiotemporal interest point detection. These approaches have largely been pure-vision based. They detect regions having a predefined set of characteristics such as complex motion or high contrast, for all kinds of videos. However, what is 'interesting' varies from one application to another. By learning features of regions that capture the attention of viewers while watching a video, this chapter aims to distinguish those that are actually salient in the given context, from the rest. The integrated approach ensures that both regions with anticipated content (top–down attention) and unanticipated content (bottom–up attention) are predicted by the proposed framework as salient. In the experiments with news videos of popular channels, the results show a significant improvement in the identification of relevant salient regions in such videos, when compared with existing approaches.

Chapter "Video-Based Human Motion Estimation by Part-Whole Gait Manifold Learning", presents a general gait representation framework for video-based human motion estimation that involves gait modeling at both the whole and part levels. The goal is to estimate the kinematics of an unknown gait from image sequences taken by a single camera. This approach involves two generative models, called the kinematic gait generative model (KGGM) and the visual gait generative model (VGGM), which represent the kinematics and appearances of a gait by a few latent variables, respectively. Particularly, the concept of gait manifold is proposed to capture the gait variability among different individuals by which KGGM and VGGM can be integrated together for gait estimation, so that a new gait with unknown kinematics can be inferred from gait appearances via KGGM and VGGM. A key issue in generating a gait manifold is the definition of the distance function that reflects the dissimilarity between two individual gaits. Specifically, three distance functions each of which leads to a specific gait manifold are investigated and compared. Moreover, this gait modeling framework from the whole level to the part level has been extended by decomposing a gait into two parts, an upper-body gait and a lower-body gait, each of which is associated with a specific gait manifold for part level gait modeling. Also, a two-stage inference algorithm is employed for whole-part gait estimation. The proposed algorithms were trained on the CMU Mocap data and tested on the HumanEva data, and the experiment results show promising results compared with the state-of-the-art algorithms with similar experimental settings.

Extremely crowded scenes present unique challenges to motion-based video analysis due to the excessive quantity of pedestrians and the large number of occlusions they produce. The interactions between pedestrians, however, collectively form a crowd that exhibits a spatially and temporally structured motion pattern within the scene. In chapter "Spatio-Temporal Motion Pattern Models of Extremely Crowded Scenes", a novel statistical framework is presented for modeling this structured motion pattern behavior, or steady state, of extremely crowded scenes. The key insight is to model the crowd by the spatial and temporal variations of the local non-uniform motion patterns generated by pedestrian interactions. The pedestrian activity is represented by modeling the rich motion information in local space-time volumes of the video. In order to capture the motion variations of the scene, a novel distribution-based hidden Markov model that encodes the temporal variations of local motion pattern is introduced. It is demonstrated that by capturing the steady-state behavior of a scene, the proposed method can naturally detect unusual activities as statistical deviations in videos with complex activities that are hard for even human observers to analyze.

Chapter "Learning Behavioral Patterns of Time Series for Video-Surveillance", deals with the problem of learning behaviors of people activities from (possibly big) sets of visual dynamic data, with a specific reference to video-surveillance applications. The study focuses mainly on devising meaningful data abstractions able to capture the intrinsic nature of the available data, and applying similarity measures appropriate to the specific representations. The methods are selected among the most promising techniques available in the literature and include classical curve fitting, string-based approaches, and hidden Markov models. The analysis considers both supervised and unsupervised settings and is based on a set of loosely labeled data acquired by a real video-surveillance system. The experiments highlight different peculiarities of the methods taken into consideration, and the final discussion guides the reader towards the most appropriate choice for a given scenario.

Part IV: Gesture and Action Recognition

Chapter "Recognition of Spatiotemporal Gestures in Sign Language Using Gesture Threshold HMMs", proposes a framework for automatic recognition of spatiotemporal gestures in sign language. An extension to the standard HMM model to develop a gesture threshold HMM (GT-HMM) framework is implemented which is specifically designed to identify inter gesture transitions. The performance of this system, and different CRF systems, is evaluated, when recognizing gestures and identifying inter gesture transitions. The evaluation of the system included testing the performance of conditional random fields (CRF), hidden CRF (HCRF) and latent-dynamic CRF (LDCRF) based systems and comparing these to the presented GT-HMM based system when recognizing motion gestures and identifying inter gesture transitions.

Learning-based approaches for human action recognition often rely on large training sets. Most of these approaches do not perform well when only a few training samples are available. Chapter "Learning Transferable Distance Functions for Human Action Recognition", considers the problem of human action recognition from a single clip per action. Each clip contains at most 25 frames. Using a patch based motion descriptor and matching scheme, promising results on three different action datasets with a single clip as the template can be achieved. The results are comparable to previously published results using much larger training sets. A method for learning a transferable distance function is also presented for these patches. The transferable distance function learning extracts generic knowledge of patch weighting from previous training sets, and can be applied to videos of new actions without further learning. Experimental results show that the transferable distance function learning not only improves the recognition accuracy of the single clip action recognition, but also significantly enhances the efficiency of the matching scheme.

Acknowledgements The publication of this book has benefited from the devotion of many people. First, we would like to thank all the authors for their tremendous effort in preparing the book chapters. We would also like to express our thanks to all of the reviewers as well as the PC members of the MLVMA Workshops, for their invaluable comments and suggestions which have helped improve the final book. Finally, we would like to thank the staff members of Springer (Wayne Wheeler, Simon Rees) for their constant supports throughout the preparation of this book.

Bath, UK — Liang Wang
Oulu, Finland — Guoying Zhao
Singapore, Singapore — Li Cheng
Oulu, Finland — Matti Pietikäinen

Contents

Part I
Manifold Learning and Clustering/Segmentation

Practical Algorithms of Spectral Clustering: Toward Large-Scale Vision-Based Motion Analysis

Tomoya Sakai and Atsushi Imiya

Abstract This chapter presents some practical algorithms of spectral clustering for large-scale data. Spectral clustering is a kernel-based method of grouping data on separate nonlinear manifolds. Reducing its computational expense without critical loss of accuracy contributes to its practical use especially in vision-based applications. The present algorithms exploit random projection and subsampling techniques for reducing dimensionality and the cost for evaluating pairwise similarities of data. The computation time is quasilinear with respect to the data cardinality, and it can be independent of data dimensionality in some appearance-based applications. The efficiency of the algorithms is demonstrated in appearance-based image/video segmentation.

1 Introduction

Clustering is a fundamental technique of machine learning for finding unknown patterns by grouping data. When one wants to detect motion segments latent in image sequences by clustering techniques, the algorithms are required to be fast and scalable. Especially for the vision-based motion analysis, one needs to handle thousands of images with an intolerably large number of pixel values. Such large-scale data often exhibit manifold structures in a feature space. For example, sequential video frames form trajectories if the frames are represented as high-dimensional feature vectors with pixel values. Clustering of large-scale data on distinct nonlinear manifolds is therefore a fundamental problem in vision-based motion analysis.

T. Sakai (✉) · A. Imiya
Nagasaki University, 1-14 Bunkyo, Nagasaki, Japan
e-mail: tsakai@ieee.org

A. Imiya
e-mail: imiya@faculty.chiba-u.jp

L. Wang et al. (eds.), *Machine Learning for Vision-Based Motion Analysis*, Advances in Pattern Recognition,
DOI 10.1007/978-0-85729-057-1_1, © Springer-Verlag London Limited 2011

Clustering methods should efficiently detect nonconvex patterns and linearly non-separable clusters in a high-dimensional space.

Spectral clustering [9, 12, 18, 22, 24, 25, 31, 33] is one of the recent approaches to nonlinear separation of data. Its algorithms were derived from graph-cut energy minimization for clustering. A popular energy function is the normalized cut (Ncut) [22, 25, 33], which yields clusters with high compactness and isolation. The major drawback of the spectral clustering algorithms is a large computational overhead. The bottleneck lies in computation of a matrix of pairwise similarities as well as its eigenvalue decomposition.

In this chapter, we address the scalability problem of the spectral clustering. We design fast and reliable computation of spectral clustering by introducing random projection and subsampling techniques. We present some practical algorithms: a time and memory efficient random projection and spectral clustering using a bipartite graph. The computational expense of our clustering algorithms is quasilinear with respect to data cardinality. We also show that the random pixel sampling in appearance-based image processing is effective for removing the dependence of data dimensionality on the computational cost. The performance of the present algorithms is demonstrated in appearance-based image and video shot segmentation.

2 Spectral Clustering

2.1 Principle

Let P be a set of n data points in a d-dimensional feature space, and let w_{ij} be the similarity between the data points $\mathbf{p}_i$ and $\mathbf{p}_j \in P$, measured by a predefined kernel function K as

$$w_{ij} = \begin{cases} K(\mathbf{p}_i, \mathbf{p}_j) & (i \neq j), \\ 0 & (i = j). \end{cases} \tag{1}$$

A similarity graph G is defined as a complete graph whose vertices represent the data points and edges are weighted by the corresponding similarities as shown in Fig. 1.

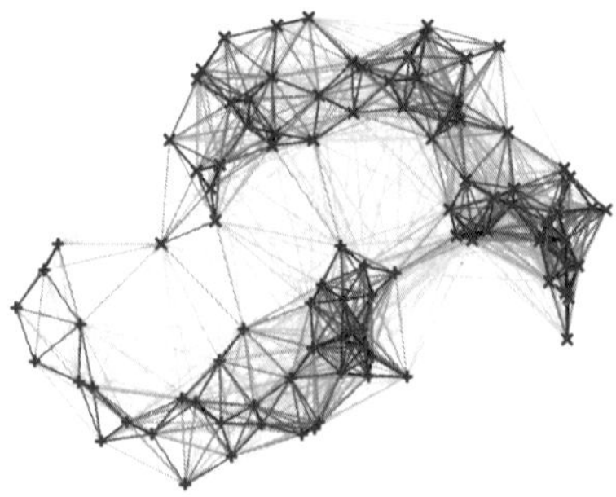

Fig. 1 Clustering by cutting complete graph G. The *nodes* represent $\mathbf{p} \in P$. The *gray level* of the edges indicates the measured similarity (*the darker* the higher)

The spectral clustering yields a set of data clusters $\{C_1, \ldots, C_k \subset P\}$ by cutting the edges of G so as to minimize a graph-cut energy function. Several energy functions have been proposed in the literature [9, 18, 25]. A popular algorithm of k-way graph cut for data clustering is the normalized cut (Ncut) [22, 25, 33], of which energy function is described as

$$\mathrm{Ncut}(C_1, \ldots, C_k) = \sum_{l=1}^{k} \frac{\mathrm{cut}(C_l, P \backslash C_l)}{\mathrm{cut}(C_l, P)}. \tag{2}$$

Here, $\mathrm{cut}(X, Y)$ is the sum of the edge weights between $\mathbf{p}_i \in X$ and $\mathbf{p}_j \in Y$. $P \backslash C_l$ is the complement of $C_l \subset P$. In (2), the numerator $\mathrm{cut}(C_l, P \backslash C_l)$ quantifies the connectivity of the cluster C_l by the sum of weights between its internal and external vertices. The denominator $\mathrm{cut}(C_l, P)$ can be regarded as the degree of C_l, measured as the total sum of edge weights related to any vertex in C_l. The minimization of Ncut therefore encourages the graph partition to yield isolated clusters with large cardinality.

Let $\mathbf{h}_l$ be an n-dimensional vector indicating the members of the cluster C_l by its discrete components such that the ith component is $1/\sqrt{\mathrm{cut}(C_l, P)}$ if $\mathbf{p}_i \in C_l$ and 0 otherwise. We define the matrix of cluster indicators $\mathbf{H}_k := [\mathbf{h}_1, \ldots, \mathbf{h}_k]$ and the affinity matrix $\mathbf{W} := [w_{ij}]$. The minimization of Ncut can be then written as a constrained trace minimization [29]:

$$\min_{C_1,\ldots,C_k} \mathrm{Ncut} \quad \Leftrightarrow \quad \min_{\mathbf{H}_k} \mathrm{tr}\, \mathbf{H}_k^\top \mathbf{L} \mathbf{H}_k \text{ subject to } \mathbf{H}_k^\top \mathbf{D} \mathbf{H}_k = \mathbf{I}. \tag{3}$$

Here, the matrix $\mathbf{D}$ is a diagonal matrix with the row sums of $\mathbf{W}$ on its diagonal:

$$\mathbf{D} := \mathrm{diag}\left(\sum_{j=1}^{n} w_{1j}, \ldots, \sum_{j=1}^{n} w_{nj} \right), \tag{4}$$

and the matrix $\mathbf{L} := \mathbf{D} - \mathbf{W}$ is known as the graph Laplacian of G.

Although finding the matrix $\mathbf{H}_k$ is NP hard, relaxation of the discreteness of $\mathbf{h}_l$ gives us a generalized eigenvalue decomposition (GEVD) problem [25, 33]

$$\mathbf{L}\mathbf{H}_k = \mathbf{D}\mathbf{H}_k \mathbf{\Lambda}, \tag{5}$$

where $\mathbf{H}_k \in \mathbb{R}^{n \times k}$ after the relaxation is composed of the eigenvectors associated with the k smallest generalized eigenvalues, and $\mathbf{\Lambda} \in \mathbb{R}^{k \times k}$ is the diagonal matrix containing the k eigenvalues. Replacing $\mathbf{H}_k = \mathbf{D}^{-1/2}\mathbf{X}_k$, one can convert the GEVD in (5) into a usual eigenvalue decomposition (EVD) problem:

$$\mathbf{S}\mathbf{X}_k = \mathbf{X}_k \mathbf{\Delta}. \tag{6}$$

Here, $\mathbf{\Delta} := \mathbf{I} - \mathbf{\Lambda}$, and

$$\mathbf{S} = \mathbf{D}^{-1/2}\mathbf{W}\mathbf{D}^{-1/2} \tag{7}$$

is the so-called normalized affinity matrix. The eigenvectors of $\mathbf{S}$ with the k largest eigenvalues are therefore calculated as the cluster indicators instead of $\mathbf{H}_k$. The clusters are obtained from $\mathbf{X}_k$: the normalized n row vectors of $\mathbf{X}_k$ exhibit the k tight clusters in k-dimensional space.

2.2 Algorithm

A spectral clustering algorithm basically consists of three procedures: (i) computation of pairwise similarities, (ii) GEVD of $(\mathbf{L}, \mathbf{D})$, or EVD of the normalized affinity matrix $\mathbf{S}$, and (iii) assignment of data points to the clusters using the eigenvectors. We present a typical algorithm of k-way spectral clustering proposed by Ng et al. [22] in Algorithm 1.

Algorithm 1: Spectral clustering (by Ng, Jordan and Weiss, 2001)

Input: $P = \{\mathbf{p}_1, \ldots, \mathbf{p}_n\}$: dataset, k: number of clusters, σ: scale

1 construct the affinity matrix $\mathbf{W} \in \mathbb{R}^{n \times n}$ whose (ij)th entry is the similarity measured as $K(\mathbf{p}_i, \mathbf{p}_j; \sigma)$ for $i \neq j$, and zero otherwise
2 compute the diagonal matrix $\mathbf{D} \in \mathbb{R}^{n \times n}$ whose (ii)th entry is the sum of the ith row of $\mathbf{W}$
3 form the normalized affinity matrix $\mathbf{S} := \mathbf{D}^{-1/2}\mathbf{W}\mathbf{D}^{-1/2}$
4 compute the top k eigenvectors $\mathbf{X}_k \in \mathbb{R}^{n \times k}$ of $\mathbf{S}$
5 form a matrix $\mathbf{Y}_k \in \mathbb{R}^{n \times k}$ by normalizing each row vector of $\mathbf{X}_k$
6 execute k-means clustering for n row vectors of $\mathbf{Y}_k$
7 assign $\mathbf{p}_i$ to the cluster C_l iff the ith row vector of $\mathbf{Y}_k$ is in the lth cluster

Output: $C = \{C_1, \ldots, C_k\}$: set of clusters

The Gaussian kernel function

$$K(\mathbf{p}_i, \mathbf{p}_j; \sigma) = \exp\left(-\frac{\|\mathbf{p}_j - \mathbf{p}_i\|_2^2}{2\sigma^2}\right) \tag{8}$$

is used in this algorithm.

If the data cardinality n and dimensionality d are both large, Algorithm 1 involves two computationally intensive tasks. One is the construction of the affinity matrix $\mathbf{W}$ at Step 1, which requires $\mathcal{O}(dn^2)$ computation time and $\mathcal{O}(n^2)$ memory space. The other task is the eigenvalue decomposition (EVD) at Step 4. The EVD of size n matrix $\mathbf{S}$ takes $\mathcal{O}(n^3)$ flops.

2.3 Related Work

Fast algorithms of spectral clustering have been proposed by different authors [10, 13, 15, 21, 26, 36]. Their algorithms are based on fast eigenvector computation of the affinity matrix. The Krylov subspace-based methods, for example, the Lanczos method [16], are the iterative algorithms for finding leading eigencomponents of a sparse matrix as used in [15, 21, 26]. However, those algorithms require $\mathcal{O}(kdn^2)$ computation time for the top k eigenvectors if the matrix is dense. Even if it is sparse, the convergence speed and accuracy highly depend on the sparsity [6, 16].

The Nyström approximation [10, 13, 32, 36] is another approach to fast computation of the eigencomponents of a large matrix. In the Nyström approximation, the eigenvectors are estimated by interpolating and orthogonalizing the eigenvectors of a randomly subsampled small submatrix. Its computation time is $\mathcal{O}(dm^3 + dnm^2)$ where m is the size of submatrix. Although the Nyström-based Ncut runs in linear time of the data cardinality n, its performance is contingent on m. Recently, Zhang and Kwok [36] improved the Nyström method using density-weighted kernel computation. Their algorithm achieves a higher accuracy without the orthogonalization, and runs in $\mathcal{O}(dm^3 + dnm)$ time if d is small. This chapter provides another improvements of the Ncut algorithm similar to these Nyström-based approaches, but with some practical advantages in simplicity, accuracy, and stability for practical uses.

The prior works mentioned above focus on large cardinality n and not on high dimensionality d of the data. The computational cost for evaluating the pairwise similarities scales as the dimensionality. For avoiding this problem, the PCA-based dimensionality reduction are unendurable because the PCA itself is an intensive EVD task. Random projection [1, 4, 7, 14, 20, 28] is an efficient dimensionality reduction technique for approximation of the affinity matrix. We will show an efficient algorithm of random projection and apply to the Ncut algorithm.

3 Dimensionality Reduction by Random Projection

In this section, we address the computational cost caused by high dimensionality. We introduce the random projection, and discuss its implication for vision-based analysis. We also present an efficient algorithm of random projection and demonstrate it.

3.1 Random Projection

Random projection is a simple technique for projecting a set of data points from a high-dimensional space to a randomly chosen low-dimensional linear subspace. Let $\mathbf{R}$ be a $\hat{d} \times d$ matrix whose $\hat{d}$ row vectors span the $\hat{d}$-dimensional subspace in $\mathbb{R}^d$ ($\hat{d} < d$). We obtain a low-dimensional representation $\hat{\mathbf{p}}_i$ of each $\mathbf{p}_i \in P$ as

$$\hat{\mathbf{p}}_i = \mathbf{R}\mathbf{p}_i. \tag{9}$$

The Johnson–Lindenstrauss lemma [20] and its proofs [1, 7, 28] ensure the following important possibility.

Theorem 1 (Johnson–Lindenstrauss embeddings) *For any* $\varepsilon > 0$, $P = \{\mathbf{p}_1, \ldots, \mathbf{p}_n \in \mathbb{R}^d\}$, *and* $\hat{d} < d$, *one can map* P *to* $\hat{P} = \{\hat{\mathbf{p}}_1, \ldots, \hat{\mathbf{p}}_n \in \mathbb{R}^{\hat{d}}\}$ *by random projection in* (9) *so that the pairwise distances are approximately preserved as*

$$(1-\varepsilon)\|\mathbf{p}_j - \mathbf{p}_i\|_2 \le \|\hat{\mathbf{p}}_j - \hat{\mathbf{p}}_i\|_2 \le (1+\varepsilon)\|\mathbf{p}_j - \mathbf{p}_i\|_2 \quad (10)$$

with probability $(1 - e^{-\mathcal{O}(\hat{d}\varepsilon^2)})$.

It has been proved that a random projection by the matrix $\mathbf{R}$ of i.i.d. normal random variables with mean zero and variance $1/\hat{d}$ satisfies the above theorem with $\hat{d} \ge \hat{d}_0 = \mathcal{O}(\varepsilon^{-2}\log n)$. The known lower bound is $\hat{d}_0 = 4(\varepsilon^2/2 - \varepsilon^3/3)^{-1}\log n$ [7]. Clearly, the dimensionality $\hat{d}$ of the low dimensional space is independent of d. The lower bound $\hat{d}_0$ is derived from the Markov inequality, and the actual approximation errors are much smaller than ε in most practical cases. Achlioptas [1] proposed an efficient random projection by the sparse matrix with (ij)th entry

$$r_{ij} = \sqrt{\frac{3}{\hat{d}}} \cdot \begin{cases} +1 & \text{with probability } 1/6, \\ 0 & \text{with probability } 2/3, \\ -1 & \text{with probability } 1/6. \end{cases} \quad (11)$$

This projection can be performed with partial sums of the vector components. Theorem 1 holds without loss of the approximation accuracy. Watanabe et al. [30] have suggested that the entries of the random matrix must be at least 4-wise independent to approximately preserve the pairwise distances. Theorem 1 can be proven even if four or more arbitrary entries in every row of the random matrix $\mathbf{R}$ are statistically independent.

The random matrix $\mathbf{R}$ defines a $\hat{d}$-dimensional random subspace independent of the dataset P. The random projection reduces the dimensionality without any consideration of data distribution. This is a major difference of the random projection from the other feature selection and reduction techniques by mining important data attributes. We would like to make some remarks on random projection for vision-based analysis.

3.1.1 Acceleration of Kernel Computation

Since the kernel function in (8) can be evaluated by pairwise distances, one can use the low-dimensional dataset $\hat{P}$ instead of the high dimensional dataset P. The $n \times n$ affinity matrix can be approximately computed in $\mathcal{O}(\varepsilon^{-2}n^2\log n)$ time. The additional cost of random projection of n data points, $\mathcal{O}(\hat{d}dn)$, is reduced to $\mathcal{O}(nd\log d)$ by an efficient algorithm we introduce later.

3.1.2 Random Sampling as Random Projection

The random projection can be regarded as random sampling of $\hat{d}$ coordinates after random rotation of the original space $\mathbb{R}^d$. The random rotation provides random orthogonal coordinates, and the random sampling picks up the $\hat{d}$-dimensional linear subspace $\mathbb{R}^{\hat{d}}$. As suggested in [1], the random rotation only serves as the randomization of coordinate directions so that all coordinates contribute equally likely to the corresponding pairwise distance. We may therefore omit the random rotation if we know in advance that the coordinates of $\mathbf{p}_i \in P$, that is, features or attributes, contribute roughly equally likely to the similarity measurements. In such case, the random sampling of coordinates takes the place of the random projection, which reduces the cost from $\mathcal{O}(\hat{d}dn)$ to $\mathcal{O}(\hat{d}n) = \mathcal{O}(\varepsilon^{-2}n \log n)$ independent of the data dimensionality d.

3.1.3 Using a Minority of Image Pixels

We suppose that the random sampling of coordinates is fairly effective for appearance-based vision techniques where pixel values compose the vector $\mathbf{p}_i$ representing an image. In fact, Sakai [23] has achieved appearance-based image classification by random sampling of image pixels. The random projection reduces the dimensionality by converting a large number of features to a small number of hash features which equally likely contribute to the similarity measurements. Presuming the pixel values to contribute roughly equally likely to the similarity measurements, one can utilize random sampling of pixels instead of random projection.

3.2 Efficient Random Projection

We offer a time and memory efficient algorithm for random projection.

Algorithm 3 is based on linear projection by a random matrix

$$\mathbf{R} := \frac{1}{\sqrt{\hat{d}}} \mathbf{\Phi} \mathbf{C} \operatorname{diag}(\mathbf{s}), \tag{12}$$

where $\operatorname{diag}(\mathbf{s}) \in \mathbb{R}^{d \times d}$ is a diagonal matrix of a random vector $\mathbf{s} = [s_1, \ldots, s_d]^\top$, and matrices $\mathbf{\Phi} \in \mathbb{R}^{\hat{d} \times d}$ and $\mathbf{C} \in \mathbb{R}^{d \times d}$ denote a random sampling matrix and a

---

Algorithm 2: Generating random vectors

Input: $d, \hat{d}$: dimensionalities

1 $\mathbf{s} = [s_1, \ldots, s_d]^\top$ where s_i $(i = 1, \ldots, d)$ are i.i.d. random variable values with mean 0 and deviation 1, or random signs $\{+1, -1\}$

2 $\mathbf{c} = [c_1, \ldots, c_d]^\top$ where c_i $(i = 1, \ldots, d)$ are i.i.d. random variable values with mean zero and deviation $1/\sqrt{\hat{d}}$

3 $\hat{\mathbf{c}} = \mathcal{F}[\mathbf{c}]$: FFT of $\mathbf{c}$

4 $L = \{l_1, \ldots, l_{\hat{d}}\} \subset \{1, \ldots, n\}$: a set of $\hat{d}$ indices without replacement

Output: $\{\hat{\mathbf{c}}, \mathbf{s}, L\}$

Algorithm 3: Efficient random projection ($\hat{\mathbf{p}}_i = \mathbf{R}\mathbf{p}_i$)

Input: $P = \{\mathbf{p}_1, \dots, \mathbf{p}_n \in \mathbb{R}^d\}$: dataset,
$\{\hat{\mathbf{w}}, \mathbf{s}, L\}$: a triplet generated by Algorithm 2

for $\forall \mathbf{p}_i \in P$ **do**
    $\boldsymbol{\xi} = \mathbf{s} \odot \mathbf{p}_i$: spectrum spreading with element-by-element multiplication
    $\eta = \mathcal{F}^{-1}[\hat{\mathbf{c}} \odot \mathcal{F}[\boldsymbol{\xi}]^*]$: circular convolution
    $\hat{\mathbf{p}}_i = [\eta_{l_1}, \dots, \eta_{l_{\hat{d}}}]^\top$
end for

Output: $\hat{P} = \{\hat{\mathbf{p}}_1, \dots, \hat{\mathbf{p}}_n \in \mathbb{R}^{\hat{d}}\}$

back-circulant matrix[1] of a random vector $\mathbf{c} = [c_1, \dots, c_d]^\top$:

$$\boldsymbol{\Phi} = \begin{bmatrix} 0 & 1 & 0 & \cdots & 0 & \cdots & 0 \\ 0 & \cdots & 0 & \cdots & 0 & 1 & 0 \\ & & \cdots & & \cdots & & \\ 0 & \cdots & 0 & 1 & 0 & \cdots & 0 \end{bmatrix} \quad \text{and}$$
$$\mathbf{C} = \begin{bmatrix} c_1 & c_2 & \cdots & c_{d-1} & c_d \\ c_2 & c_3 & \cdots & c_d & c_1 \\ \vdots & \ddots & \cdots & \ddots & \vdots \\ c_d & c_1 & \cdots & c_{d-2} & c_{d-1} \end{bmatrix}. \tag{13}$$

One can also set $\boldsymbol{\Phi} = [\mathbf{I}\ \mathbf{O}]$. We define that the vector components $\forall c_i$ and $\forall s_i$ are statistically independent. If we set $\mathrm{E}[\mathbf{c}] = \mathrm{E}[\mathbf{d}] = \mathbf{0}$ and $\mathrm{E}[\mathbf{c}\mathbf{c}^\top] = \mathrm{E}[\mathbf{s}\mathbf{s}^\top] = \mathbf{I}$, then we have $\mathrm{E}[\|\mathbf{R}\mathbf{p}\|_2^2] = \mathrm{E}[\|\mathbf{p}\|_2^2]$. The d-dimensional $\hat{d}$ row vectors of the matrix $\boldsymbol{\Phi}\mathbf{C}$ act as the orthogonal coordinates of a random subspace in expectation. The diagonal matrix diag($\mathbf{s}$) enhances the statistical independence of the entries r_{ij} enough for $\mathbf{R}$ to satisfy Theorem 1.

For the random projection of $\mathbf{p}_i$, the multiplication by diag($\mathbf{s}$) costs $\mathcal{O}(d)$. This process is a modulation technique called the direct sequence spread spectrum, which densifies the spectrum of $\mathbf{p}_i$ so that all the Fourier components of the random vector $\mathbf{c}$ are involved in the subsequent convolution. The multiplication by $\mathbf{C}$, or the circular convolution with $\mathbf{c}$, costs $\mathcal{O}(d \log d)$ using a fast algorithm of orthogonal transform such as the fast Fourier transform (FFT) or the Walsh-Hadamard transform. Consequently, we do not have to generate and store a huge $\hat{d} \times d$ random matrix. Altogether, Algorithm 3 requires $\mathcal{O}(nd \log d)$ time and $\mathcal{O}(d)$ space for n data points. We would note that Algorithm 3 is more efficient than the random projection of the sparse matrix as (11). We also remark that Algorithm 3 is quite practical because it can be easily implemented by FFT using GPGPU (general-purpose computing on graphics processing units), although we do not evaluate the performance on GPU in this work.

[1] $\mathbf{C}$ can be a forward circulant matrix. We prefer the back-circulant matrix just because it is symmetric.

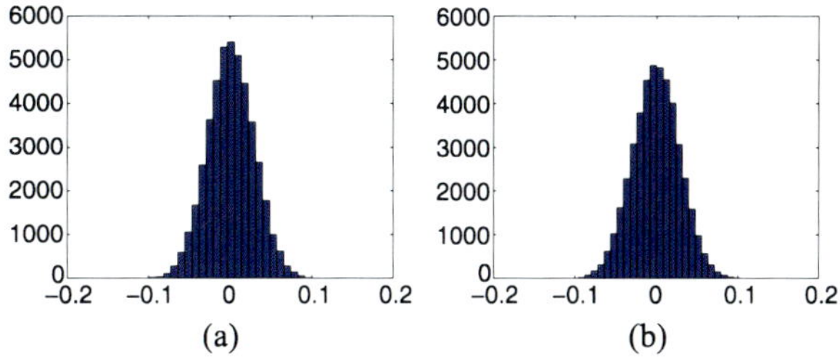

Fig. 2 Distance preservation. The *histograms* show the distributions of relative errors in pairwise distances after random projection by (**a**) a random matrix and by (**b**) Algorithm 3

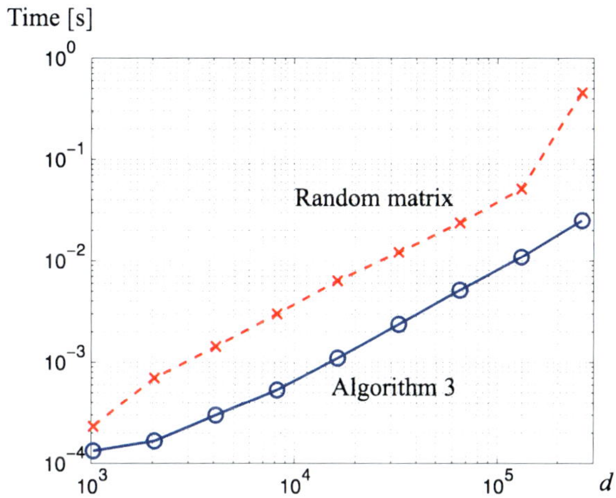

Fig. 3 Computation time (per sample) with respect to original dimensionality d

We demonstrate the performance of Algorithm 3. We set $n = 300$, $d = 65{,}536$, and $\hat{d} = 634$ ($\varepsilon = 0.3$), and generated a dataset P using Gaussian random variables. Figure 2 compares the distributions of relative errors in the pairwise distances. We confirmed that both projections by (a) a random matrix with i.i.d. Gaussian entries and (b) Algorithm 3 could approximately preserve all the distances to within $\pm\varepsilon$ or a much smaller interval.

We also evaluate the computation time on Core i7 2.93 GHz with a single core running. We set $n = 300$, $\hat{d} = 634$ ($\varepsilon = 0.3$), and $d = 1{,}024$ to $262{,}144$. It is impractical to generate and store the random matrix with a higher dimension beyond the range studied in this experiment. On the other hand, Algorithm 3 can afford to run at higher dimensions without generating and storing the random matrix. Algorithm 3 outperforms at any dimension as shown in Fig. 3. It would run in real-time for videos whose frames are represented as $\mathbf{p}_i$ containing pixel values.

4 Size Reduction of Affinity Matrix by Sampling

The computational cost of the spectral clustering quadratically grows with respect to the cardinality n as long as we compute all the entries of $n \times n$ affinity matrix. In this section, we aim to compute the eigenvectors of the normalized affinity matrix from a part of it. The key idea is to estimate the clusters not from the complete graph G but from another graph with less edges. We introduce two methods of obtaining such graphs: random subsampling for a high-dimensional dataset and pre-clustering for a low-dimensional dataset.

4.1 Random Subsampling

We will prune the edges of the similarity graph G by random subsampling. Let $\hat{Q}$ be a set of m subsamples randomly chosen from the low-dimensional dataset $\hat{P}$. We measure the similarities only between $\hat{\mathbf{p}}_i \in \hat{P}$ and $\hat{\mathbf{q}}_j \in \hat{Q}$. Figure 4 illustrates an incomplete similarity graph whose vertices represent $\hat{\mathbf{p}}_i \in \hat{P}$ and edges are weighted by the measured similarities. One can observe in Fig. 4(a) that there are two clusters, and the data points in each cluster are strongly connected each other through the randomly chosen points. If $\hat{Q}$ roughly exhibits the data clusters, one can extract the clusters from $\hat{P}$ by cutting this incomplete graph in the same manner as spectral clustering.

The incomplete graph can be converted into a complete bipartite graph between $\hat{P}$ and $\hat{Q}$ as shown in Fig. 4(b). The affinity matrix of this bipartite graph with $(n+m)$ nodes is described as

$$\mathbf{W}_{\hat{P}\cup\hat{Q}} = \begin{bmatrix} \mathbf{O} & \mathbf{W}_{\hat{P}\hat{Q}} \\ \mathbf{W}_{\hat{P}\hat{Q}}^{\top} & \mathbf{O} \end{bmatrix} \in \mathbb{R}^{(n+m)\times(n+m)}. \tag{14}$$

Here, $\mathbf{W}_{\hat{P}\hat{Q}}$ is an $n \times m$ rectangular affinity matrix whose (ij)th entry is calculated as $K(\hat{\mathbf{p}}_i, \hat{\mathbf{q}}_j)$ where $\hat{\mathbf{p}}_i \in \hat{P}$ and $\hat{\mathbf{q}}_j \in \hat{Q}$ for $\hat{\mathbf{p}}_i \neq \hat{\mathbf{q}}_j$, and zero otherwise. The eigenvalue decomposition of $\mathbf{W}_{\hat{P}\cap\hat{Q}}$ after normalization as in (7) can be performed by the singular value decomposition (SVD) of the normalized rectangular matrix

$$\mathbf{S}_{\hat{P}\hat{Q}} = \mathbf{D}_{\hat{P}}^{-1/2}\mathbf{W}_{\hat{P}\hat{Q}}\mathbf{D}_{\hat{Q}}^{-1/2} \in \mathbb{R}^{n\times m}. \tag{15}$$

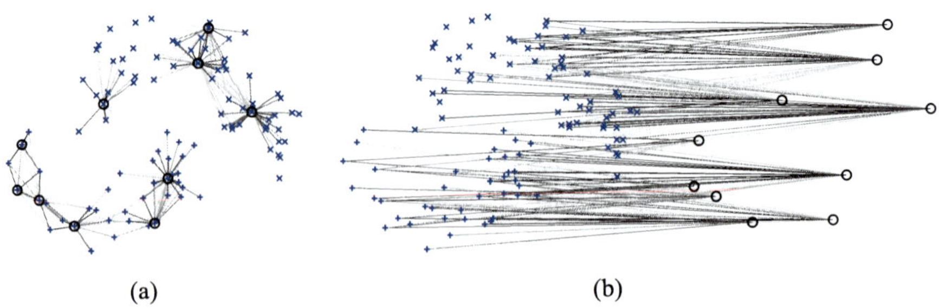

Fig. 4 Approximation of the similarity graph G. (**a**) Incomplete similarity graph. *Open circles* indicate the randomly chosen data points $\hat{\mathbf{q}} \in \hat{Q}$. (**b**) Complete bipartite graph equivalent to (**a**)

Here, $\mathbf{D}_{\hat{P}} \in \mathbb{R}^{n\times n}$ and $\mathbf{D}_{\hat{Q}} \in \mathbb{R}^{m\times m}$ are diagonal matrices with the row and column sums of $\mathbf{W}_{\hat{P}\hat{Q}}$ on the diagonals, respectively

$$\mathbf{D}_{\hat{P}} = \mathrm{diag}(\mathbf{W}_{\hat{P}\hat{Q}}\mathbf{1}_r), \tag{16}$$

$$\mathbf{D}_{\hat{Q}} = \mathrm{diag}\big(\mathbf{1}_n^\top \mathbf{W}_{\hat{P}\hat{Q}}\big). \tag{17}$$

We denote the SVD of $\mathbf{S}_{\hat{P}\hat{Q}}$ as

$$\mathbf{S}_{\hat{P}\hat{Q}} = \mathbf{U}\boldsymbol{\Delta}\mathbf{V}^\top. \tag{18}$$

The n row vectors of the matrix of the top k left singular vectors, $\mathbf{U}_k$, provide the k clusters in $\hat{P}$. The right singular vectors indicate the clusters in $\hat{Q}$. This clustering method using a bipartite graph is known as the co-clustering [8].

Note that the number of randomly chosen subsamples, m, is independent of the cardinality n, because the role of $\hat{Q}$ is to sketch roughly the k clusters. One may set $m \geq m_0 = \alpha k$ where α is constant independent of n if the clusters are well detached from each other. The full SVD of $n \times m$ matrix $\mathbf{S}_{\hat{P}\hat{Q}}$ needs $\mathcal{O}((n+m)\min^2(n,m)) \approx \mathcal{O}(\alpha^2 k^2 n)$ computation time and $\mathcal{O}(nm) = \mathcal{O}(\alpha k n)$ memory space [16, 17]. That is, the cost of computing the eigenvectors is linear with respect to the data cardinality n even if we calculate all m left singular vectors.

4.2 Pre-clustering

Pre-clustering is another approach to obtaining an incomplete graph. We divide $\hat{P}$ into $m \gg k$ groups as pre-clusters, and use the m pre-cluster centers as $\hat{Q}$ in place of the random subsamples. The pre-cluster centers can be thought of as local centers found by importance sampling because the over-clustering of $\hat{P}$ estimates local density of $\hat{P}$. This approach can therefore achieve better performance than random subsampling.

According to the recent work by Zhang and Kwok [36], the affinity matrix can be well approximated by using m-means cluster centers of $\hat{P}$ as $\hat{Q}$ if the similarity is defined as a radial-based function such as the Gaussian. We can adopt the m-means pre-clustering, and approximate the Ncut spectral clustering using a complete bipartite graph as shown in Fig. 5. Let ω_j $(j = 1, \ldots, m)$ denote the cardinality of the jth pre-cluster. Then, $n \times m$ rectangular affinity matrix $W_{\hat{P}\hat{Q}}$ is calculated as

$$\mathbf{W}_{\hat{P}\hat{Q}} = \big[\omega_j K(\hat{\mathbf{p}}_i, \hat{\mathbf{q}}_j)\big]. \tag{19}$$

In the same manner as described in Sect. 4.1, the EVD of the affinity matrix of the bipartite graph between $\hat{P}$ and the m-means pre-cluster centers $\hat{Q}$ boils down to the SVD of $\mathbf{S}_{\hat{P}\hat{Q}}$ as in (15).

Note that the m-means pre-clustering is effective only if the dimensionality of $\hat{P}$ is low ($\hat{d} < \mathcal{O}(10^1)$). The pre-clustering generally suffers from the so-called curse

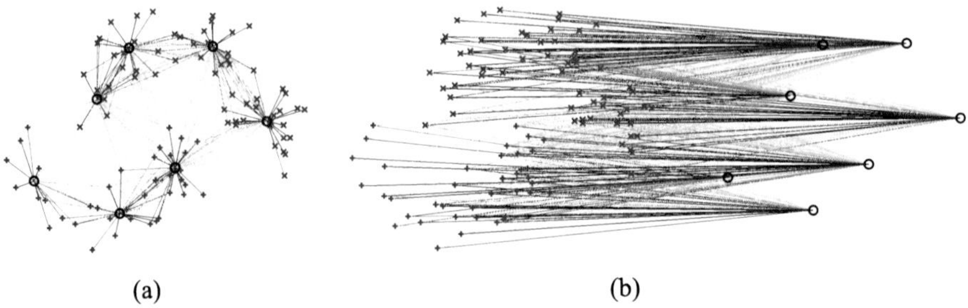

Fig. 5 Approximation of the similarity graph G by pre-clustering. (a) Complete bipartite graph between $\hat{P}$ and pre-cluster centers. *Open circles* indicate the m-means pre-cluster centers. (b) Complete bipartite graph equivalent to (a)

of dimensionality. We also remark that the advantages of the use of SVD derived from a bipartite graph are in the clustering accuracy and stability. The singular vectors are always orthogonal while the Nyström methods have to orthogonalize the eigenvectors to improve the accuracy. Besides, we conjugate that the left singular vectors of the $n \times m$ normalized affinity matrix are more stable than the eigenvectors estimated from a $m \times m$ affinity matrix by the Nyström methods. The stability is experimentally confirmed in Sect. 6.

5 Practical Ncut Algorithms

Using the reduction techniques shown in the previous sections, we present time and memory efficient algorithms of Ncut spectral clustering for large-scale data. We utilize random projection and random subsampling techniques to reduce the computational expense owing to the high dimensionality and large cardinality. The resulting algorithm is Algorithm 4, which uses an efficient algorithm of the random projection as shown in Algorithm 3. We also present an algorithm specialized for low-dimensional data in Algorithm 5.

5.1 Randomized Ncut Algorithm

The spectral clustering using random projection and subsampling is shown in Algorithm 4. This algorithm first reduces the dimensionality of the given dataset P at Step 1, which costs $\mathcal{O}(nd \log d)$ flops using Algorithm 3 or $\mathcal{O}(\varepsilon^{-2} n \log n)$ by random sampling of coordinates. The construction of the affinity matrix at Step 3 costs $\mathcal{O}(\varepsilon^{-2} \alpha k n \log n)$ flops. The computation of left singular vectors at Step 6 can be done in $\mathcal{O}(\alpha^2 k^2 n)$ flops. The cost of k-means clustering is negligibly small. Consequently, Algorithm 4 works in quasilinear time and space with respect to the data cardinality n for fixed k. The dimensionality d concerns only the cost of random

Algorithm 4: Randomized Ncut algorithm

Input: $P = \{\mathbf{p}_1, \ldots, \mathbf{p}_n \in \mathbb{R}^d\}$: dataset, k: number of clusters, σ: scale, ε: distortion; α: constant factor

1 perform random projection by Algorithm 3 or random sampling to obtain $\hat{d} = \mathcal{O}(\varepsilon^{-2} \log n)$-dimensional dataset $\hat{P}$
2 set $\hat{Q}$ to be a set of $m = \alpha k$ data points randomly chosen from $\hat{P}$
3 construct the affinity matrix $\mathbf{W}_{\hat{P}\hat{Q}} \in \mathbb{R}^{n\times m}$ whose (ij)th entry is $K(\hat{\mathbf{p}}_i \in \hat{P}, \hat{\mathbf{q}}_j \in \hat{Q}; \sigma)$ for $\hat{\mathbf{p}}_i \neq \hat{\mathbf{q}}_j$, and zero otherwise
4 compute the diagonal matrix $\mathbf{D}_{\hat{P}}$ and $\mathbf{D}_{\hat{Q}}$ in (16) and (17), respectively
5 form the matrix $\mathbf{S}_{\hat{P}\hat{Q}}$ in (15)
6 compute the k leading left singular vectors $\mathbf{U}_k \in \mathbb{R}^{n\times k}$
7 form the matrix $\mathbf{Y}_k \in \mathbb{R}^{n\times k}$ by normalizing each row vector of $\mathbf{U}_k$
8 execute k-means clustering for n row vectors of $\mathbf{Y}_k$
9 assign $\mathbf{p}_i$ to the cluster C_l iff the ith row vector of $\mathbf{Y}_k$ is in the lth cluster

Output: $C = \{C_1, \ldots, C_k\}$: set of clusters

projection, which is avoidable by the random sampling in appearance-based applications, expectedly.

We give some remarks on our Algorithm 4.

5.1.1 Invocation of Dimensionality Reduction

The dimensionality reduction at Step 1 should be performed only when the data dimensionality d is intractably high. Otherwise, use P as $\hat{P}$. The time complexity becomes linear to n. For a low-dimensional dataset, Algorithm 5 is effective unless cursed by dimensionality.

5.1.2 Relation to the Original Algorithm

Algorithm 4 is equivalent to Algorithm 1 if $\hat{Q} = \hat{P} = P$. That is, if we skip the dimensionality reduction at Step 1 and choose all data points as $\hat{Q}$ at Step 2, Algorithm 4 provides the same clustering result as the original Ncut algorithm without any disadvantage.

5.1.3 Scale Selection

The scale parameter σ can be automatically tuned by local scaling [35]. The local scaling can be easily incorporated in the algorithm as

$$\mathbf{W}_{\hat{P}\hat{Q}} = \left[K(\hat{\mathbf{p}}_i, \hat{\mathbf{q}}_j; \sqrt{\sigma_i \sigma_j})\right] \tag{20}$$

at Step 3. Here, σ_i and σ_j are ν-nearest neighbor distances ($\nu = 7$ in [35]) of $\hat{\mathbf{p}}_i$ and $\hat{\mathbf{q}}_j$ for $\hat{P}$, respectively. The average distance of ν-nearest neighbors is also useful when we assume the clusters have almost the same spatial densities. The nearest neighbor search, however, could be the most intensive task for large $\hat{d}$.

5.1.4 Number of Clusters

In substitution for the k-means at Step 8, it is possible to classify the samples by binarization of the entries of $\mathbf{Y}_k$ after orthogonal transformation. The cluster number k can be automatically found by a method of recovering the aligning rotation of eigenvectors [35]. The number k can also be found by observing the compactness of the k clusters exhibited by the n row vectors of $\mathbf{Y}_k$ with increasing $k > 1$. One can plug the algorithm into these methods of finding k.

5.2 Ncut Algorithm with Pre-clustering

The spectral clustering using pre-clustering is shown in Algorithm 5. For low-dimensional datasets, this algorithm effectively works with a smaller constant α compared to Algorithm 4. Time complexity is linear with respect to the cardinality n. For the parameters σ and k, the same remarks as on Algorithm 4 are applicable to Algorithm 5.

Algorithm 5: Ncut algorithm with pre-clustering

Input: $P = \{\mathbf{p}_1, \ldots, \mathbf{p}_n \in \mathbb{R}^d\}$: dataset, k: number of clusters, σ: scale, ε: distortion; α: constant factor

1 set $\hat{P} = P$ or perform random projection if necessary
2 set $\hat{Q}$ and $\{\omega_1, \ldots, \omega_m\}$ to respectively be the sets of the centers and cardinalities of the m-means clusters of $\hat{P}$ where $m = \alpha k$
3 construct the affinity matrix $\mathbf{W}_{\hat{P}\hat{Q}} = [\omega_j K(\hat{\mathbf{p}}_i \in \hat{P}, \hat{\mathbf{q}}_j \in \hat{Q}; \sigma)] \in \mathbb{R}^{n\times m}$ where K is the kernel function in (8)
4 compute the diagonal matrix $\mathbf{D}_{\hat{P}}$ and $\mathbf{D}_{\hat{Q}}$ in (16) and (17), respectively
5 form the matrix $\mathbf{S}_{\hat{P}\hat{Q}}$ in (15)
6 compute the k leading left singular vectors $\mathbf{U}_k \in \mathbb{R}^{n\times k}$
7 form the matrix $\mathbf{Y}_k \in \mathbb{R}^{n\times k}$ by normalizing each row vector of $\mathbf{U}_k$
8 execute k-means clustering for n row vectors of $\mathbf{Y}_k$
9 assign $\mathbf{p}_i$ to the cluster C_l iff the ith row vector of $\mathbf{Y}_k$ is in the lth cluster

Output: $C = \{C_1, \ldots, C_k\}$: set of clusters

6 Experiments

We experimentally evaluate the performance of the spectral clustering algorithms. We also show some applications of our algorithms to appearance-based motion segmentation and video shot segmentation. We implemented Algorithms 1, 4, and 5 using MATLAB. Note that we used SVD that computes all singular vectors of a full matrix for numerical reliability. The computation time would be further reduced by computing only the top k singular vectors in practice.

6.1 Performance Tests

6.1.1 Error Analysis

We experimentally compared the clustering errors of our Algorithms 4, 5, and Ncut with Nyström approximation [13], referring to the clustering result of Algorithm 1 as optimal. Figure 6 shows the result of Algorithm 1 with $k = 3$, $\sigma = 3$ applied to a synthetic dataset P ($n = 3{,}000$) used for the evaluation.

We examined two clustering measures: normalized mutual information (NMI) and conditional entropy (CE) for different numbers m of subsamples. See Appendix for the definitions of these measures. Figure 7 shows the average scores over one hundred trials. It reveals slightly better performance of Algorithm 4 than the Nyström-based Ncut. The deviation of NMI indicates the stability of each algorithm. While the Nyström-based Ncut estimates the eigenvectors from those of a $m \times m$ affinity matrix, Algorithm 4 obtains the eigenvectors from a $n \times m$ rectangular affinity matrix, which results in more stable performance. We can also observe that Algorithm 5 achieves accurate and stable clustering with a significantly small number of pre-cluster centers. The pre-clustering approach can therefore drastically reduce the computational cost if the pre-clustering itself is inexpensive. Since SVD

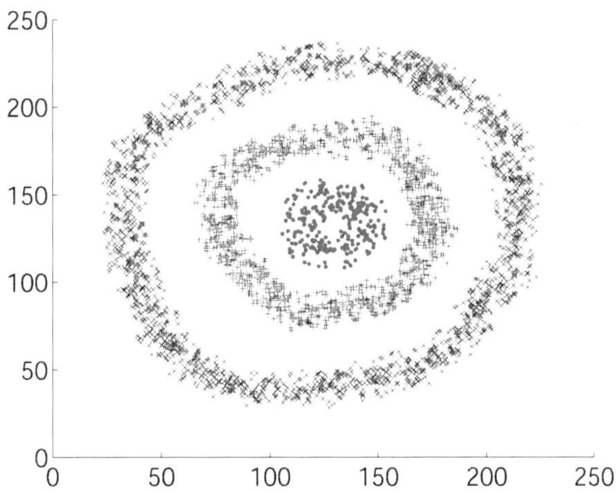

Fig. 6 Test dataset P clustered by Algorithm 1

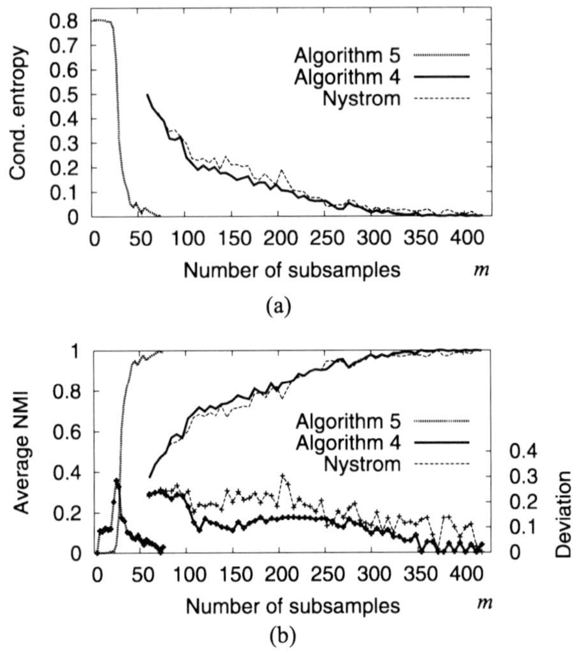

Fig. 7 Clustering scores with respect to subsample number m. (**a**) Conditional entropy (CE), (**b**) normalized mutual information (NMI) and deviation. The smaller the CE is, or the larger the NMI is, the better the clustering solution is. For Algorithm 5, m indicates the number of pre-cluster centers

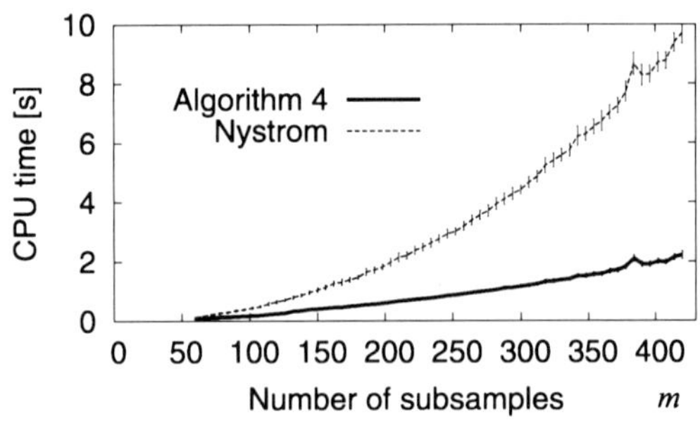

Fig. 8 Computation time with respect to m

requires quadratic time in m, Algorithm 5 is about a hundred times faster than Algorithm 4 to achieve a comparable accuracy in this experiment.

6.1.2 Computational Cost

We show, in Fig. 8, the computation time with respect to the subsample number m under the same condition as described in Sect. 6.1.1. Algorithms 4, 5, and Nyström-based Ncut run in linear time $\mathcal{O}(n)$, but the Nyström-based Ncut requires cubic time in m. We remark that Algorithm 4 runs always faster than the Nyström-based Ncut, which implies Algorithm 4 can afford to use more subsamples to improve the accuracy.

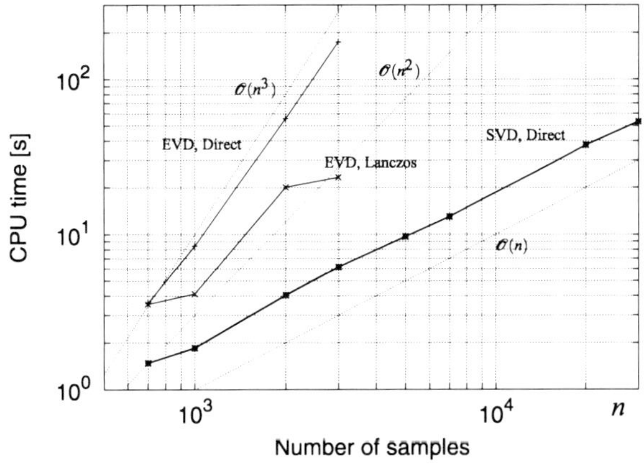

Fig. 9 Computation time with respect to cardinality n

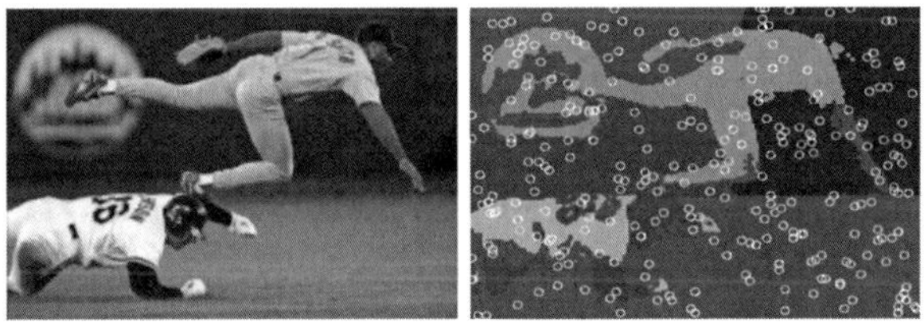

Fig. 10 A baseball scene (216×147 *gray image*), and the segmentation result by Algorithm 4. *Open circles in the right image* indicate the pixels corresponding to the randomly chosen data points

We also confirmed the computation time with respect to the cardinality n. By random sampling of coordinates, Algorithm 4 actually runs in $\mathcal{O}(n)$ time as shown in Fig. 9. Even if we compute all the eigenvectors by the direct method of SVD for a full $n \times m$ matrix, our algorithm is faster than Algorithm 1 implemented by a Krylov subspace-based method (e.g., Lanczos method of EVD). This outperformance is apparent unless the matrix $\mathbf{S}_{\hat{P}\hat{Q}}$ is sparse. If $\mathbf{S}_{\hat{P}\hat{Q}}$ is sparse, the state-of-the-art SVD driver routines for a sparse matrix, for example, [3], would be helpful for further acceleration.

Another advantage of Algorithm 4 is the low memory consumption. The experiments could be performed on large datasets with up to $\mathcal{O}(10^5)$ cardinality even on a 32 bit system. Such system can hardly run the EVD of a full matrix of $\mathcal{O}(10^4)$ size by in-memory algorithms.

6.2 Image Segmentation

We perform image segmentation by clustering the image pixels. The purpose of this experiment is to confirm that sparse pixel sampling has potential to achieve the segmentation. For more practical approach, see [6]. Figure 10 shows an example of the image segmentation by Algorithm 4. The left image is the classical *baseball* image introduced in [25]. We expressed this image as a set of $n = 216 \times 147 = 31{,}752$ data points in $d = 3$-dimensional space $(i/\sigma_X, j/\sigma_X, I(i, j)/\sigma_I)$. Here,

(i, j) and $I(i, j)$ are the pixel position and corresponding gray value, respectively. We set the scales $\sigma_X = 0.6\sqrt{n}$ and $\sigma_I = 0.3 \max I(i, j)$. Algorithm 4 with $k = 7$, $\sigma = 1$, and $\alpha = 40$ extracts the seven major segments: three parts of backwall around the top player, two grassy regions around the bottom player, and each of the two players' uniforms. The top and bottom players' undershirts are respectively detected as disjoint parts of the ground and backwall. The execution time was about ten seconds. It is impossible to run Algorithm 1 on a 32 bit system.

6.3 Motion Segmentation

Motion segmentation is a fundamental task for computer vision and video analysis. Clustering image pixels using optical flow vectors is one of the common approaches to the motion segmentation. As evaluated in [11], spectral clustering has a potential for grouping dominant optical flow fields according to the motion of individual objects.

We show in Fig. 11 an example of the motion segmentation by spectral clustering using optical flow. The left image is a sample frame of a 480×360 image sequence *PedCross* taken from the UCF crowd dataset [2]. For each pair of subsequent frames of this sequence, we computed the dense optical-flow field by a TV-L^1 approach presented in [34]. Using the flow vectors with norm larger than 0.5 (a half pixel size), we represented the flow field as a set of data points in $d = 4$-dimensional space $(i/\sigma_X, j/\sigma_X, u_{ij}/\sigma_F, v_{ij}/\sigma_F)$. Here, (i, j) and (u_{ij}, v_{ij}) are the pixel position and corresponding flow vector, respectively. We set the scales $\sigma_X = 0.6\sqrt{\#\text{pixels}}$ and $\sigma_F = 0.3 \max \sqrt{u_{ij}^2 + v_{ij}^2}$ in the same way as the image segmentation. As shown in the right panel of Fig. 11, Algorithm 4 with $k = 3$, $\sigma = 1$, and $\alpha = 100$ classifies the regions of moving objects with three dominant flows: two spatially mixed flows of pedestrians in opposite directions and a flow associated with car arrival at the crossing. The execution time was about four seconds. We have confirmed that Algorithm 5 with $\alpha = 10$ runs about twice as fast as Algorithm 4, and provides the segmentation result with a comparable quality. It is intolerable to run Algorithm 1 due to the large cardinality $n = \mathcal{O}(10^4)$ in this experiment.

Fig. 11 A sample frame of *PedCross* sequence and the segmentation result by Algorithm 5

6.4 Video Shot Segmentation

We show two examples of appearance-based video shot segmentation by spectral clustering. We used video sequences *Mountain Skywater* and *level3*, publicly available on the websites of [5] and [19], respectively. Although both sequences are in color, we used the grayscale values for simplicity.

6.4.1 Segmentation Using Appearance-Based Similarities

The sequence *Mountain Skywater* consists of six shots with dynamic objects and camera motions. Figure 12(a) shows typical video frame of each shot. Each pair of successive shots are connected with cross-fade transition (except an abrupt transition between the 2nd and 3rd shots).

We represented each frame of this video as a $d = 352 \times 240 = 84{,}480$ dimensional data point storing the grayscale values, and treated the set of all $n = 1{,}188$ frames as the dataset P. We applied Algorithm 1 to this dataset P with $k = 6$ and $\sigma = 3 \times 10^4$. The resulting six clusters were sets of sequential frames as shown in Fig. 12(b). We could stably extract almost the same clusters by Algorithm 4 with $\varepsilon = 0.3$ ($\hat{d} = 786$, random sampling) and $\alpha = 100$ ($m = 600$). The differences of the clustering results are a few frames during the transitions. The computation time is reduced from about four minutes to five seconds, excluding time of loading data and estimating σ.

We have also confirmed that the dimensionality reduction down to $\hat{d} = 300$ is effective for the segmentation of this dataset. We conclude from this fact that the similarities between frames are efficiently estimated from a minority of pixels. The random sampling of pixels capably plays the role of random projection because the pixels contribute roughly equally likely to the similarity measure between frames.

In this experiment, we achieved the video shot segmentation by clustering frames without relying on the frame numbers or the order of the frames. Figure 12(c) plots the six clusters on randomly chosen 2D spaces. One can observe that the clusters are detected as curved manifolds. This implies that the manifold structures are preserved in $\hat{d}$ dimensional space, and are recognized as the groups of similar images by the spectral clustering with random projection.

6.4.2 Segmentation with Local Scaling

If a video sequence contains a variety of transition effects or fast camera motions, the shot segmentation using the similarities measured with a fixed scale σ may fail to detect the correct shot boundaries. In that case, the local scaling technique introduced in Sect. 5.1 improves the robustness of the shot segmentation. The video frame numbers (sequential frame IDs) can also be used for incorporating temporal distances in the similarity measure as follows: We represent the ith frame as a d-dimensional vector $\mathbf{p}_i$ with d pixel values, and compute the $\hat{d}$-dimensional

(a)

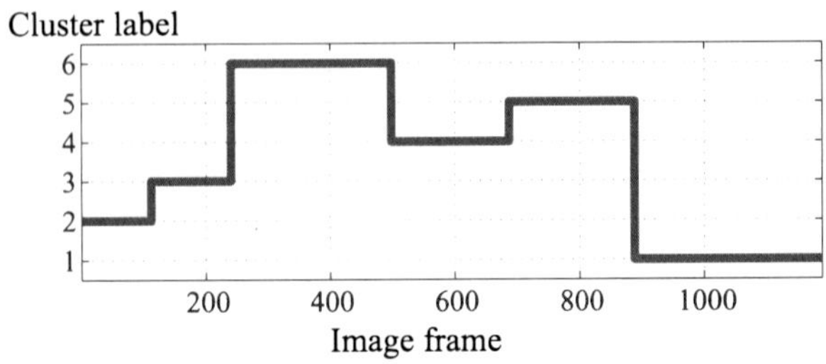

(b)

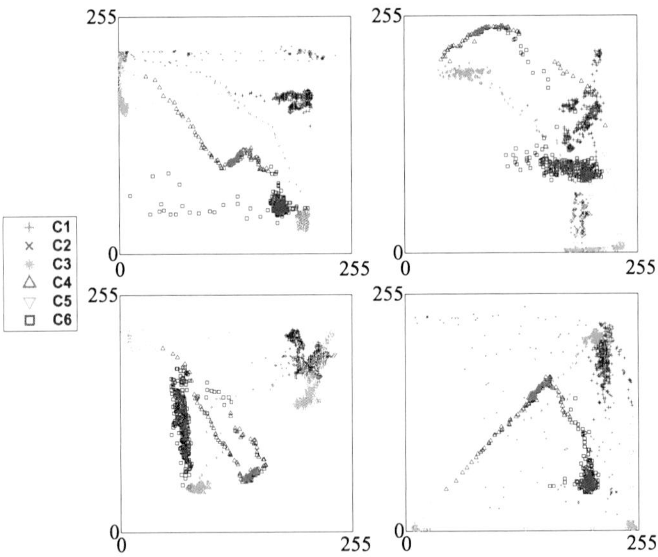

(c)

Fig. 12 Video shot segmentation of *Mountain Skywater* sequence. (**a**) Typical six frames: the 100, 200, 400, 600, 800, and 1,000th frame *from left to right, top to bottom*. (**b**) Segmentation result. The video frames were classified into six shots: frames 1–122, 123–239, 240–497, 498–686, 687–887, and 888–1188. (**c**) Scatter plots of P on random 2D spaces (random sampling)

vector $\hat{\mathbf{p}}_i$ by random projection. Using the νth nearest neighbor distance σ_i in the $\hat{d}$-dimensional space, we append the $(\hat{d}+1)$th component as a temporal position

$$\hat{\mathbf{p}}_{i+1}(\hat{d}+1) = \hat{\mathbf{p}}_i(\hat{d}+1) + \sqrt{\frac{\sigma_{i+1}\sigma_i}{\hat{d}+1}}. \tag{21}$$

We deal with the resulting dataset as $\hat{P}$ at Step 1 in Algorithms 4 or 5.

We performed the shot segmentation of *level3* video sequence using the local scaling. This sequence consists of seven scenery shots with wide panning and long zooming out. They are connected by fade, dissolve, and wipe transition effects. For the dataset of $d = 720 \times 480 = 345{,}600$ pixels and $n = 3{,}207$ frames of this sequence, we set $\varepsilon = 0.3$ $(\hat{d} = 897)$ and $\alpha = 100$ $(m = 700)$ for the dimensionality and cardinality reduction, and $\nu = 7$ for the nearest neighbor search. Figure 13(a) shows the clustering result by Algorithm 4 with the local scaling. A tiny fragment of the frames 2184–2186 does not cause any problem because it is in a transition. We can find a long shot with successive frames in each cluster. Thus, all of the shots were detected, correctly. The computation time was twelve seconds while Algorithm 4 without any reduction or Algorithm 1 took about 10 minutes. We would remark that the shot segmentation of *Mountain Skywater* was also successful with the local scaling in the same way.

For comparision, we show a k-means clustering result of the *level3* dataset in Fig. 13(b). As observed in Fig. 13(c), the clusters in this dataset are nonconvex, and k-means clustering cannot detect them.

7 Conclusions

We designed practical Ncut algorithms of spectral clustering. The data dimensionality is reduced by random projection, and the spectral analysis is accelerated by approximation of the similarity graph by incomplete one. The random projection and random subsampling in Algorithm 4 resolve the two major problems of spectral clustering: intensive computations of the affinity matrix and its eigenvectors. Present spectral clustering algorithms run at most in quasilinear time with respect to the number of samples. The algorithms are highly scalable, simple to implement, fairly stable and accurate because it utilizes SVD, and not orthogonalization or interpolation of the eigenvectors as the Ncut with Nyström approximation.

In the applications to appearance-based image and video shot segmentation, a minority of pixels provide enough information for similarity measurement. The random projection explains and ensures the performance. The random sampling of pixels plays the role of dimensionality reduction in the video shot segmentation, and of cardinality reduction in the image segmentation. Since the random projection of a feature vector is the random sampling of features after random rotation of the feature space, the random sampling of pixels can be considered as the random projection without random rotation. If it is presumable that the pixels contribute equally likely to the similarity measure used for the clustering, hundreds of pixels are enough to

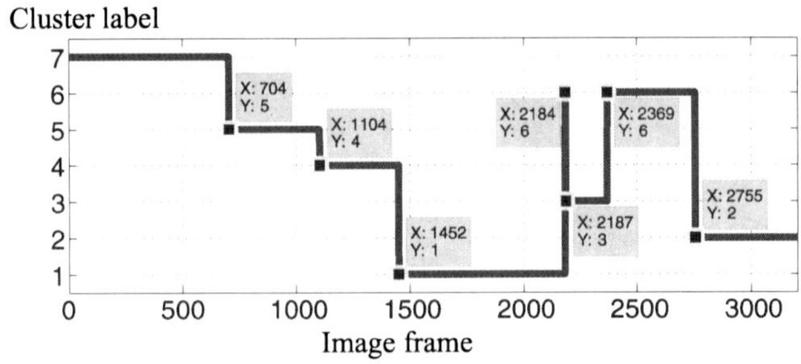

(a)

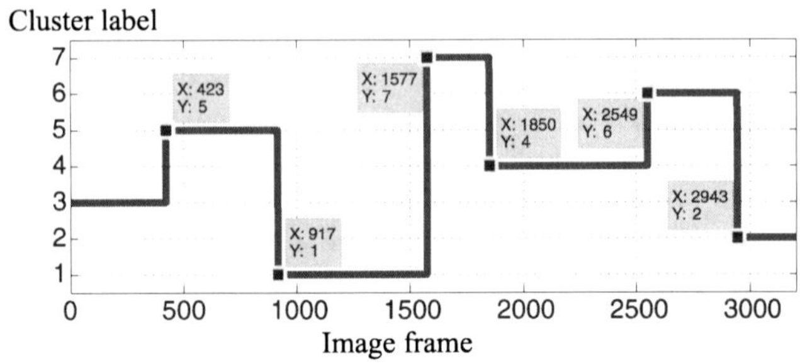

(b)

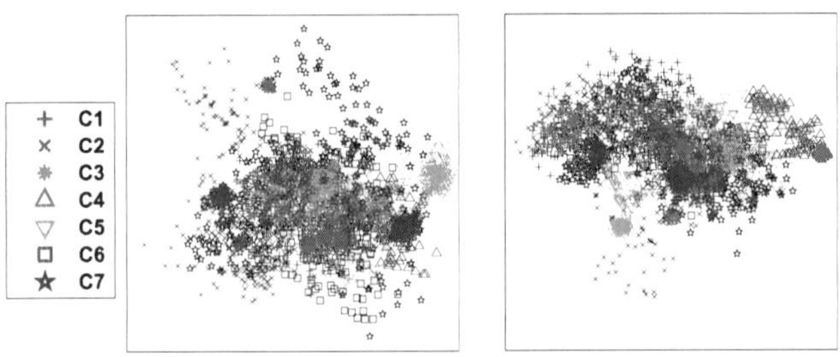

(c)

Fig. 13 Video shot segmentation of *level3* sequence. (**a**) Segmentation result using local scaling and the frame numbers. The detected shot boundaries are all in the transitions (697–729, 1086–1118, 1443–1475, 2156–2188, 2369, and 2732–2764). (**b**) Segmentation result by k-means clustering of $\hat{P}$. (**c**) Scatter plots of P on random 2D spaces (random projection)

evaluate the similarity because its accuracy is guaranteed in the same fashion as the Johnson–Lindenstrauss lemma.

This chapter focused on the acceleration of the spectral clustering. Of course, the selection of the cluster number and design of the similarity measure still remain to be addressed. These issues should be carefully resolved according to the applications, and the prior work introduced in this chapter would be helpful. We have no doubts that vision-based motion analyses require nonlinear machine learning methods that can handle large-scale data. We expect that the present clustering algorithms will contribute to the scalability for computer vision applications in general, and the motion analyses in particular.

Acknowledgement The first author was partially supported by the Grant-in-Aid for Young Scientists, from the Ministry of Education, Culture, Sports, Science and Technology of Japan under MEXT KAKEN 22700163.

Appendix: Clustering Scores

The conditional entropy (CE) and the normalized mutual information (NMI) [27] are defined as follows.

$$\mathrm{CE} = \sum_{i=1}^{k} \frac{|C_i|}{-n \log k} \sum_{j=1}^{k} \frac{|X_{ij}|}{|C_i|} \log \frac{|X_{ij}|}{|C_i|},$$

$$\mathrm{NMI} = \frac{\sum_{i=1}^{k} \sum_{j=1}^{k} \frac{|X_{ij}|}{n} \log \frac{n|X_{ij}|}{|C_i||A_j|}}{\sqrt{(\sum_{i=1}^{k} \frac{|C_i|}{n} \log \frac{|C_i|}{n})(\sum_{j=1}^{k} \frac{|A_j|}{n} \log \frac{|A_j|}{n})}}.$$

Here, $|C_i|$ and $|A_j|$ are the numbers of samples in the estimated cluster C_i and the optimal cluster A_j, respectively. $X_{ij} = C_i \cap A_j$ is the set of common samples. The smaller the CE is, or the larger the NMI is, the better the clustering result is. The NMI takes a value between 0 and 1.

References

1. Achlioptas, D.: Database-friendly random projections: Johnson–Lindenstrauss with binary coins. J. Comput. Syst. Sci. **66**, 671–687 (2003)
2. Ali, S., Shah, M.: A Lagrangian particle dynamics approach for crowd flow segmentation and stability analysis. In: CVPR (2007)
3. Berry, M.W.: Large scale sparse singular value computations. Int. J. Supercomput. Appl. **6**, 13–49 (1992)
4. Brigham, E., Maninila, H.: Random projection in dimensionality reduction: applications to image and text data. In: ACM SIGKDD ICKDDM, pp. 245–250. ACM, New York (2001)
5. Carnegie Mellon University Informedia Project: Mountain Skywater, segment 11 of 12. http://www.open-video.org/ (1996)
6. Cour, T., Benezit, F., Shi, J.: Spectral segmentation with multiscale graph decomposition. In: CVPR Proc., vol. 2, pp. 1124–1131. IEEE Comput. Soc., Washington (2005)
7. Dasgupta, S., Gupta, A.: An elementary proof of the Johnson–Lindenstrauss lemma. Technical Report, UC Berkeley (1999)
8. Dhillon, I.S.: Co-clustering documents and words using bipartite spectral graph partitioning. In: ACM SIGKDD, pp. 269–274. ACM, New York (2001)
9. Ding, C.H.Q., He, X., Zha, H., Gu, M., Simon, H.D.: A min-max cut algorithm for graph partitioning and data clustering. In: Proc. of ICDM 2001, pp. 107–114 (2001)
10. Drineas, P., Mahoney, M.W.: On the Nyström method for approximating a Gram matrix for improved kernel-based learning. J. Mach. Learn. Res. **6**, 2153–2175 (2005)

11. Eibl, G., Brändle, N.: Evaluation of clustering methods for finding dominant optical flow fields in crowded scenes. In: ICPR08 (2008)
12. Fiedler, M.: A property of eigenvectors of nonnegative symmetric matrices and its application to graph theory. Czechoslov. Math. J. **25**, 619–633 (1975)
13. Fowlkes, C., Belongie, S., Chung, F., Malik, J.: Spectral grouping using the Nyström method. IEEE Trans. Pattern Anal. Mach. Intell. **26**(2), 214–225 (2004)
14. Fradkin, D., Madigan, D.: Experiments with random projections for machine learning. In: ACM SIGKDD ICKDDM, pp. 517–522. ACM, New York (2003)
15. Freitas, N.D., Wang, Y., Mahdaviani, M., Lang, D.: Fast Krylov methods for N-body learning. In: Advances in Neural Information Processing Systems, vol. 18, pp. 251–258. MIT Press, Cambridge (2006)
16. Golub, G.H., Loan, C.F.V.: Matrix Computations, 3rd edn. Johns Hopkins University Press, Baltimore (1996)
17. Gu, M., Eisenstat, S.C.: A stable and fast algorithm for updating the singular value decomposition. Technical Report YALEU/DCS/RR-966, Yale University (1994)
18. Hagen, L., Kahng, A.: New spectral methods for ratio cut partitioning and clustering. IEEE Trans. Comput. Aided Des. **11**(9), 1074–1085 (1992)
19. IEICE Information and System Society: PRMU algorithm contest 2006: Shot boundary detection from image sequence, sample data. http://www-sens.sys.es.osaka-u.ac.jp/alcon/data/level3.avi (2006) (in Japanese)
20. Johnson, W., Lindenstrauss, J.: Extensions of Lipschitz maps into a Hilbert space. Contemp. Math. **26**, 189–206 (1984)
21. Mahadevan, S.: Fast spectral learning using Lanczos eigenspace projections. In: AAAI, pp. 1472–1475 (2008)
22. Ng, A.Y., Jordan, M.I., Weiss, Y.: On spectral clustering: analysis and an algorithm. In: Advances in Neural Information Processing Systems, vol. 14, pp. 849–856. MIT Press, Cambridge (2002)
23. Sakai, T.: Monte Carlo subspace method: an incremental approach to high-dimensional data classification. In: International Conference on Pattern Recognition (2008)
24. Scott, G.L., Longuet-Higgins, H.C.: Feature grouping by relocalisation of eigenvectors of the proximity matrix. In: British Machine Vision Conference, pp. 103–108 (1990)
25. Shi, J., Malik, J.: Normalized cuts and image segmentation. IEEE Trans. Pattern Anal. Mach. Intell. **22**(8), 888–905 (2000)
26. Song, Y., Chen, W.Y., Bai, H., Lin, C.J., Chang, E.Y.: Parallel spectral clustering. In: ECML PKDD. Lecture Notes in Computer Science, vol. 5212, pp. 374–389. Springer, Berlin (2008)
27. Strehl, A., Ghosh, J.: Cluster ensembles—a knowledge reuse framework for combining multiple partitions. J. Mach. Learn. Res. **3**, 583–617 (2002)
28. Vempala, S.S.: The Random Projection Method. Series in Discrete Mathematics and Theoretical Computer Science, vol. 65. American Mathematical Society, Providence (2004)
29. Von Luxburg, U.: A tutorial on spectral clustering. Stat. Comput. **17**(4), 395–416 (2007)
30. Watanabe, T., Takimoto, E., Amano, K., Maruoka, A.: Random projection and its application to learning. In: Proc. 2005 Workshop on Randomness and Computation, pp. 3–4 (2005)
31. Weiss, Y.: Segmentation using eigenvectors: a unifying view. In: International Conference on Computer Vision, pp. 975–982 (1999)
32. Williams, C.K.I., Seeger, M.: Using the Nyström method to speed up kernel machines. In: Advances in Neural Information Processing Systems, vol. 13, pp. 682–688. MIT Press, Cambridge (2001)
33. Yu, S.X., Shi, J.: Multiclass spectral clustering. In: International Conference on Computer Vision, pp. 313–319 (2003)
34. Zach, C., Pock, T., Bischof, H.: A duality based approach for realtime TV-L1 optical flow. In: Pattern Recognition, Proc. DAGM, Heidelberg, Germany, pp. 214–223 (2007)
35. Zelnik-Manor, L., Perona, P.: Self-tuning spectral clustering. In: Advances in Neural Information Processing Systems, vol. 17, pp. 1601–1608. MIT Press, Cambridge (2004)
36. Zhang, K., Kwok, J.T.: Density-weighted Nyström method for computing large kernel eigensystems. Neural Comput. **21**(1), 121–146 (2009). doi:10.1162/neco.2009.11-07-651

Riemannian Manifold Clustering and Dimensionality Reduction for Vision-Based Analysis

Alvina Goh

Abstract Segmentation is one fundamental aspect of vision-based motion analysis, thus it has been extensively studied. Its goal is to group the data into clusters based upon image properties such as intensity, color, texture, or motion. Most existing segmentation algorithms proceed by associating a feature vector to each pixel in the image or video and then segmenting the data by clustering these feature vectors. This process can be phrased as a manifold learning and clustering problem, where the objective is to learn a low-dimensional representation of the underlying data structure and to segment the data points into different groups. Over the past few years, various techniques have been developed for learning a low-dimensional representation of a nonlinear manifold embedded in a high-dimensional space. Unfortunately, most of these techniques are limited to the analysis of a single connected nonlinear manifold. In addition, all these manifold learning algorithms assume that the feature vectors are embedded in a Euclidean space and make use of (at least locally) the Euclidean metric or a variation of it to perform dimensionality reduction. While this may be appropriate in some cases, there are several computer vision problems where it is more natural to consider features that live in a Riemannian space. To address these problems, algorithms for performing simultaneous nonlinear dimensionality reduction and clustering of data sampled from multiple submanifolds of a Riemannian manifold have been recently proposed. In this book chapter, we give a summary of these newly developed algorithms as described in Goh and Vidal (Conference on Computer Vision and Pattern Recognition, 2007 and 2008; European Conference on Machine Learning, 2008; and European Conference on Computer Vision, 2008) and demonstrate their applications to vision-based analysis.

A. Goh (✉)
National University of Singapore, Singapore, Singapore
e-mail: alvinagoh@gmail.com

L. Wang et al. (eds.), *Machine Learning for Vision-Based Motion Analysis*,
Advances in Pattern Recognition,
DOI 10.1007/978-0-85729-057-1_2,

1 Introduction

Nonlinear dimensionality reduction (NLDR) refers to the problem of finding a low-dimensional representation for a set of points lying in a nonlinear manifold embedded in a high-dimensional space. This question of how to detect and represent low-dimensional structure in high-dimensional data is fundamental to many disciplines and several attempts have been made in different areas to address this question. For example, the number of pixels in an image can be rather large, yet most computer vision models use only a few parameters to describe the geometry, photometry and dynamics of the scene. Since most datasets often have fewer degrees of freedom than the dimension of the ambient space, NLDR is fundamental to many problems in computer vision, machine learning, and pattern recognition.

When the data lives in a low-dimensional linear subspace of a high-dimensional space, simple linear methods such as Principal Component Analysis (PCA) [26] and metric Multi-Dimensional Scaling (MDS) [11] can be used to learn the subspace and its dimension. However, when the data lies in a low-dimensional submanifold, its structure may be highly nonlinear, hence linear dimensionality reduction methods are likely to fail. This has motivated extensive efforts toward developing NLDR algorithms for computing low-dimensional embeddings.

A huge family of such algorithms computes a low-dimensional representation from the eigenvectors of a matrix constructed from the local geometry of the manifold. Such algorithms include ISOMAP [42], Kernel PCA (KPCA) [38], locally linear embedding (LLE) [35], and its variants such as Laplacian Eigenmaps (LE) [5], Hessian LLE [14], Local Tangent Space Alignment (LTSA) [50], maximum variance unfolding [47], and conformal eigenmaps [39]. A recent survey of many of these algorithms can be found in [8]. Most of these NLDR techniques can be categorized into two main groups: global and local techniques.

Global techniques attempt to preserve global properties of the data lying in a submanifold, similar to what PCA attempts to preserve for data lying in a linear subspace. In addition, they are also capable of constructing a nonlinear transformation between the high-dimensional data and the low-dimensional representation. Two of the best-known examples of this family of algorithms are ISOMAP and KPCA. ISOMAP attempts to preserve the geodesic distance between two data points on a manifold by approximating it as the length of the shortest path in the graph connecting the two points. KPCA reformulates linear PCA in a high-dimensional space via the kernel trick. Rather than considering the covariance matrix, KPCA computes the principal components of the kernel matrix. The kernel function allows KPCA to construct nonlinear mappings from the high-dimensional space to the low-dimensional space.

Local techniques are based on the preservation of local properties obtained from small neighborhoods around the data points. The key idea of such techniques is that by preserving local properties of the data, one can also retain global properties of the data. LLE, LE, Hessian LLE and LTSA fall under this category of algorithms. It has been proven that LLE is a special form of kernel principal component analysis

(KPCA) [23]. However, unlike conventional KPCA where one defines a kernel function in order to map the input space to a higher dimensional feature space, deriving the analytic form for the LLE kernel is not straightforward.

Although the goals of dimensionality reduction, classification and segmentation have always been intertwined with each other, considerably less work has been done on extending NLDR techniques for the purpose of clustering data living on different manifolds. For linear manifolds, there are many existing subspace clustering methods including K-subspaces [25], Local Subspace Affinity [49], Mixtures of Probabilistic PCA (MPPCA) [43], Generalized Principal Component Analysis (GPCA) [45], Agglomerative Lossy Compression (ALC) [48], and spectral clustering [1, 10, 22]. K-subspaces [25] proceeds similarly to K-means: it is initialized with a collection of K subspace bases of dimension d, and then it alternates between assigning points to their nearest subspace, and computing a subspace that minimizes the sum-of-the-squares distance to all points in each cluster. MPPCA [43] applies Expectation Maximization (EM) to a mixture of probabilistic PCAs. It assumes that the distribution of the data inside each subspace is Gaussian and uses EM to learn the parameters of the mixture model. GPCA [45] is an algebraic solution to subspace clustering based on fitting a union of m subspaces with a polynomial of degree m. The gradient of this polynomial at a point gives a vector normal to the subspace containing that point. The subspace clustering problem is then equivalent to fitting and differentiating a set of homogeneous polynomials. ALC [48] models each subspace with a degenerate Gaussian, and iteratively merges pairs of points so as to minimize the coding length needed to encode these points with a mixture of Gaussians. The multi-way spectral clustering algorithm [10] applies spectral clustering to a multi-way similarity that captures the curvature of a collection of points within an affine subspace.

All the aforementioned subspace clustering methods are formulated specifically for mixtures of linear manifolds, and thus they do not work in the presence of nonlinear manifolds. Existing works that extend NLDR techniques to clustering nonlinear manifolds include [7, 33, 40]. The work of [40] develops an EM-like extension of ISOMAP for clustering multiple nonlinear manifolds. However, this method is very sensitive to good initialization and is not a principled EM method as it uses heuristics in the E-step to assign points to manifolds. The work of [33] applies LLE to a manifold with m connected components. It shows that m eigenvalues of the matrix M are zero and that the clustering of the data can be obtained from the corresponding eigenvectors. However, this LLE clustering algorithm suffers from degeneracies in the presence of linear subspaces. In [7], the authors proved that in spectral clustering, it is possible to obtain a lower dimensional hypersphere representation via an eigendecomposition of the affinity matrix. In particular, [7] shows that it is possible to find a low-dimensional embedding in spaces having a mixture of linear and cyclic axes, and to cluster the data by repeated projection. However, the embedding algorithm maps the data only to a mixed vector and toric space, with the linear or cyclic nature of each axis determined from statistical tests. Also, the clustering is done via an iterative method that requires several projections.

A significant amount of work [4, 17, 24, 29, 31] has also been done on clustering data according to the dimensionality of the manifolds that contain the data rather than the manifolds themselves. For example, in human activity recognition, the videos describing different activities such as walking, jumping, and running can be described with a different number of parameters. Barbará and Chen [4] proposes a new clustering algorithm that is based on the use of the fractal dimension and clusters the data so that points in the same cluster are more self-affine among themselves than points in different clusters. The dimension of a manifold is estimated using a tensor voting scheme [31]. In [17], the local correlation dimension and density of a point is estimated and used as the input to standard clustering techniques. In [29], a maximum likelihood estimator of the intrinsic dimension of a dataset is derived by a Poisson process approximation. This idea is extended in [24] by modeling the high-dimensional sample points as a mixtures of Poisson processes, with regularizing restrictions and spatial continuity constraints. By proceeding in a EM-like manner, it is shown that it is possible to simultaneously estimate the soft clustering and the intrinsic dimension and density of each cluster. Even though such methods have proven to be efficient in clustering when the manifolds are of different dimensions, it is common in computer vision problems that this assumption is violated. In motion segmentation, for example, the manifolds for two translational motions are of the same dimension.

Chapter summary In this book chapter, we give a summary of the newly developed algorithms for performing simultaneous nonlinear dimensionality reduction and clustering of data sampled from multiple submanifolds of a Riemannian manifold and demonstrate their applications to vision-based analysis. More specifically, this chapter first reviews how to perform locally linear manifold clustering and dimensionality reduction for data sampled from nonlinear manifolds using the Euclidean metric as the distance between feature vectors and its limitations [18]. Unfortunately, such manifold clustering algorithms assume that the feature vectors are embedded in a Euclidean space and use (at least locally) the Euclidean metric or a variation of it to perform clustering. While this may be appropriate in some cases, there are several computer vision problems where it is more natural to consider features that live in a non-Euclidean space. For example, Grassmann manifolds and Lie groups are used for motion segmentation and multibody factorization; symmetric positive semi-definite matrices are common in diffusion tensor imaging and structure tensor analysis; and the statistical manifold, that is, the space of probability density functions, is found in texture analysis. The second part of this chapter shows a novel algorithm for clustering data sampled from multiple submanifolds of a Riemannian manifold [19–21]. First, a representation of the data using generalizations of local nonlinear dimensionality reduction algorithms from Euclidean to Riemannian spaces is learnt. Such generalizations exploit geometric properties of the Riemannian space, particularly its Riemannian metric. Then, assuming that the data points from different groups are separated, it is shown that the null space of a matrix built from the local representation gives the segmentation of the data. Finally, the applications to various vision problems are shown.

2 Review of Local Nonlinear Dimensionality Reduction Methods in Euclidean Spaces

In this section, we review three local nonlinear dimensionality reduction algorithms for data lying in a single manifold. Section 2.1 reviews the classical LLE, Laplacian Eigenmaps, and Hessian LLE algorithms for a nonlinear manifold. Section 2.2 shows that it is possible that such NLDR algorithms become degenerate when the data lies in a linear subspace.

2.1 *NLDR for a Nonlinear Manifold*

In this section, we review three local NLDR algorithms. Let $X = \{\mathbf{x}_i \in \mathcal{M}\}_{i=1}^n$ be a set of n data points sampled from a d-dimensional manifold $\mathcal{M}$ embedded in $\mathbb{R}^D$, $d \ll D$. We assume that the n points are *k-connected*, that is, for any two points $\mathbf{x}_i, \mathbf{x}_j \in X$ there is an ordered sequence of points in X having $\mathbf{x}_i$ and $\mathbf{x}_j$ as endpoints, such that any two consecutive points in the sequence have at least one k-nearest neighbor in common. The goal of dimensionality reduction is to find a set of vectors $\{\mathbf{y}_i \in \mathbb{R}^d\}_{i=1}^n$, such that nearby points remain close and distant points remain far.

Locally Linear Embedding (*LLE*) [36] assumes that the local neighborhood of a point in the manifold can be well approximated by the affine subspace spanned by the k-nearest neighbors of the point, and finds a low-dimensional embedding of the data based on these affine approximations. *Laplacian Eigenmaps* (*LE*) [6] are based on computing the low dimensional representation that best preserves *locality* instead of *local linearity* in LLE. *Hessian LLE* (*HLLE*) [14] bears substantial resemblance to LLE and LE, with the main difference being that the local neighborhood is represented by the tangent space at each point and the Laplacian matrix is replaced by the Hessian matrix. The main steps of these local NLDR algorithms are as follows:

1. *Nearest neighbor search*: for each data point $\mathbf{x}_i \in X$, find its k-nearest neighbors (kNN) $\{\mathbf{x}_{i_j}\}_{j=1}^k$ according to the Euclidean distance.
2. *Construction of similarity matrix*: construct a weighted graph whose elements encode the local geometry of the data. Define a similarity matrix M based on these weights. M is symmetric and positive semidefinite.
3. *Sparse eigenvalue problem*: obtain the embedding coordinates, that is, the columns of $Y = [\mathbf{y}_1, \ldots, \mathbf{y}_n]^\top \in \mathbb{R}^{n\times d}$, from the d (generalized) eigenvectors of the matrix M associated with its second to $(d+1)$th smallest (generalized) eigenvalues. The vector of all ones, $\mathbf{1} \in \mathbb{R}^n$, is a eigenvector of M associated with eigenvalue 0.

We now describe the construction of M for each NLDR algorithm in greater detail.

2.1.1 Calculation of M in LLE

1. *Weight matrix*: find a matrix of weights $W \in \mathbb{R}^{n\times n}$ whose entries W_{ij} minimize the reconstruction error

$$\varepsilon(W) = \sum_{i=1}^{n} \left\| \sum_{j=1}^{n} W_{ij}\mathbf{x}_j - \mathbf{x}_i \right\|^2 = \sum_{i=1}^{n} \mathrm{dist}^2(\hat{\mathbf{x}}_i, \mathbf{x}_i), \tag{1}$$

subject to the constraints (i) $W_{ij} = 0$ if $\mathbf{x}_j$ is not a k-nearest neighbor of $\mathbf{x}_i$ and (ii) $\sum_{j=1}^{n} W_{ij} = 1$. In (1), $\hat{\mathbf{x}}_i = \mathbf{x}_i + \sum_{j=1}^{n} W_{ij}\overrightarrow{\mathbf{x}_i\mathbf{x}_j}$ is the linear interpolation of $\mathbf{x}_i$ and its kNN. The solution to this problem can be computed as

$$\begin{bmatrix} W_{i\,i_1} & W_{i\,i_2} & \dots & W_{i\,i_k} \end{bmatrix} = \frac{\mathbf{1}^\top C_i^{-1}}{\mathbf{1}^\top C_i^{-1}\mathbf{1}} \in \mathbb{R}^{1\times k}, \tag{2}$$

where $\mathbf{1} \in \mathbb{R}^n$ is the vector of all ones, and $C_i \in \mathbb{R}^{k\times k}$ is the local Gram matrix at $\mathbf{x}_i$, that is, $C_i(j,l) = (\mathbf{x}_j - \mathbf{x}_i)\cdot(\mathbf{x}_l - \mathbf{x}_i)$.

2. *Objective function*: find vectors $\{\mathbf{y}_i \in \mathbb{R}^d\}_{i=1}^n$ that minimize the error

$$\phi(Y) = \sum_{i=1}^{n} \left\| \mathbf{y}_i - \sum_{j=1}^{n} W_{ij}\mathbf{y}_j \right\|^2 = \mathrm{trace}\big(Y^\top M Y\big), \tag{3}$$

subject to the constraints (i) $\sum_{i=1}^{n} \mathbf{y}_i = \mathbf{0}$ and (ii) $\frac{1}{n}\sum_{i=1}^{n} \mathbf{y}_i\mathbf{y}_i^\top = I$. The solution to this optimization problem is given by the d eigenvectors of $M = (I - W)^\top \times (I - W)$ associated with its second to $(d+1)$th smallest eigenvalues.

2.1.2 Calculation of M in LE

1. *Weight matrix*: construct a matrix of weights $W \in \mathbb{R}^{n\times n}$ where the entries of W, W_{ij}, measure the proximity between two points $\mathbf{x}_i$ and $\mathbf{x}_j$ subject to the constraint $W_{ij} = 0$ if $\mathbf{x}_j$ is not a k-nearest neighbor of $\mathbf{x}_i$. A possible weight of generating W is to use the heat kernel

$$W_{ij} = \exp\big(-\|\mathbf{x}_i - \mathbf{x}_j\|^2/\sigma^2\big). \tag{4}$$

2. *Objective function*: find vectors $\{\mathbf{y}_i \in \mathbb{R}^d\}_{i=1}^n$ that minimize the error

$$\phi(Y) = \sum_{i,j} \|\mathbf{y}_i - \mathbf{y}_j\|^2 W_{ij} = \mathrm{trace}\big(Y^\top M Y\big), \tag{5}$$

subject to the constraints (i) $Y^\top D\mathbf{1} = \sum_{i=1}^{n} D_{ii}\mathbf{y}_i = \mathbf{0}$ (weighted low-dimensional coordinates centered at the origin) and (ii) $Y^\top D Y = I$ (weighted low-dimensional coordinates having unit covariance). In (5), $M = D - W$ is the graph Laplacian matrix and D is a diagonal matrix whose entries are given by $D_{ii} = \sum_j W_{ij}$. The solution to this optimization problem is given by the d generalized eigenvectors of (M, D) associated with its second to $(d+1)$th smallest generalized eigenvalues.

2.1.3 Calculation of M in HLLE

1. *Tangent coordinates*: for each data point $\mathbf{x}_i$, let $\{\mathbf{x}_{i_j}\}_{j=1}^k$ be its kNN. Form the D by D covariance matrix $\mathrm{cov}(\mathbf{x}_i) = \frac{1}{k}\sum_{j=1}^k (\mathbf{x}_{i_j} - \bar{\mathbf{x}}_i)(\mathbf{x}_{i_j} - \bar{\mathbf{x}}_i)^\top$, where $\bar{\mathbf{x}}_i$ is the mean of the kNN. Perform an eigenanalysis of the matrix $\mathrm{cov}(\mathbf{x}_i)$ to obtain the d eigenvectors $\{\mathbf{u}_q \in \mathbb{R}^D\}_{q=1}^d$. The tangent coordinates of the kNN are given by the d columns of the $k \times d$ matrix V given below, where $p = 1, \ldots, k$ and $q = 1, \ldots, d$

$$V_{pq} = (\mathbf{x}_{i_p} - \bar{\mathbf{x}}_i)^\top \mathbf{u}_q = \langle \mathbf{x}_{i_p} - \bar{\mathbf{x}}_i, \mathbf{u}_q \rangle. \tag{6}$$

2. *Objective function*: the embedding vectors are obtained based on the null vectors of a matrix M that indicates the Hessian quadratic cost. While we refer the reader to [14] for details on the estimation of M, the basic principle is as follows. We first locally estimate a Hessian operator h^i at each point $\mathbf{x}_i$ in the manifold in a least squares sense. In particular, consider a smooth function $f : \mathcal{M} \to \mathbb{R}$. We evaluate the function at all kNN of a point $\mathbf{x}_i$ in the manifold and stack these entries into a vector $\mathbf{f}_i$. It can be shown that $h^i \mathbf{f}_i$ approximates the entries of the Hessian, whose (p, q)th entry is given by $\frac{\partial^2 f}{\partial V_p \delta V_q}$. These local estimates are then used to obtain an empirical estimate of the (i, j)th entry of M as

$$M_{i,j} = \sum_l \sum_r ((h^l)_{r,i} (h^l)_{r,j}). \tag{7}$$

The embedding coordinates are then found by selecting a basis for the space spanned by d eigenvectors of M associated with its second to $(d+1)$th smallest eigenvalues with the restriction that it provides an orthonormal basis to a specific fixed neighborhood $\mathbf{N}$. Let U denote the $n \times d$ matrix associated with the second to $(d+1)$th smallest eigenvectors where $U_{l,r}$ is the lth entry in the rth eigenvector of M. The embedding coordinates is obtained as $UR^{-\frac{1}{2}}$, where $R_{r,s} = \sum_{j \in \mathbf{N}} U_{j,r} U_{j,s}$.

2.2 NLDR for a Single Subspace

Consider now the application of NLDR to a single k-connected linear manifold. As the goal of NLDR is to unfold a low-dimensional manifold of a high-dimensional space into a low-dimensional linear subspace, intuitively one would expect that if

NLDR is applied to a dataset that is already a subspace of dimension d, the output representation should again be a subspace of the same dimension.

We will first illustrate that one of the NLDR algorithms, namely LLE, becomes degenerate when the data lies in a linear subspace. Proposition 1 [33] below shows that when d is known, the low-dimensional representation is indeed a subspace of dimension d, which is contained in the null space of the matrix M representing the local geometry of the manifold. However, since the vector $\mathbf{1}$ is also in $\ker(M)$, some degeneracies can show up when applying LLE to data lying in a linear subspace, as shown by the following proposition in [33].

Proposition 1 *Assume that the data points $\mathbf{x}_i \in \mathbb{R}^D$ lie in a subspace of $\mathbb{R}^D$ of dimension $d < k - 1$. Then the dimension of the null space of M is at least $d + 1$.*

Proof Since the data lie in a subspace of $\mathbb{R}^D$ of dimension $d < k - 1$, each point $\{\mathbf{x}_i\}$ can be reconstructed with zero error in (1), that is, for all $i = 1, \dots, n$, there are W_{ij} such that $\mathbf{x}_i = \sum_{j=1}^n W_{ij}\mathbf{x}_j$. If we let $X \in \mathbb{R}^{n\times D}$ be the matrix whose rows are the data points, then we have that $WX = X$, hence $MX = 0$. In other words, the D n-dimensional vectors formed by taking each one of the D coordinates of the n given data points are in the null space of M. Therefore, the null space of M is at least d-dimensional, because $\operatorname{rank}(X) = d$. On the other hand, since the data points live in a subspace of dimension d, there exist a matrix $T \in \mathbb{R}^{D\times d}$, $T^\top T = I$ and vectors $\{\mathbf{y}_i\}$ such that $\mathbf{x}_i = T\mathbf{y}_i + \boldsymbol{m}$ and $\sum_{i=1}^n \mathbf{y}_i = 0$, where $\boldsymbol{m} = \frac{1}{n}\sum_{i=1}^n \mathbf{x}_i \in \mathbb{R}^D$ is the mean of the data. Now, by construction, the vector of all ones $\mathbf{1}$ is also in $\ker(M)$, because $\sum_{j=1}^n W_{ij} = 1$. This implies that $X = YT^\top + \mathbf{1}\boldsymbol{m}^\top \Rightarrow MX = MYT^\top + M\mathbf{1}\boldsymbol{m}^\top = 0 \Rightarrow MYT^\top = 0 \Rightarrow MYT^\top T = 0$; hence $MY = 0$, where $Y \in \mathbb{R}^{n\times d}$ is a matrix whose rows are the $\{\mathbf{y}_i\}$ vectors. Since in addition $\sum_{i=1}^n \mathbf{y}_i = 0$, we have that $\mathbf{1}^\top Y = \mathbf{0}^\top$, hence the vector $\mathbf{1}$ is linearly independent from the columns of Y. Therefore, the null space of M is at least $(d + 1)$-dimensional. □

From Proposition 1, we see that if we apply LLE to data lying in a subspace of dimension d and choose the second to $(d + 1)$th smallest eigenvectors of M for dimensionality reduction, we might not get the correct subspace reconstruction. This is because the embedding eigenvectors (the columns of Y) may be mixed with the vector $\mathbf{1}$, which is also a null vector of M and therefore, we cannot guarantee that $\mathbf{1}$ is the first smallest eigenvector given.

We will now illustrate how Proposition 1 is applicable for Laplacian eigenmaps and HLLE as well. For the Laplacian eigenmaps, it is shown in [6] that the solution that LLE finds is an approximation of the eigenfunctions of the iterated Laplace Beltrami operator $\mathbf{L}^2$ whereas LE attempts to find the eigenfunctions of the Laplace Beltrami operator $\mathbf{L}$. Note that eigenfunctions of $\mathbf{L}^2$ coincide with those of $\mathbf{L}$. Therefore, depending on the approximation that is used for the graph Laplacian, it is also possible that LE suffers from the same degeneracy. Finally, for Hessian LLE, as the entries of the Hessian approximates $\frac{\partial^2 f}{\partial V_p \delta V_q}$, it is easy to see that when the function f is linear, it becomes a null eigenfunction as well. Therefore, it is possible that the embedding vectors are mixed with the vector $\mathbf{1}$ when we have linear manifolds for LE and HLLE as well.

3 Manifold Clustering and Dimensionality Reduction Using the Euclidean Metric

This section presents an algorithm for simultaneous NLDR and manifold clustering. In Sect. 3.1, we review the algorithm for clustering nonlinear manifolds using NLDR algorithms. This algorithm is based on the fact that the null vectors of M are actually m membership vectors indicating the grouping of the data. In Sect. 3.2, we show that these clustering algorithms based on NLDR can become degenerate when applied to data lying in multiple linear subspaces. More specifically, we show that if the data live in a k-separated union of m connected manifolds, of which $m_1 \leq m$ are linear subspaces of dimensions $\{d_i\}_{i=1}^{m_1}$, then, depending on the NLDR algorithm used, the null space of M might contain m eigenvectors that give the segmentation of the data and $\sum_{i=1}^{m_1} d_i$ eigenvectors that give the embedding coordinates for the subspaces.

3.1 Manifold Clustering and Dimensionality Reduction for a k-Separated Union of k-Connected Nonlinear Manifolds

In this section, we review an extension of the NLDR algorithm for clustering a union of m k-connected manifolds under the assumption that the manifolds are k-separated, that is, *no kNN of a data point in a manifold lies in one of the other* $(m-1)$ *manifolds*. The following proposition shows how NLDR (LLE, LE, and HLLE) can be extended for clustering a k-separated union of m k-connected nonlinear manifolds. The proposition follows from the block-diagonal properties of M in the presence of multiple k-separated nonlinear manifolds. Polito and Perona [33] illustrates this proposition for LLE only and make use of it to cluster different groups. Notice that, contrary to intuition, the case of nonlinear manifolds is simpler than the case of linear subspaces, as we will see in Sect. 3.2.

Proposition 2 *Let* $\{\mathbf{x}_i\}_{i=1}^{n}$ *be a set of points drawn from a k-separated union of m k-connected nonlinear manifolds of dimension* $d < k-1$. *There exist m vectors* $\{\mathbf{v}_j\}_{j=1}^{m}$ *in the null space of M such that* $\mathbf{v}_j$ *corresponds to the jth group of points, that is,* $\mathbf{v}_{j,i} = 1$ *if the ith data point is in the jth manifold* ($\mathbf{x}_i \in \mathcal{M}_j$), *and* $\mathbf{v}_{j,i} = 0$ *otherwise* ($\mathbf{x}_i \notin \mathcal{M}_j$).

Proof If the data can be partitioned into m k-connected groups, then the matrix M is block-diagonal with m blocks. This is because if points $\mathbf{x}_i$ and $\mathbf{x}_j$ belong to different groups, then they cannot be kNN of each other, hence $M_{ij} = 0$. Therefore, the matrix M is also block diagonal, and we can write it as $M = \operatorname{diag}(M_j)$, where $M_j \in \mathbb{R}^{n_j \times n_j}$ is the matrix for the jth group. Now, from the properties of the local NLDR algorithms reviewed in Sect. 2, we know that each one of the m blocks of M, has the vector $\mathbf{1} \in \mathbb{R}^{n_j}$ in its null space. Therefore, there are m vectors $\{\mathbf{v}_j\}$ in

Algorithm 1: Unsupervised clustering and dimensionality reduction on nonlinear manifolds

Given data points $\mathbf{x}_1, \ldots, \mathbf{x}_n \in \mathcal{M}$

1. *nearest neighbors*: find the kNN of each data point $\mathbf{x}_i$
2. *construction of* M: for each NLDR algorithm, construct the appropriate M on the entire data and compute a basis B for the null space of M
3. *clustering*: compute the segmentation of the data by applying K-means to the rows of B
4. *low-dimensional embedding*: apply NLDR as described in Sect. 2.1 to each group to obtain a low-dimensional embedding for each submanifold

$\ker(M)$, with each $\mathbf{v}_j$ taking the values 1 and 0, indicating the group membership, as claimed. □

Notice that when computing a basis $B \in \mathbb{R}^{n\times m}$ for $\ker(M)$, we do not necessarily obtain the set $\{\mathbf{v}_j\}_{j=1}^m$, but rather linear combinations of them, including the constant vector. Nevertheless, a generic linear combination of these membership vectors will still contain the segmentation of the data. Hence, we can cluster the data into m groups by applying a central clustering algorithm to the rows of B, for example, K-means. Therefore, this algorithm can be seen as a spectral clustering algorithm where the similarity matrix is obtained from the M matrix of NLDR. Algorithm 1 summarizes the dimensionality reduction and clustering algorithm for a union of k-connected nonlinear manifolds.

3.2 Degeneracies for a k-Separated Union of k-Connected Linear Manifolds

In this section, we illustrate the limitations of Algorithm 1 [18]. More precisely, we will show that NLDR becomes degenerate when the data points $\{\mathbf{x}_i\}_{i=1}^n$ are drawn from a union of m k-connected subspaces $\{\mathcal{M}_j\}_{j=1}^m$ of $\mathbb{R}^D$ with dimensions $\{d_j\}_{j=1}^m$ [18]. We first define two different types of vectors, given as follows:

Definition 1 Membership vectors $\mathbf{v}_j$ indicate the membership of each point for manifold $\mathcal{M}_j$. That is, $\mathbf{v}_{j,i} = 1$ if $\mathbf{x}_i \in \mathcal{M}_j$, and $\mathbf{v}_{j,i} = 0$ otherwise. Note that $\mathbf{v}_j$ is an $n \times 1$ vector.

Definition 2 Embedding vectors $\mathbf{e}_j$ give the embedding coordinates for each manifold $\mathcal{M}_j$. We define $\mathbf{e}_j$ as the $n \times d_j$ matrix that contains the low-dimensional coordinates of each manifold. That is,

$$\mathbf{e}_j = \left[\mathbf{0}_{(d_j \times \sum_{i=1}^{j-1} n_i)}, Y_j^\top, \mathbf{0}_{(d_j \times \sum_{i=j+1}^{m} n_i)}\right]^\top. \tag{8}$$

From Propositions 1–2, we know that there are two types of vectors in the null space of M: the embedding vectors coming from the coordinates and the membership vectors coming from each one of the m connected components. However, it is unclear if these vectors are linearly independent, and if one can recover the segmentation of the data and a nonlinear embedding for each group from $\ker(M)$. That is because an arbitrary vector in $\ker(M)$ is a linear combination of the embedding and membership vectors. The following proposition addresses these issues in detail.

Proposition 3 *Let $\{\mathbf{x}_i\}_{i=1}^n$ be a set of points drawn from a k-separated union of m k-connected subspaces of dimensions $d_j < k-1$, $j = 1, \dots, m$. The null space of M is of dimension at least $m + \sum_{j=1}^m d_j$ and contains orthonormal zero-padded vectors formed from the individual embedding $\{\mathbf{e}_j\}$ and membership $\{\mathbf{v}_j\}$ vectors.*

Proof From Proposition 2, we know that M is block diagonal and can be written as $M = \operatorname{diag}(M_j)$, where $M_j \in \mathbb{R}^{n_j \times n_j}$ is the matrix for the jth group and n_j is the number of points in the jth group. From Proposition 1, we also know that the matrix M_j has $d_j + 1$ vectors in the null space: the vector of all ones and the d_j linearly independent columns of the matrix of coordinates $Y_j \in \mathbb{R}^{n_j \times d_j}$. That is $M_j\,[Y_j\ \mathbf{1}] = 0$. Therefore, the matrix $Y = \operatorname{diag}\left([Y_j\ \mathbf{1}_{(n_j \times 1)}]\right) \in \mathbb{R}^{n \times (\sum_{j=1}^m d_j + m)}$ is such that $MY = 0$. Furthermore, as Y is block diagonal and $\operatorname{rank}\left([Y_j\ \mathbf{1}]\right) = d_j + 1$, we have that Y is of rank $\sum_{j=1}^m d_j + m$, and so the dimension of $\ker(M)$ is at least $\sum_{j=1}^m d_j + m$. Also, the matrix of embedding vectors of the jth group $\mathbf{e}_j = [\mathbf{0}, Y_j^\top, \mathbf{0}]^\top \in \mathbb{R}^{n \times d_j}$ is orthogonal to its membership vector $\mathbf{v}_j = [\mathbf{0}, \mathbf{1}_{(1 \times n_j)}, \mathbf{0}]^\top \in \mathbb{R}^{n \times 1}$. Since $\mathbf{e}_j$ and $\mathbf{v}_j$ are zero-padded, they are always orthogonal to $\mathbf{e}_i$ and $\mathbf{v}_i$ for $i \neq j$. In addition, one can choose the embedding vectors $\mathbf{e}_j$ to be orthogonal to each other, because the matrix M is symmetric. Therefore, we can assume that the vectors $\{\mathbf{v}_1, \mathbf{e}_1, \dots, \mathbf{v}_m, \mathbf{e}_m\}$ are orthonormal. $\square$

Given a set of points $\{\mathbf{x}_i \in \mathbb{R}^D\}_{i=1}^n$, it follows from the proof of Proposition 3, $\ker(M)$ contains the orthonormal set of embedding vectors $\mathbf{e}_j$ and membership vectors $\mathbf{v}_j$. More precisely, when the points $\{\mathbf{x}_i\}_{i=1}^n$ are drawn from a k-separated union of m k-connected manifolds, we have,

1. for m nonlinear manifolds:

$$\{\mathbf{v}_j\}_{j=1}^m \in \ker(M) \quad \text{and} \quad \dim\bigl(\ker(M)\bigr) = m$$

2. for m linear manifolds of dimensions $\{d_j\}_{j=1}^m$:

$$\{\mathbf{v}_j\}_{j=1}^m, \{\mathbf{e}_j\}_{j=1}^m \in \ker(M) \quad \text{and} \quad \dim\bigl(\ker(M)\bigr) = m + \sum_{j=1}^m d_j$$

3. for $m - m_1$ nonlinear manifolds, and m_1 linear manifolds of dimensions $\{d_j\}_{j=1}^{m_1}$:

$$\{\mathbf{v}_j\}_{j=1}^{m}, \{\mathbf{e}_j\}_{j=1}^{m_1} \in \ker(M) \quad \text{and} \quad \dim\big(\ker(M)\big) = m + \sum_{j=1}^{m_1} d_j$$

Therefore, in the presence of linear manifolds, we cannot directly obtain the segmentation of the data or an embedding for each one of the subspaces from $\ker(M)$, since an arbitrary vector in $\ker(M)$ is a linear combination of both membership and embedding vectors. This is a limitation of Algorithm 1; for this rest of this chapter, we assume that the data does not lie on linear manifolds.

4 Manifold Clustering and Dimensionality Reduction Using the Riemannian Metric

The NLDR techniques presented in Sects. 2 and 3.1 are applicable only in the presence of manifolds with *unknown structure*. Every operation is approximated by the corresponding Euclidean operation as the metric is unknown. However, for Riemannian manifolds with well-studied geometries, closed-form formulae for Riemannian operations are often available. The question now is to extend NLDR techniques for Riemannian manifolds in a way that takes into consideration the appropriate Riemannian structure. In this section, we present an algorithm for clustering and dimensionality reduction on Riemannian manifolds. We first present a brief summary of the theory of Riemannian manifolds in Sect. 4.1. For a more complete description, we refer the reader to [13]. Next, in Sect. 4.2, we illustrate how to extend manifold clustering and dimensionality reduction algorithms to Riemannian manifolds by adopting the framework in [21]. This framework has been applied to motion segmentation in [21], diffusion tensor images in [19] and probability density functions in [20].

4.1 Review of Riemannian Manifolds

In this section, we will give an overview of Riemannian theory and show how the various operations such as interpolation on the manifold and computation of the mean and principal components are carried out.

A smooth manifold is a topological space that is locally diffeomorphic to a Euclidean space smooth function $\gamma(t) : \mathbb{R} \to \mathcal{M}$. Figure 1 shows an example of a two-dimensional manifold. The *tangent space* of $\mathcal{M}$ at a point $\mathbf{x} \in \mathcal{M}$, denoted as $T_{\mathbf{x}}\mathcal{M}$, is then defined as the span of the tangent vectors for all the possible curves γ passing through $\mathbf{x}$. A *Riemannian metric* is a continuous collection of dot products $\langle \cdot, \cdot \rangle_{\mathbf{x}}$. Using this metric, we define the length of a curve between two points $\mathbf{x}_i, \mathbf{x}_j \in \mathcal{M}$ as

$$\mathcal{L}_a^b(\gamma) = \int_a^b \langle \dot{\gamma}(t) | \dot{\gamma}(t) \rangle_{\gamma(t)}^{\frac{1}{2}} \, dt, \tag{9}$$

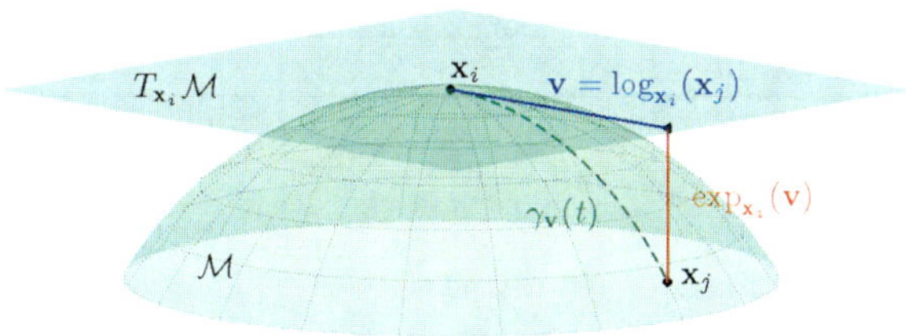

Fig. 1 An example of a two-dimensional manifold. The tangent plane at $\mathbf{x}_i$, together with the exponential and logarithm maps relating $\mathbf{x}_i$ and $\mathbf{x}_j$, are also shown

where $\gamma(a) = \mathbf{x}_i$ and $\gamma(b) = \mathbf{x}_j$. A curve between $\mathbf{x}_i$ and $\mathbf{x}_j$ with minimum length is called a geodesic. The distance between two points in the manifold is subsequently defined as the length of the geodesic curve between them,

$$\mathrm{dist}(\mathbf{x}_i, \mathbf{x}_j) = \mathcal{L}_0^1(\gamma), \quad \gamma(0) = \mathbf{x}_i,\ \gamma(1) = \mathbf{x}_j. \tag{10}$$

Let $\mathbf{v}$ be a unit-length tangent vector in $T_{\mathbf{x}}\mathcal{M}$, that is, $\|\mathbf{v}\|_{\mathbf{x}} = \langle \mathbf{v}, \mathbf{v}\rangle_{\mathbf{x}}^{\frac{1}{2}} = 1$. We can then define the *exponential map* $\exp_{\mathbf{x}} : T_{\mathbf{x}}\mathcal{M} \to \mathcal{M}$, which maps each tangent vector $t\mathbf{v} \in T_{\mathbf{x}}\mathcal{M}$ to the point $\gamma(t) \in \mathcal{M}$ obtained by following the geodesic $\gamma(t)$ (parametrized with arc-length) passing through $\mathbf{x}$ with direction $\mathbf{v}$ for a distance t. Define the set $C_{\mathbf{x}} \subset \mathcal{M}$ as the set of all the points $\mathbf{x}_i = \exp(t_0\mathbf{v})$ such that $\gamma = \exp(t\mathbf{v})$ is a length-minimizing geodesic for $t \in [0, t_0]$. The boundary of $C_{\mathbf{x}}$ is called the *cut locus*, and, intuitively, it is the set of points for which the distance from $\mathbf{x}$ stops to be a differentiable function of t. The exponential map is therefore invertible for all the points in the interior of $C_{\mathbf{x}}$ and we can define the *logarithm map* $\log_{\mathbf{x}} : C_{\mathbf{x}} \to T_{\mathbf{x}}\mathcal{M}$ as $\log_{\mathbf{x}} = \exp_{\mathbf{x}}^{-1}$. Note that for any $\mathbf{x}_i \in C_{\mathbf{x}}$ and $\mathbf{x}_i = \exp_{\mathbf{x}}(t\mathbf{v})$ we have that (by definition),

$$\left\| \log_{\mathbf{x}}(\mathbf{x}_i) \right\|_{\mathbf{x}} = \|t\mathbf{v}\|_{\mathbf{x}} = \mathrm{dist}(\mathbf{x}, \mathbf{x}_i) = t. \tag{11}$$

Linear geodesic interpolation makes use of the exponential and logarithm maps and is defined as $\hat{\mathbf{x}} = \exp_{\mathbf{x}_i}(w \log_{\mathbf{x}_i}(\mathbf{x}_j))$, $w \in [0, 1]$. $\hat{\mathbf{x}}$ is the linear interpolation at $t = w$ of $\mathbf{x}_i$ and $\mathbf{x}_j$. Finally, we recall that the Riemannian metric, exponential and logarithm maps depend on the point $\mathbf{x}$ under consideration, hence the subscripts reflecting this dependency.

We will now briefly summarize how to calculate the mean and principal components of data points lying in a manifold. As defined by Fréchet in [16] and used in several recent works [15, 32], the intrinsic mean $\bar{\mathbf{x}}$ is defined as the solution to the following minimization problem

$$\bar{\mathbf{x}} = \underset{\mathbf{x}\in\mathcal{M}}{\arg\min} \sum_{i=1}^{n} \mathrm{dist}(\mathbf{x}, \mathbf{x}_i)^2 = \underset{\mathbf{x}\in\mathcal{M}}{\arg\min} \sum_{i=1}^{n} \left\| \log_{\mathbf{x}}(\mathbf{x}_i) \right\|_{\mathbf{x}}^2. \tag{12}$$

Note that, unlike in the Euclidean case, in general there is no closed form for $\bar{\mathbf{x}}$. Moreover, there is no guarantee that $\bar{\mathbf{x}}$ exists or is unique. However, one can show the existence and uniqueness of $\bar{\mathbf{x}}$ [27] by assuming that the data lie in a small enough neighborhood, that is, the maximum distance between any $\mathbf{x}_i$ and $\mathbf{x}_j$ is small enough. Furthermore, $\bar{\mathbf{x}}$ can be computed as shown in Algorithm 2.

Algorithm 2: Intrinsic mean

Given data points $\mathbf{x}_1, \ldots, \mathbf{x}_n \in \mathcal{M}$, a predefined threshold ϵ, maximum number of iterations T

1. initialize $t = 1$, $\bar{\mathbf{x}}_1 = \mathbf{x}_i$ for a random i, $\mathbf{v} \neq 0 \in T_{\bar{\mathbf{x}}_1}\mathcal{M}$
2. while $t \leq T$ or $\|\mathbf{v}\|_{\bar{\mathbf{x}}} \geq \epsilon$
   (a) compute tangent vector $\mathbf{v} = \frac{1}{n}\sum_{i=1}^{n} \log_{\bar{\mathbf{x}}_t}(\mathbf{x}_i)$
   (b) set $\bar{\mathbf{x}}_{t+1} = \exp_{\bar{\mathbf{x}}_t}(\mathbf{v})$

Algorithm 3: Principal geodesic analysis

Given data points $\mathbf{x}_1, \ldots, \mathbf{x}_n \in \mathcal{M}$

1. compute intrinsic mean $\bar{\mathbf{x}}$ as in Algorithm 2
2. calculate the tangent vectors $\mathbf{v}_i = \log_{\bar{\mathbf{x}}}(\mathbf{x}_i)$about $\bar{\mathbf{x}}$
3. construct the sample covariance matrix $\text{cov}(\mathbf{x}) = \frac{1}{n}\sum_{i=1}^{n} \mathbf{v}_i\mathbf{v}_i^{\top}$
4. perform eigenanalysis of the matrix $\text{cov}(\mathbf{x})$, with the eigenvectors $\{\mathbf{u}_i\}_{i=1}^{d}$ giving the principal directions. $\{\mathbf{u}_i\}_{i=1}^{d}$ forms an orthonormal basis for $T_{\bar{\mathbf{x}}}\mathcal{M}$

Table 1 Comparison of Euclidean and Riemannian operations, where $\{\mathbf{x}_i\}_{i=1}^{n}$, are data points

Operation	Euclidean	Riemannian
Subtraction $\overrightarrow{\mathbf{x}_i\mathbf{x}_j}$	$\mathbf{x}_j - \mathbf{x}_i$	$\log_{\mathbf{x}_i}(\mathbf{x}_j)$
Addition $\mathbf{x}_j$	$\mathbf{x}_i + \overrightarrow{\mathbf{x}_i\mathbf{x}_j}$	$\exp_{\mathbf{x}_i}(\overrightarrow{\mathbf{x}_i\mathbf{x}_j})$
Distance $\text{dist}(\mathbf{x}_i, \mathbf{x}_j)$	$\|\overrightarrow{\mathbf{x}_i\mathbf{x}_j}\| = \|\mathbf{x}_j - \mathbf{x}_i\|$	$\|\log_{\mathbf{x}_i}(\mathbf{x}_j)\|_{\mathbf{x}_i} = \sqrt{\langle \log_{\mathbf{x}_i}(\mathbf{x}_j), \log_{\mathbf{x}_i}(\mathbf{x}_j)\rangle_{\mathbf{x}_i}}$
Linear interpolation $\hat{\mathbf{x}}$	$\mathbf{x}_i + w\overrightarrow{\mathbf{x}_i\mathbf{x}_j}$	$\exp_{\mathbf{x}_i}(w\overrightarrow{\mathbf{x}_i\mathbf{x}_j})$
Mean $\bar{\mathbf{x}}$	$\sum_{i=1}^{n} \overrightarrow{\bar{\mathbf{x}}\mathbf{x}_i} = 0$	$\sum_{i=1}^{n} \log_{\bar{\mathbf{x}}}(\mathbf{x}_i) = 0$
Covariance $\text{cov}(\mathbf{x})$	$\frac{1}{n}\sum_{i=1}^{n}(\overrightarrow{\bar{\mathbf{x}}\mathbf{x}_i})(\overrightarrow{\bar{\mathbf{x}}\mathbf{x}_i})^{\top}$	$\frac{1}{n}\sum_{i=1}^{n}(\log_{\bar{\mathbf{x}}}(\mathbf{x}_i))(\log_{\bar{\mathbf{x}}}(\mathbf{x}_i))^{\top}$

Given $\bar{\mathbf{x}}$, the calculation of principal components on a Riemannian manifold is not as straightforward as in the Euclidean case. It involves projecting a point onto a geodesic curve, which is also defined as a minimization problem for which existence and uniqueness are not ensured [15]. Again, by making the assumptions that the data lie in a small neighborhood, the projection can be shown to be unique. In [15], it is shown that finding principal components boils down to doing PCA in the tangent vectors $\log_{\bar{\mathbf{x}}}(\mathbf{x}_i) \in T_{\bar{\mathbf{x}}}\mathcal{M}$ about the mean $\bar{\mathbf{x}}$. This algorithm, known as Principal Geodesic Analysis (PGA), is summarized in Algorithm 3. Table 1 compares the standard operations in Euclidean and Riemannian spaces.

4.2 Extending Manifold Clustering and Dimensionality Reduction to Riemannian Manifolds

In this section, we will show how to extend manifold clustering and dimensionality reduction to Riemannian manifolds. We assume here that the various Riemannian operations are known and closed-form. First of all, notice that the information about the local geometry of the manifold is essential only in the first two steps of each algorithm and therefore, modifications are made only to these two stages. The key issues are how to select the kNN and how to compute the matrix M representing the local geometry. As shown in [21], the former is straightforward, while the latter requires some thought. Given M, calculating the low-dimensional representation remains the same as in the Euclidean case. We let $X = \{\mathbf{x}_i \in \mathbb{R}^D\}_{i=1}^n$ be a set of n data points sampled from a known Riemannian manifold.

4.2.1 Selection of the Riemannian kNN

The first step of any NLDR algorithm is the computation of the kNN associated with each data point. We define the kNN of $\mathbf{x}_i$ by incorporating the Riemannian distance, that is, *the kNN of* $\mathbf{x}_i$ *are the k data points* $\mathbf{x}_j$ *that minimize* $\|\log_{\mathbf{x}_i}(\mathbf{x}_j)\|_{\mathbf{x}_i}$.

4.2.2 Riemannian Calculation of M for LLE

The second step of LLE is to compute the matrix of weights $W \in \mathbb{R}^{n\times n}$. In order to do so, we will answer two main questions: (1) how does one express a point as a linear combination of its neighbors? and (2) what is the reconstruction cost? First of all, we know that from Sect. 4.1 that

$$\hat{\mathbf{x}}_{\mathrm{Riem},i} = \exp_{\mathbf{x}_i}\left(\sum_{j=1}^n W_{ij}\log_{\mathbf{x}_i}(\mathbf{x}_j)\right), \tag{13}$$

is the geodesic linear interpolation of $\mathbf{x}_i$ by $\{\mathbf{x}_j\}_{j=1}^n$. Now, instead of minimizing the Euclidean error, we rewrite (1) to minimize the Riemannian reconstruction error and make use of the fact that exp and log are inverse mappings. Therefore, we have

$$\varepsilon_{\mathrm{Riem}}(W) = \sum_{i=1}^n \|\log_{\mathbf{x}_i}(\hat{\mathbf{x}}_{\mathrm{Riem},i})\|_{\mathbf{x}_i}^2 = \sum_{i=1}^n \left\|\sum_{j=1}^n W_{ij}\log_{\mathbf{x}_i}(\mathbf{x}_j)\right\|_{\mathbf{x}_i}^2, \tag{14}$$

subject to $W_{ij} = 0$ if $\mathbf{x}_j$ is not a kNN of $\mathbf{x}_i$ and $\sum_j W_{ij} = 1$. Using similar manipulations as in the Euclidean case, the optimal weights are obtained as in (2), with the local Gram matrix $C_i \in \mathbb{R}^{k\times k}$ defined as

$$C_i(j,l) = \left\langle \log_{\mathbf{x}_i}(\mathbf{x}_j), \log_{\mathbf{x}_i}(\mathbf{x}_l)\right\rangle_{\mathbf{x}_i}. \tag{15}$$

M is then $(I - W)^\top(I - W)$.

4.2.3 Riemannian Calculation of M for LE

Here, instead of attempting to write each data point as a linear combination of its kNN, we find a matrix of weights $W \in \mathbb{R}^{n\times n}$ whose entries W_{ij} measure the proximity between two points $\mathbf{x}_i$ and $\mathbf{x}_j$ as in (4). Therefore, modifying LE for Riemannian manifolds is less involved than in the case of LLE. Instead of using $\exp(-\|\mathbf{x}_i - \mathbf{x}_j\|^2/\sigma^2)$ as in (4), we construct the weight matrix W using the Riemannian distance as

$$W_{ij} = \exp\left(-\frac{\mathrm{dist}_{\mathrm{Riem}}(\mathbf{x}_i, \mathbf{x}_j)^2}{\sigma^2}\right) = \exp\left(-\frac{\|\log_{\mathbf{x}_i}(\mathbf{x}_j)\|^2_{\mathbf{x}_i}}{\sigma^2}\right), \tag{16}$$

subject to the constraint $W_{ij} = 0$ if $\mathbf{x}_j$ is not a k-nearest neighbor of $\mathbf{x}_i$. As before, $M = D - W$ and D is a diagonal matrix, where $D_{ii} = \sum_j W_{ij}$.

4.2.4 Riemannian Calculation of M for HLLE

The second step of HLLE involves computing the tangent coordinates for each $\mathbf{x}_i$ by applying Euclidean PCA to its neighbors. This implicitly assumes that these local points lie in a subspace. This assumption is no longer valid if $\mathbf{x}_i$ and its kNN lie in a Riemannian manifold. From Sect. 4.1, we know that in this case, calculating the principal geodesic components and the projection coordinates is not as simple as doing Euclidean PCA. There is a need to incorporate the correct Riemannian metric, mean and covariance matrix.

Again, let $\{\mathbf{x}_{i,j}\}_{j=1}^k$ denote the set of k-nearest neighbors of $\mathbf{x}_i$. First, we calculate the intrinsic mean $\bar{\mathbf{x}}_i$ of the kNN (Algorithm 2). Next, we find the tangent vectors $\mathbf{v}_j = \log_{\bar{\mathbf{x}}_i}(\mathbf{x}_{i,j})$ about $\bar{\mathbf{x}}_i$ and the geodesic principal directions $\{\mathbf{u}_q\}_{q=1}^d$ using PGA (Algorithm 3). Since $\{\mathbf{u}_q \in \mathbb{R}^D\}_{q=1}^d$ is an orthonormal basis for $T_{\bar{\mathbf{x}}_i}\mathcal{M}$, we will rewrite the projection operator in (6) using the Riemannian metric. Thus, the tangent coordinates of the kNN are given by the $k \times d$ matrix V, where

$$V_{pq} = \left\langle \log_{\bar{\mathbf{x}}_i}(\mathbf{x}_{i,p}), \mathbf{u}_q \right\rangle_{\bar{\mathbf{x}}_i}, \quad p = 1, \ldots, k,\ q = 1, \ldots, d. \tag{17}$$

Once the tangent coordinates are found, the estimation of the Hessian matrix M is the same as in the Euclidean case (7).

4.2.5 Calculation of the Embedding Coordinates

The last step of NLDR is to find a Euclidean low-dimensional representation of the data points. As this step is independent of the Riemannian structure, one can find the embedding coordinates as described in Sect. 2. That is, the embedding coordinates are obtained based on the d (generalized) eigenvectors of the matrix M associated with its second to $(d+1)$th smallest (generalized) eigenvalues. Finally, notice that if the Riemannian operations are available in closed-form, then extending NLDR to Riemannian manifolds do not require significant additional computational complexity.

Algorithm 4: Unsupervised clustering and dimensionality reduction on Riemannian manifolds

Given data points $\mathbf{x}_1, \ldots, \mathbf{x}_n \in \mathcal{M}$

1. *nearest neighbors*: find the kNN of each data point $\mathbf{x}_i$ according to the Riemannian distance
2. *construction of M*: for each NLDR algorithm, construct the appropriate M described in Sect. 4.2 on the entire data and compute a basis B for the null space of M
3. *clustering*: compute the segmentation of the data by applying K-means to the rows of B
4. *low-dimensional embedding*: apply NLDR as described in Sect. 4.2 to each group to obtain a low-dimensional embedding for each submanifold

4.2.6 Extending Manifold Clustering to Riemannian Manifolds

We will now consider the case when we have data lying in m submanifolds of a Riemannian space. Similar to what is done in the Euclidean space, we assume that the data is distributed in a k-disconnected union of m k-connected submanifolds of $\mathcal{M}$. Notice that just as done in the calculation of the embedding coordinate, Algorithm 1 is independent of the Riemannian structure. Therefore, it is immediate to see that they are applicable to m submanifolds, both linear and nonlinear, of a Riemannian space. Algorithm 4 summarizes the dimensionality reduction and clustering algorithm for m submanifolds of a Riemannian space.

5 Experiments

5.1 Application and Experiments on SPSD(3) [19, 21]

This section present an application of the theory developed in Sect. 4 to the space of 3 by 3 symmetric positive semi-definite matrices SPSD(3). It is well known [3, 15, 32, 46] that the traditional Euclidean distance is not the most appropriate metric for SPSD matrices as they lie on a Riemannian symmetric space. An example of such data is the well-known structure tensor found in direct 2-D motion segmentation from the image intensities *without* extracting features such as optical flow or point correspondences. Under the assumption that all surfaces are Lambertian, the optical flow (u, v) between two images of a sequence is related to the image partial derivatives $\nabla I = (I_x, I_y, I_t)$ by $I_x u + I_y v + I_t = 0 \Rightarrow \nabla I^\top (u, v, 1) = 0$, where (x, y) denotes pixel location and t denotes time. Premultiplying by ∇I gives an equation of the form $(\nabla I \nabla I^\top)(u, v, 1) = 0$, This system of linear equations involves the spatial-temporal structure tensor $(\nabla I \nabla I^\top)$. SPSD matrices also play an important role in Diffusion Tensor Imaging (DTI). DTI is a 3-D imaging technique that measures the diffusion of water molecules in living tissues. Water diffusion is represented mathematically with a symmetric positive semi-definite tensor field $\mathbf{T} : \mathbb{R}^3 \to \mathrm{SPSD}(3) \subset \mathbb{R}^{3\times 3}$ that measures the diffusion in a direction $\mathbf{d} \in \mathbb{R}^3$ as $\mathbf{d}^\top \mathbf{T}\mathbf{d}$.

The goal is to automatically segment a set of SPSD matrices $\{\mathbf{T}_j \in \mathrm{SPSD}(r)\}_{j=1}^n$ into different clusters, where different groups correspond to different 2-D motions in a video or to different fiber bundles in DTI. There are many possible metrics in $\mathrm{SPSD}(r)$ [3, 15, 28, 32, 46]. Each metric is derived from different geometrical, statistical or information-theoretic considerations. The question of which one is the best metric remains an active research area. The Riemannian metric proposed in [32] $\mathrm{dist}_{\mathrm{Riem}}(\mathbf{T}_i, \mathbf{T}_j) = \| \log(\mathbf{T}_i^{-\frac{1}{2}} \mathbf{T}_j \mathbf{T}_i^{-\frac{1}{2}}) \|_F$, where $\| \cdot \|_F$ is the Frobenius norm and $\log(\cdot)$ is the matrix logarithm, is used here. For this metric, the exponential map is defined as $\exp_{\mathbf{T}_i}(\mathbf{V}) = \mathbf{T}_i^{\frac{1}{2}} \exp(\mathbf{T}_i^{-\frac{1}{2}} \mathbf{V} \mathbf{T}_i^{-\frac{1}{2}}) \mathbf{T}_i^{\frac{1}{2}}$, where $\exp(\cdot)$ is the matrix exponential and $\mathbf{V} \in T_{\mathbf{T}_i}\mathrm{SPSD}(r)$. The logarithm map is $\overrightarrow{\mathbf{T}_i\mathbf{T}_j} = \log_{\mathbf{T}_i}(\mathbf{T}_j) = \mathbf{T}_i^{\frac{1}{2}} \log(\mathbf{T}_i^{-\frac{1}{2}} \mathbf{T}_j \mathbf{T}_i^{-\frac{1}{2}}) \mathbf{T}_i^{\frac{1}{2}}$. The Gram matrix is $C_i(j, l) = \langle \log_{\mathbf{T}_i}(\mathbf{T}_j), \log_{\mathbf{T}_i}(\mathbf{T}_l) \rangle_{\mathbf{x}_i} = \mathrm{trace}(\log(\mathbf{T}_i^{-\frac{1}{2}} \mathbf{T}_j \mathbf{T}_i^{-\frac{1}{2}}) \log(\mathbf{T}_i^{-\frac{1}{2}} \mathbf{T}_l \mathbf{T}_i^{-\frac{1}{2}}))$. The geodesic linear interpolation $\hat{\mathbf{T}}_{\mathrm{Riem},i}$ of $\{\mathbf{T}_j \in \mathrm{SPSD}(r)\}_{j=1}^n$ about $\mathbf{T}_i$ with weights $W_{i1}, \ldots, W_{in}$ is given by $\mathbf{T}_i^{\frac{1}{2}} \exp(\sum_{j=1}^n W_{ij} \log(\mathbf{T}_i^{-\frac{1}{2}} \mathbf{T}_j \mathbf{T}_i^{-\frac{1}{2}})) \mathbf{T}_i^{\frac{1}{2}}$.

Our algorithm is tested on 2-D motion segmentation from two consecutive frames of video sequences [21]. The spatial-temporal structure tensor is $\mathbf{T} = K * (\nabla I \nabla I^\top)$, where $*$ is the convolution operator, K is a smoothing kernel (the Gaussian kernel is commonly used), and $\nabla I = (I_x, I_y, I_t)$ is the spatial-temporal image gradient. We use the same data set as in [12]. Figure 2 shows two examples of moving patches of homogeneously textured wallpaper in which the different regions cannot be distinguished on the sole basis of appearance. This is because the input frames contain regions with the same intensity and texture with no clear edges or corners. Thus, all results are obtained exclusively from the motion information. Figure 2(a) contains two regions, the text region with the word "UCLA" and the background. In Fig. 2(c),

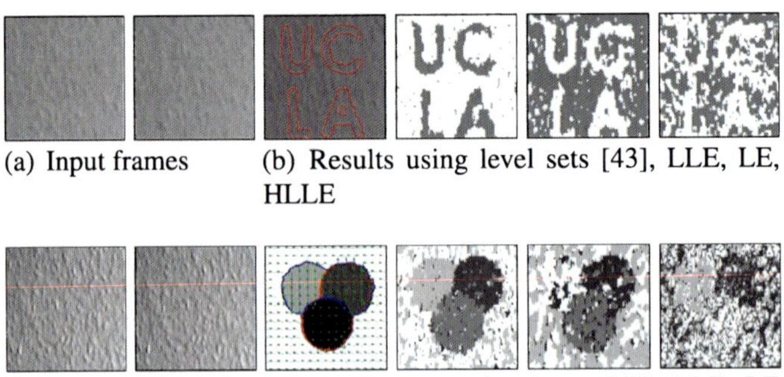

(a) Input frames (b) Results using level sets [43], LLE, LE, HLLE

(c) Input frames (d) Results using level sets [43], LLE, LE, HLLE

Fig. 2 2-D motion segmentation using the structure tensor [21]

there are three overlapping circles, each with its own motion, and the background. As shown in Figs. 2(b) and 2(d), LLE yields the best results among all the algebraic methods, distinguishing the text "UCLA" and the three circles. As none of the NLDR methods incorporates a smoothness constraint (as done in level set methods), it is of no surprise that the level set method produces a cleaner segmentation. Nevertheless, it is immediate that our algorithm provides a very good initialization for iterative techniques such as level sets.

The next set of experiments is done on real video sequences [21]. The first video involves a camera tracking a car going along a road, as shown in Fig. 3(a). There are three different motion groups found in the two consecutive frames. The first group contains mostly the pixels of the car, the second group contains the background pixels where the camera movement is apparent (e.g., edges and corners), and the last group contains the background pixels with the aperture problem (e.g., middle of the road). The second video of a car is taken with a stationary camera, as shown in Fig. 3(c). There are two different motion groups in this case, the first group being the car and the second group being the background. The last video, shown in Fig. 3(e), is taken from the Hamburg Taxi sequence. In this dataset, the moving taxi forms the first group and the stationary background forms the second group. From Figs. 3(b), 3(d), and 3(f), it is clear that LLE is able to segment the different groups, LE gives a reasonable segmentation but suffers from artifacts, whereas the performance of HLLE performance is poor.

Our algorithm is also tested in the segmentation of the corpus callosum and the cingulum from real DTI data [19]. The size of the entire DTI volume of the brain is

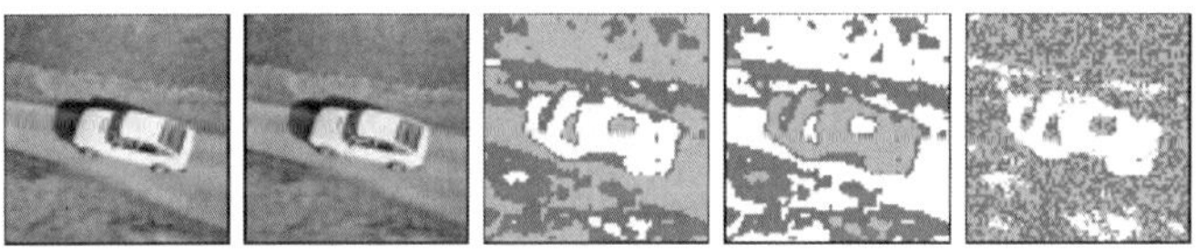

(a) Input frames (b) Segmentation results: LLE, LE, HLLE

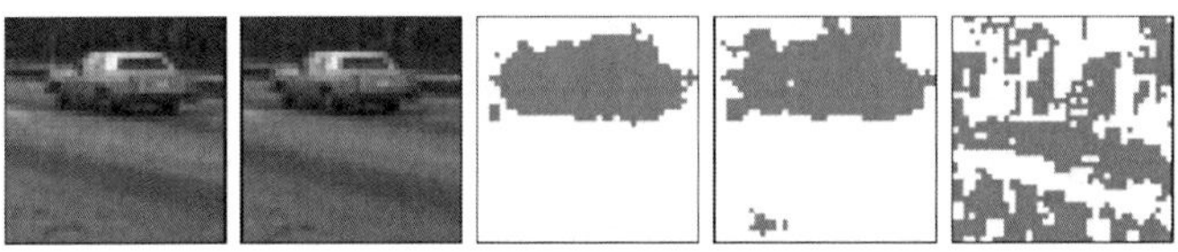

(c) Input frames (d) Segmentation results: LLE, LE, HLLE

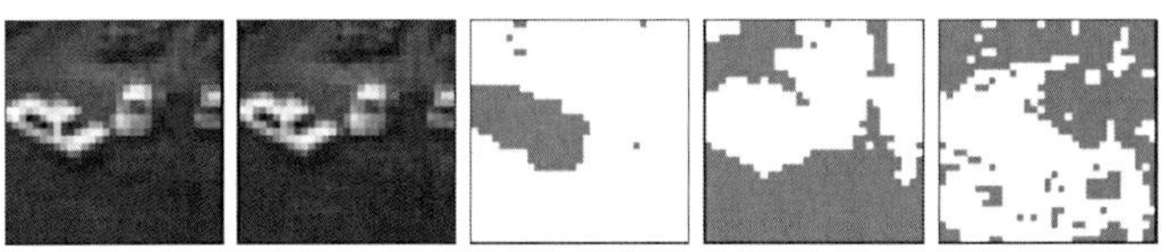

(e) Input frames (f) Segmentation results: LLE, LE, HLLE

Fig. 3 2-D motion segmentation on real video sequences [21]

$128 \times 128 \times 58$ voxels and the voxel size is $2 \times 2 \times 2$ mm. From the visualization of the tensor data, we know the approximate location of each cingulum bundle in the left and right hemispheres. Hence, we reduce the input volume to the algorithm by focusing in this location. In addition, we also mask out voxels with fractional anisotropy below a threshold of 0.2 in order to separate white matter from the rest of the brain. Also, tensors at adjacent voxels within a fiber bundle are similar (similar eigenvectors and eigenvalues), while tensors at distant voxels could be very different, even if they lie in the same bundle. In order to account for this fact, the kNN $\{\mathbf{D}(x_j)\}_{j=1}^k$ of a tensor $\mathbf{D}(x_i)$ at x_i subject to $\|x_j - x_i\| \leq R$ is chosen. The value of the spatial radius is set to $R = 10$ and the number of nearest neighbors to $k = 30$.

Figure 4 shows the results of the left hemisphere. Figure 4(a) shows the sagittal slices used and the ellipsoid visualization of the tensors. The corpus callosum is the bundle with the red tensors pointing out of the plane and resembles the letter 'C'. The cingulum, which is significantly smaller, is the bundle left to the corpus callosum with the green tensors oriented vertically. The corpus callosum and the cingulum are clustered around different centers. Figure 4(b) shows the results of LLE. The corpus callosum forms a distinct cluster (in red). Figure 4(c) shows the results of LE. Even though it appears that the cingulum forms a distinct group, the corpus callosum is merged into the same group as the tensors in the background. HLLE (not shown) failed to produce any reasonable segmentation of the fiber bundles.

As our algorithm does not incorporate any smoothness constraint, the segmentation is noisier compared to energy minimization methods such as in [30]. However, for the segmentation of the cingulum bundle in [30], a significant effort was required to manually remove voxels in the corpus callosum before running their respective algorithms. Our algorithm, on the other hand, is automatic. Hence, an immediate use of our algorithm is that the output could be used as an automatic initialization for such algorithms.

5.2 Application and Experiments on the Space of Probability Density Functions

This section present an application of the theory developed in Sect. 4 to the space of probability density functions. The class of constrained non-negative continuous functions under study here is the set of pdfs defined below. Without loss of generality, we can assume that these functions are defined on the interval $[0, T]$. Therefore, the set $\mathcal{P}$ of pdfs is given by

$$\mathcal{P} = \left\{ p : [0, T] \to \mathbb{R} | \forall s, p(s) \geq 0, \int_0^T p(s)\, ds = 1 \right\}. \tag{18}$$

The question of how to regard the space of pdfs as a differential manifold endowed with a Riemannian metric and a family of affine connections has a long history behind it. Nevertheless, it remains an active and important research area. Treating statistical structures as geometric structures has the advantage that geometric

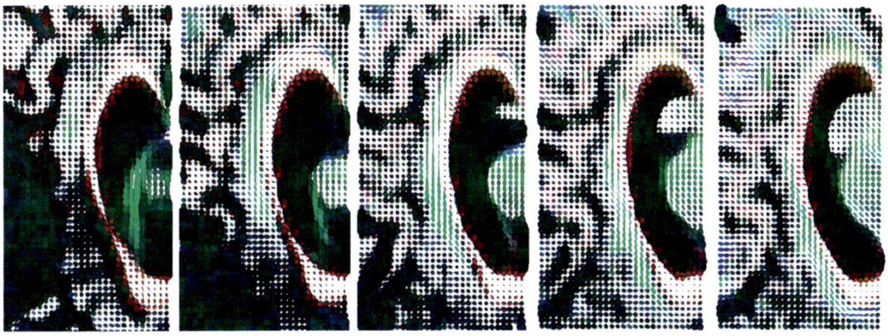

(a) Ellipsoid visualization of the tensors in five sagittal slices

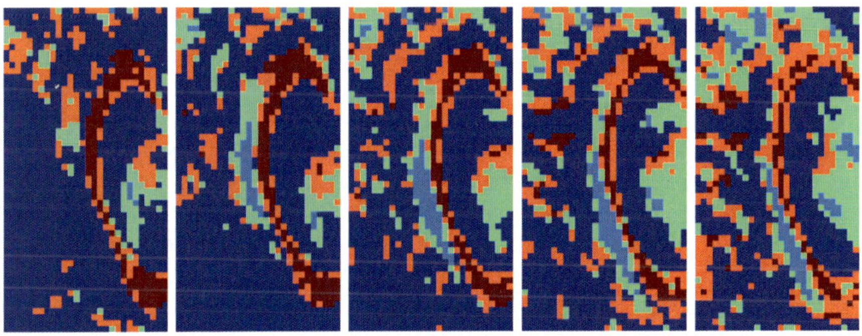

(b) Segmentation results of the five sagittal slices using LLE

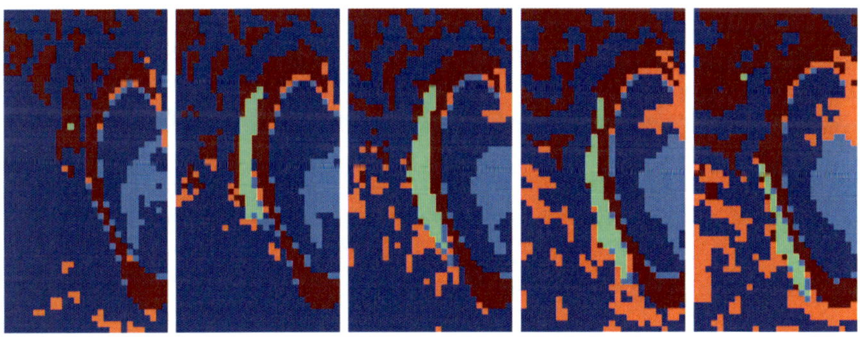

(c) Segmentation results of the five sagittal slices using LE

Fig. 4 Segmenting the corpus callosum and the cingulum [19]. The *first row* shows the visualization of the data in five sagittal slices and the tensor at each voxel is represented by an ellipsoid. The *second row* shows the clustering result given by LLE. The *third row* shows the clustering result given by LE

structures remain invariant under coordinate transforms. Rao [34] first introduces the Riemannian structure formed by the statistical manifold where each point in the manifold denotes a pdf. In addition, [34] also shows that the *Fisher–Rao* metric determines a Riemannian metric. The Fisher–Rao metric is later shown to be the *unique intrinsic metric* on the statistical manifold in [9]. This study of probability and information via differential geometry is known as *information geometry*. The reader is referred to the seminal work of [2] for a complete description.

We will consider the manifold $\mathcal{P}$ of pdfs on the interval $[0, T]$. For any point $\boldsymbol{p}_i \in \mathcal{P}$, the Fisher–Rao metric is defined as

$$\langle \boldsymbol{q}_j, \boldsymbol{q}_k \rangle_{\boldsymbol{p}_i} = \int_0^T \boldsymbol{q}_j(s) \boldsymbol{q}_k(s) \frac{1}{\boldsymbol{p}_i(s)} ds, \tag{19}$$

where $\boldsymbol{q}_j, \boldsymbol{q}_k \in T_{\boldsymbol{p}_i}(\mathcal{P})$ are tangent vectors and $T_{\boldsymbol{p}_i}(\mathcal{P})$ is the set containing the functions tangent to $\mathcal{P}$ at the point $\boldsymbol{p}_i$. This representation turns out to be extremely difficult to work with as ensuring the geodesic between two elements lies on $\mathcal{P}$ is not easy [41].

Even though the space $\mathcal{P}$ turns out to be difficult to work with, we know that it is not the only possible representation for pdfs and in addition, we also know that the Fisher–Rao metric is the only metric that is invariant to re-parameterizations (essentially coordinate transforms) of the functions [9]. There are many different re-parameterizations of pdfs that are equivalent representations. Depending on the representation, the resulting Riemannian structure can have varying degrees of complexity and numerical techniques may be required to compute geodesics on the manifold. Therefore, the natural question to ask now is, is it possible to use a re-parameterization such that the resulting manifold is simple and the Riemannian operations are easy, preferably closed-form, to compute? Once an efficient representation is found, the corresponding Fisher–Rao metric, which depends on the tangent vector, will then be used as the Riemannian metric. In a recent work [41], it is proved that by using the square-root representation, the resulting manifold is a unit sphere in a Hilbert space with the Fisher–Rao metric being the usual $\mathbb{L}^2$ metric. Therefore, the various Riemannian operations such as geodesics, exponential maps, logarithmic maps are available in closed form. This is one of the most efficient representation found to date.

The square-root density function is defined as $\boldsymbol{\psi} = \sqrt{\boldsymbol{p}}$, where $\boldsymbol{\psi}$ is assumed to be nonnegative to ensure uniqueness. The space of such functions is defined as:

$$\boldsymbol{\Psi} = \left\{ \boldsymbol{\psi} : [0, T] \to \mathbb{R} | \forall s, \boldsymbol{\psi}(s) \geq 0, \int_0^T \boldsymbol{\psi}^2(s)\, ds = 1 \right\}. \tag{20}$$

From (20), it is easy to see that the functions $\boldsymbol{\psi}$ lie on a unit sphere. In addition, $\boldsymbol{\Psi}$ forms a convex subset of the unit sphere. The advantage of choosing the square-root density becomes immediately obvious, as many of the Riemannian expressions for the unit sphere are well-known and closed-form. By making use of the representation in (20), we can rewrite (19) and obtain the Fisher–Rao metric as

$$\langle \mathbf{v}_j, \mathbf{v}_k \rangle_{\boldsymbol{\psi}_i} = \int_0^T \mathbf{v}_j(s) \mathbf{v}_k(s)\, ds, \tag{21}$$

where $\mathbf{v}_j, \mathbf{v}_k \in T_{\boldsymbol{\psi}_i} \boldsymbol{\Psi}$ are tangent vectors. Now, for any two functions $\boldsymbol{\psi}_i, \boldsymbol{\psi}_j \in \boldsymbol{\Psi}$, the geodesic distance between these two points on a unit sphere is simply the angle between them, that is,

$$\mathrm{dist}(\boldsymbol{\psi}_i, \boldsymbol{\psi}_j) = \cos^{-1}\langle \boldsymbol{\psi}_i, \boldsymbol{\psi}_j \rangle = \cos^{-1}\left(\int_0^T \boldsymbol{\psi}_i(s)\boldsymbol{\psi}_j(s)\, ds \right), \tag{22}$$

where $\langle \cdot, \cdot \rangle$ is the normal dot product between points in the sphere under the $\mathbb{L}^2$ metric.

From the differential geometry of the sphere, the exponential map is defined as

$$\exp_{\boldsymbol{\psi}_i}(\mathbf{v}) = \cos\big(\|\mathbf{v}\|_{\boldsymbol{\psi}_i}\big)\boldsymbol{\psi}_i + \sin\big(\|\mathbf{v}\|_{\boldsymbol{\psi}_i}\big)\frac{\mathbf{v}}{\|\mathbf{v}\|_{\boldsymbol{\psi}_i}}, \tag{23}$$

where $\mathbf{v} \in T_{\boldsymbol{\psi}_i}(\boldsymbol{\Psi})$ is a tangent vector at $\boldsymbol{\psi}_i$ and $\|\mathbf{v}\|_{\boldsymbol{\psi}_i} = \sqrt{\langle \mathbf{v}, \mathbf{v} \rangle_{\boldsymbol{\psi}_i}} = (\int_0^T \mathbf{v}(s)\mathbf{v}(s)\, ds)^{\frac{1}{2}}$. In order to ensure that the exponential map is a bijective function, we restrict $\|\mathbf{v}\|_{\boldsymbol{\psi}_i} \in [0, \pi]$. The logarithm map from $\boldsymbol{\psi}_i$ to $\boldsymbol{\psi}_j$ is then given by

$$\overrightarrow{\boldsymbol{\psi}_i \boldsymbol{\psi}_j} = \log_{\boldsymbol{\psi}_i}(\boldsymbol{\psi}_j) = \frac{\mathbf{u}}{(\int_0^T \mathbf{u}(s)\mathbf{u}(s)\, ds)^{\frac{1}{2}}} \cos^{-1}\langle \boldsymbol{\psi}_i, \boldsymbol{\psi}_j \rangle, \tag{24}$$

with $\mathbf{u} = \boldsymbol{\psi}_j - \langle \boldsymbol{\psi}_i, \boldsymbol{\psi}_j \rangle \boldsymbol{\psi}_i$.

We test the algorithm in the segmentation of different textures [20]. From the Columbia-Utrecht Reflectance and Texture Database (CUReT) found at http://www1.cs.columbia.edu/CAVE//software/curet/, we obtain samples of different textures and each grayscale image contains only one texture. In order to construct a histogram that reflects the texture statistics in an image, we will calculate what is commonly known as *textons* [44]. This is done by first applying a filter bank to all images in the training set. We use the Schmid [37] filter banks shown in Fig. 5. This will provide us with a feature vector $f(x, y)$ of dimension 13 at each pixel. Next, we apply k-means to all the vectors in the entire dataset to get 30 cluster centers, also known as the textons. For each image in the dataset, we then compute a histogram that contains the number of pixels corresponding to each one of these 30 bins. This is done by assigning a pixel (x, y) to bin i if the feature vector $f(x, y)$ is closest to cluster center $i = 1, \ldots, 30$, according to the Euclidean distance in $\mathbb{R}^{13}$.

We test our algorithm on 4 sets of data containing 2 different textures each. There are 92 images in each texture class. In these experiments, the number of nearest neighbors is set to 10. Figure 6 shows these 4 sets with a typical example of the 2 different textures and the corresponding histograms in each set. Table 2 shows the misclustering percentage of LLE and LE for each set.

Finally, we test our algorithm on a set of data containing 3 different textures. Figure 7 shows a typical example of the different textures and the corresponding histograms in each set. The error produced by LLE in clustering is 5.43% whereas LE is significantly higher at 30.07%.

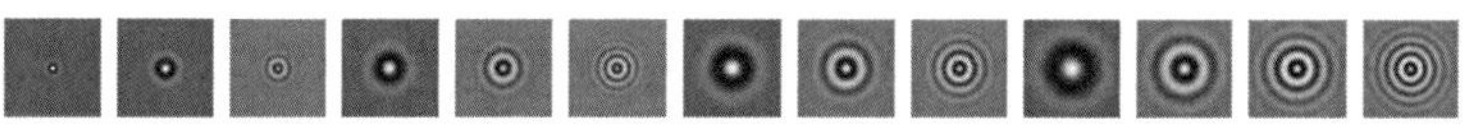

Fig. 5 Schmid filter bank that we use to generate the textons and in turn the histograms

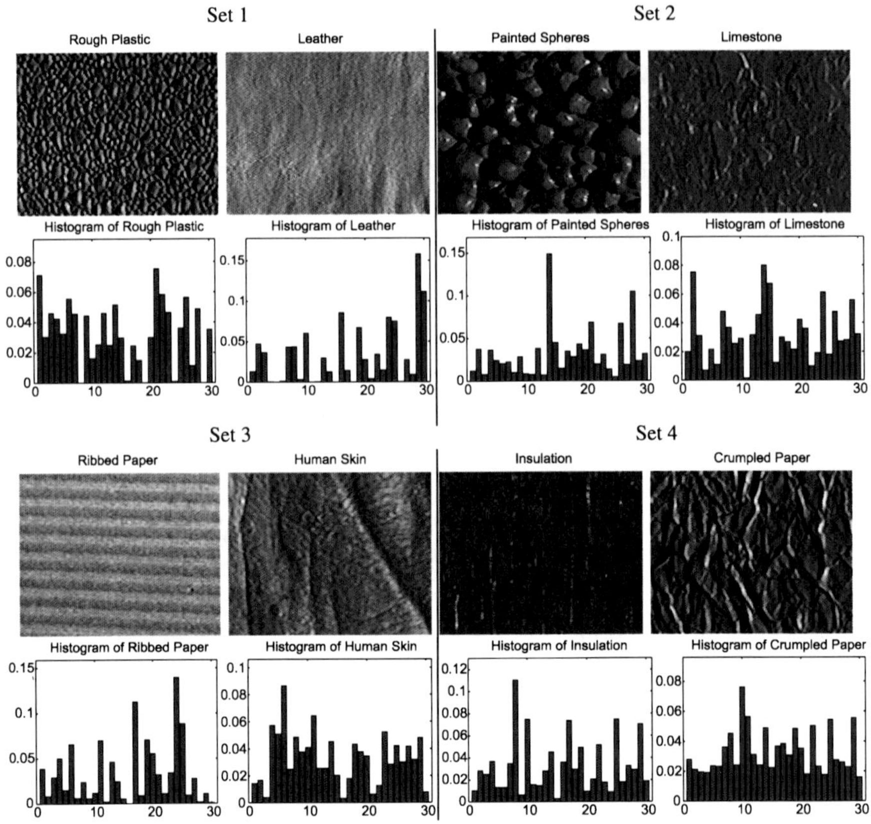

Fig. 6 Textures and corresponding histograms used in the two-class clustering experiments [20]

Table 2 Misclustering rates in % for two-class segmentation

Algorithm	Set 1	Set 2	Set 3	Set 4
Riemannian LLE	0	0	1.63	0
Riemannian LE	0	0	19.68	22.9

6 Conclusion and Open Research Problems

An algorithm for simultaneous NLDR and clustering of data sampled from multiple submanifolds of a Riemannian manifold is presented. We focused our investigation on the three NLDR algorithms, which computes a low-dimensional embedding from the eigenvectors of a matrix M that depends on the local properties of the data. It is important for the user to note that, it is possible that these algorithms become degenerate if the construction of a matrix M, which captures the local geometry of the data, is done in the presence of linear manifolds. Presently, there are several open research problems. First of all, notice that the various Riemmanian operations

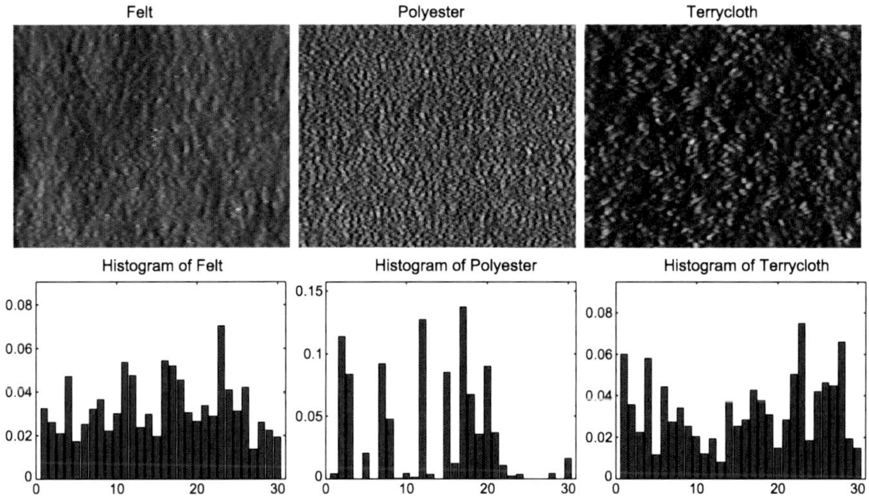

Fig. 7 Textures and corresponding histograms used in the three-class clustering experiment [20]

are assumed to be known and closed-form. This is not true for generic cases and there is a question of how one would be able to perform a similar analysis when the Riemannian operations need to be solved in an iterative manner. The next open problem that future research efforts are focusing on is to address the assumption that data lying on different manifolds do not intersect each other. Finally, there is also a need to construct a large-scale version of the algorithms presented in order to handle large datasets.

References

1. Agarwal, S., Lim, J., Zelnik-Manor, L., Perona, P., Kriegman, D., Belongie, S.: Beyond pairwise clustering. In: Computer Vision and Pattern Recognition, vol. 2, pp. 838–845 (2005)
2. Amari, S.: Differential-Geometrical Methods in Statistics. Springer, Berlin (1985)
3. Arsigny, V., Fillard, P., Pennec, X., Ayache, N.: Log-Euclidean metrics for fast and simple calculus on diffusion tensors. Magn. Reson. Med. **56**, 411–421 (2006)
4. Barbará, D., Chen, P.: Using the fractal dimension to cluster datasets. In: KDD '00: Proceedings of the 6th ACM SIGKDD International Conference on Knowledge Discovery and Data Mining, pp. 260–264. ACM, New York (2000)
5. Belkin, M., Niyogi, P.: Laplacian eigenmaps and spectral techniques for embedding and clustering. In: Advances in Neural Information Processing Systems, pp. 585–591. MIT Press, Cambridge (2002)
6. Belkin, M., Niyogi, P.: Laplacian eigenmaps for dimensionality reduction and data representation. Neural Comput. **15**(6), 1373–1396 (2003)
7. Brand, M., Huang, K.: A unifying theorem for spectral embedding and clustering. In: International Workshop on Artificial Intelligence and Statistics (2003)
8. Burges, C.: Geometric methods for feature extraction and dimensional reduction—a guided tour. In: The Data Mining and Knowledge Discovery Handbook, pp. 59–92. Kluwer Academic, Norwell (2005)

9. Cencov, N.N.: Statistical decision rules and optimal inference. In: Translations of Mathematical Monographs, vol. 53. AMS, Providence (1982)
10. Chen, G., Lerman, G.: Spectral curvature clustering, SCC. Int. J. Comput. Vis. **81**(3), 317–330 (2009)
11. Cox, T.F., Cox, M.A.A.: Multidimensional Scaling. Chapman & Hall, London (1994)
12. Cremers, D., Soatto, S.: Motion competition: a variational framework for piecewise parametric motion segmentation. Int. J. Comput. Vis. **62**(3), 249–265 (2005)
13. do Carmo, M.P.: Riemannian Geometry. Birkhäuser, Boston (1992)
14. Donoho, D., Grimes, C.: Hessian eigenmaps: locally linear embedding techniques for high-dimensional data. Natl. Acad. Sci. **100**(10), 5591–5596 (2003)
15. Fletcher, P.T., Joshi, S.: Riemannian geometry for the statistical analysis of diffusion tensor data. Signal Process. **87**(2), 250–262 (2007)
16. Frechet, M.: Les elements aleatoires de nature quelconque dans un espace distancie. Ann. Inst. Henri Poincare **10**, 235–310 (1948)
17. Gionis, A., Hinneburg, A., Papadimitriou, S., Tsaparas, P.: Dimension induced clustering. In: KDD '05: Proceeding of the 11th ACM SIGKDD International Conference on Knowledge Discovery in Data Mining, pp. 51–60. ACM, New York (2005)
18. Goh, A., Vidal, R.: Segmenting motions of different types by unsupervised manifold clustering. In: Conference on Computer Vision and Pattern Recognition (2007)
19. Goh, A., Vidal, R.: Segmenting fiber bundles in diffusion tensor images. In: European Conference on Computer Vision (2008)
20. Goh, A., Vidal, R.: Unsupervised Riemannian clustering of probability density functions. In: European Conference on Machine Learning (2008)
21. Goh, A., Vidal, R.: Clustering and dimensionality reduction on Riemannian manifolds. In: Conference on Computer Vision and Pattern Recognition, pp. 238–250 (2008)
22. Govindu, V.: A tensor decomposition for geometric grouping and segmentation. In: Computer Vision and Pattern Recognition, vol. 1, pp. 1150–1157 (2005)
23. Ham, J., Lee, D.D., Mika, S., Schölkopf, B.: A kernel view of the dimensionality reduction of manifolds. In: International Conference on Machine learning, vol. 69, p. 47 (2004)
24. Haro, G., Randall, G., Sapiro, G.: Translated poisson mixture model for stratification learning. Int. J. Comput. Vis. **80**, 358–374 (2008)
25. Ho, J., Yang, M.H., Lim, J., Lee, K.C., Kriegman, D.: Clustering appearances of objects under varying illumination conditions. In: IEEE Conference on Computer Vision and Pattern Recognition, vol. 1, pp. 11–18 (2003)
26. Hotelling, H.: Analysis of a complex of statistical variables into principal components. J. Educ. Psychol. **24**, 417–441 (1933)
27. Karcher, H.: Riemannian center of mass and mollifier smoothing. Commun. Pure Appl. Math. **30**(5), 509–541 (1977)
28. Kindlmann, G., Estepar, R.S.J., Niethammer, M., Haker, S., Westin, C.F.: Geodesic-loxodromes for diffusion tensor interpolation and difference measurement. In: Medical Image Computing and Computer-Assisted Intervention (2007)
29. Levina, E., Bickel, P.J.: Maximum likelihood estimation of intrinsic dimension. In: NIPS (2004)
30. Melonakos, J., Mohan, V., Niethammer, M., Smith, K., Kubicki, M., Tannenbaum, A.: Finsler tractography for white matter connectivity analysis of the cingulum bundle. In: Medical Image Computing and Computer-Assisted Intervention (2007)
31. Mordohai, P., Medioni, G.G.: Unsupervised dimensionality estimation and manifold learning in high-dimensional spaces by tensor voting. In: International Joint Conference on Artificial Intelligence, pp. 798–803 (2005)
32. Pennec, X., Fillard, P., Ayache, N.: A Riemannian framework for tensor computing. Int. J. Comput. Vis. **66**(1), 41–46 (2006)
33. Polito, M., Perona, P.: Grouping and dimensionality reduction by locally linear embedding. In: Advances in Neural Information Processing Systems. MIT Press, Cambridge (2002)
34. Rao, C.R.: Information and accuracy attainable in the estimation of statistical parameters. Bull. Calcutta Math. Soc. **37**, 81–89 (1945)

35. Roweis, S., Saul, L.: Nonlinear dimensionality reduction by locally linear embedding. Science **290**(5500), 2323–2326 (2000)
36. Roweis, S., Saul, L.: Think globally, fit locally: unsupervised learning of low dimensional manifolds. J. Mach. Learn. Res. **4**, 119–155 (2003)
37. Schmid, C.: Constructing models for content-based image retrieval. In: IEEE Conference on Computer Vision and Pattern Recognition (2001)
38. Schölkopf, B., Smola, A.: Learning with Kernels. MIT Press, Cambridge (2002)
39. Sha, F., Saul, L.: Analysis and extension of spectral methods for nonlinear dimensionality reduction. In: International Conference on Machine learning, pp. 784–791 (2005)
40. Souvenir, R., Pless, R.: Manifold clustering. In: IEEE International Conference on Computer Vision, vol. I, pp. 648–653 (2005)
41. Srivastava, A., Jermyn, I., Joshi, S.: Riemannian analysis of probability density functions with applications in vision. In: IEEE Conference on Computer Vision and Pattern Recognition (2007)
42. Tenenbaum, J.B., de Silva, V., Langford, J.C.: A global geometric framework for nonlinear dimensionality reduction. Science **290**(5500), 2319–2323 (2000)
43. Tipping, M., Bishop, C.: Mixtures of probabilistic principal component analyzers. Neural Comput. **11**(2), 443–482 (1999)
44. Varma, M., Zisserman, A.: A statistical approach to texture classification from single images. Int. J. Comput. Vis. **62**(1–2), 61–81 (2005)
45. Vidal, R., Ma, Y., Sastry, S.: Generalized principal component analysis, GPCA. IEEE Trans. Pattern Anal. Mach. Intell. **27**(12), 1–15 (2005)
46. Wang, Z., Vemuri, B.: An affine invariant tensor dissimilarity measure and its applications to tensor-valued image segmentation. In: IEEE Conference on Computer Vision and Pattern Recognition, vol. 1, pp. 228–233 (2004)
47. Weinberger, K.Q., Saul, L.: Unsupervised learning of image manifolds by semidefinite programming. In: IEEE Conference on Computer Vision and Pattern Recognition, pp. 988–955 (2004)
48. Wright, J., Ma, Y.: Dense error correction via l1-minimization. IEEE Trans. Inf. Theory **56**(7), 3540–3560 (2010). doi:10.1109/TIT.2010.2048473
49. Yan, J., Pollefeys, M.: A factorization approach to articulated motion recovery. In: IEEE Conference on Computer Vision and Pattern Recognition, vol. II, pp. 815–821 (2005)
50. Zhang, Z., Zha, H.: Principal manifolds and nonlinear dimensionality reduction via tangent space alignment. SIAM J. Sci. Comput. **26**(1), 313–338 (2005)

Manifold Learning for Multi-dimensional Auto-regressive Dynamical Models

Fabio Cuzzolin

Abstract We present a general differential-geometric framework for learning distance functions for dynamical models. Given a training set of models, the optimal metric is selected among a family of pullback metrics induced by the Fisher information tensor through a parameterized automorphism. The problem of classifying motions, encoded as dynamical models of a certain class, can then be posed on the learnt manifold. In particular, we consider the class of multidimensional autoregressive models of order 2. Experimental results concerning identity recognition are shown that prove how such optimal pullback Fisher metrics greatly improve classification performances.

1 Introduction

Manifold learning [2, 5, 12, 30, 34, 39] has become a popular topic in machine learning and computer vision in the last few years, as many objects of interests (like natural images, or sequences representing walking persons), in spite of their apparent high dimensionality, live in a nonlinear space of usually limited dimension. Many unsupervised algorithms (e.g., locally linear embedding [29]) take an input dataset and embed it into some other space, implicitly learning a metric. Extensions learning a full metric for the whole input space have been recently formulated [4]. Extremely interesting is the case in which the objects of interest (whose lower dimensional manifold we would like to identify) are not simple collections of real numbers (vectors), but are complex structures such as, for instance, graphs, trees, or graphical models. The branch of machine learning dealing with the representation and classification of such complex objects goes under the name of "structured learning", and has achieved widespread popularity in the last few years.

F. Cuzzolin (✉)
Department of Computing, Oxford Brookes University, Oxford, OX33 1HX, UK
e-mail: fabio.cuzzolin@brookes.ac.uk

L. Wang et al. (eds.), *Machine Learning for Vision-Based Motion Analysis*,
Advances in Pattern Recognition,
DOI 10.1007/978-0-85729-057-1_3,

In particular, videos or image sequences are often represented as realizations of some sort of dynamical model, either stochastic (e.g., HMMs) or deterministic (e.g., ARMA). Such an approach has proven to be effective in problems such as video coding, action recognition, or identity recognition from gait [33].

Recognizing human activities from video is indeed a significant example of such applications. Even though the formulation of the problem is simple and intuitive, activity recognition is a much harder problem than it may look. Motions inherently possess an extremely high degree of variability. Methods which neglect to take into account action dynamics for recognition have recently proven very effective. Typically, these approaches extract spatiotemporal features from the 3D volume associated with a video [19, 38].

However, encoding the dynamics of videos or image sequences by means of some sort of dynamical model can be useful in situations in which the dynamics is critically discriminative. Chaudry et al. [8], for instance, have used nonlinear dynamical systems (NLDS) to model times series of histograms of oriented optical flow, measuring distances between NLDS by means of Cauchy kernels. Wang and Mori [35], instead, have proposed to use sophisticated max-margin conditional random fields to address locality by recognizing actions modeled as constellations of local motion patterns. Generally speaking, dynamical representations are very effective in coping with time warping or helping with the crucial issue of action segmentation. Dynamic textures [11] based on deterministic linear dynamical systems (LDS) have proven to be effective in video coding.

In all these scenarios, action (or identity) recognition reduces to classifying dynamical models. One way of doing this is to learn a new model for each test image sequence, measure its distance from the old models, and attribute to the new sequence the label of the closest model. A number of distance functions between linear systems have indeed been introduced [6, 23, 32], and a vast literature about dissimilarity measures between Markov models also exists [10, 36], mostly about variants of the Kullback–Leibler divergence [20].

Consider, though, the general problem of classifying a dynamical model (as the representative of an input image sequence). Since models (or sequences) can be endowed with different labeling, while maintaining the same geometrical structure, no single distance function can possibly outperform all the others in each and every classification problem. A reasonable approach, when possessing some a-priori information, consists therefore in trying to *learn* in a supervised fashion the "best" distance function for a specific classification problem [2, 5, 12, 30, 34, 39]. A natural optimization criterion seeks to maximize the classification performance achieved by means of the learnt metric. Efforts have been done in this sense in the linear case [31, 37]. However, as even linear dynamical models live in a nonlinear space, the need for a principled way of learning Riemannian metrics from such data naturally arises.

Such a tool is provided by the formalism of *pullback metrics*. If the models belong to a Riemannian manifold $\mathcal{M}$, any diffeomorphism of $\mathcal{M}$ onto itself (or *automorphism*) induces such a metric on $\mathcal{M}$. By designing a suitable parameterized family of automorphisms, we obtain a family of pullback metrics on $\mathcal{M}$ that we can optimize upon.

In this chapter, we propose a general framework for learning the optimal pullback metric for a data-set D of dynamical models. Assume each input observation sequence is mapped to a model of a certain class by parameter identification. If such models belong to a Riemannian manifold (for instance endowed with the Fisher metric [1]), we can design a parametric family of automorphisms which induce a family of pullback metrics. If the training set of models is labeled, we can then find the parameter of the metric which optimizes classification performance by cross-validation [7]. Otherwise, the metric which optimizes some purely geometric objective function can be sought (like, for instance, the inverse volume of the manifold around the data-points in D [21]).

In particular, we consider here the class $\mathcal{AR}(2)$ of multidimensional autoregressive models of order 2. We study the Riemannian structure of their manifold, and design a number of automorphisms inducing families of parameterized pullback metrics on $\mathcal{AR}(2)$. We apply this framework to identity recognition from gait. We use the video sequences of the Mobo database [15] to prove that classifiers based on an optimal pullback Fisher metric between stochastic models significantly improve classification scores with respect to what obtained by standard, a-priori distance functions.

2 Learning Pullback Metrics for Linear Models

Let us suppose a data-set of dynamical models is available. Suppose also that such models live on a Riemannian manifold $\mathcal{M}$ of some sort, that is, a Riemannian *metric* is defined in any point of the manifold. Any automorphism (a differentiable map) from $\mathcal{M}$ to itself induces a new metric, called "pullback metric". The use of pullback metrics has been recently proposed by Lebanon [21] in the context of document retrieval. However, pullback metrics are a well studied notion of differential geometry [17], which has found several applications in computer vision [18].

2.1 Pullback Metrics

Formally, consider a family of automorphisms between the Riemannian manifold $\mathcal{M}$ in which the data-set $D = \{m_1, \dots, m_N\} \subset \mathcal{M}$ resides and itself:

$$\begin{aligned} F_p : \mathcal{M} &\to \mathcal{M} \\ m \in \mathcal{M} &\mapsto F_p(m). \end{aligned}$$

Let us denote by $T_m\mathcal{M}$ the tangent space to $\mathcal{M}$ in m. Any such automorphism F is associated with a "push-forward" map

$$\begin{aligned} F_* : T_m\mathcal{M} &\to T_{F(m)}\mathcal{M} \\ v \in T_m\mathcal{M} &\mapsto F_* v \in T_{F(m)}\mathcal{M} \end{aligned}$$

defined as $F_* v(f) = v(f \circ F)$ for all smooth functions f on $\mathcal{M}$ (see Fig. 1). Consider now a Riemannian metric

$$g : T\mathcal{M} \times T\mathcal{M} \to \mathbb{R}$$

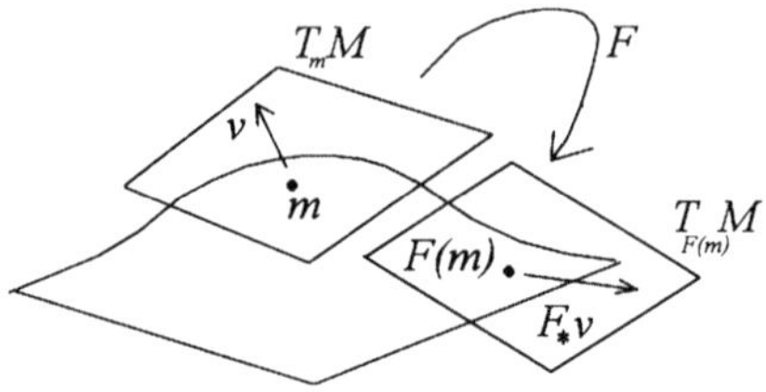

Fig. 1 The push-forward map F_* associated with an automorphism F on a Riemannian manifold $\mathcal{M}$

on $\mathcal{M}$. Roughly speaking, g determines how to compute scalar products of tangent vectors $v \in T_m\mathcal{M}$. The map F induces a *pullback* metric g_* on $\mathcal{M}$:

$$g_{*m}(u, v) \doteq g_{F(m)}(F_*u, F_*v). \tag{1}$$

The scalar product of two vectors u, v of $T_m\mathcal{M}$ according to g_* is computed as the scalar product with respect to the *original* metric g of the images F_*u, F_*v of the vectors u, v under the push-forward map F_*. The pullback geodesic between any two points m_1, m_2 of the manifold $\mathcal{M}$ is the geodesic connecting their images with respect to the original metric. If we manage to define an entire class of such automorphisms depending on some parameter λ, we get a corresponding family of pullback metrics on $\mathcal{M}$, also depending on λ. We can then define an optimization problem over such a family in order to select an "optimal" metric, which in turn determines the desired manifold. The nature of this manifold will depend on the objective function we choose to optimize.

2.2 Fisher Metric for Linear Models

To apply the pullback metric framework to dynamical models, we first need to define a structure of Riemannian manifold on them. The study of the geometrical structure of the space formed by a family of probability distribution is due to Rao, and has been developed by Nagaoka and Amari [1]. A family S of probability distributions $p(x, \xi)$ depending on a n-dimensional parameter ξ can be regarded in fact as an n-dimensional manifold. If the Fisher information matrix

$$g_{ij} \doteq E\left[\frac{\partial \log p(x,\xi)}{\partial \xi_i}\frac{\partial \log p(x,\xi)}{\partial \xi_j}\right]$$

is nondegenerate, $G = [g_{ij}]$ is a Riemannian metric, and S is a Riemannian manifold. The Fisher information matrix for several manifolds of linear MIMO systems has been computed in [16].

2.3 General Framework

As linear dynamical models do live in a Riemannian space, we can apply to them the pullback metric formalism and obtain a family of metrics on which to optimize. Itoh

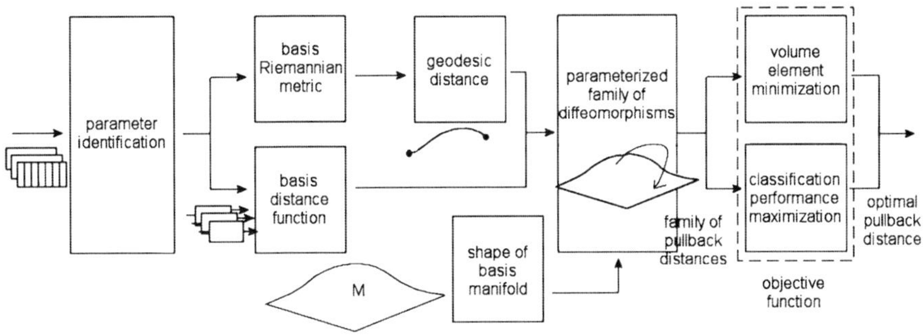

Fig. 2 A bird's eye view of the proposed framework for learning pullback metrics on dynamical models

et al. [17] have indeed recently done some work on pullbacks of Fisher information metrics.

Here, we design a general framework for learning an optimal pullback metric from a training set of dynamical models, as depicted in Fig. 2.

1. A data-set Y of observation sequences $\{y_k = [y_k(t), t = 1, \ldots, T_k]$, $k = 1, \ldots, N\}$ of variable length T_k is available.
2. From each sequence, a dynamical model m_i of a certain class $\mathcal{C}$ is estimated by parameter identification, yielding a data-set of such models $D = \{m_1, \ldots, m_N\}$.
3. Models of a certain class $\mathcal{C}$ belong to a manifold $\mathcal{M}_\mathcal{C}$; its atlas of coordinate charts has to be known.[1]
4. To measure distances between pairs of models on the manifold $\mathcal{M}_\mathcal{C}$, either a distance function $d_\mathcal{M}$ or a proper Riemannian metric $g_\mathcal{M}$ have to be defined on it.
5. In the case of a Riemannian metric, it is necessary to know the geodesic path between two models in order to compute the associated geodesic distance.
6. A family F_λ of automorphisms from $\mathcal{M}_\mathcal{C}$ onto itself, parameterized by a vector λ, is designed to provide a search space of metrics from which to select the optimal one.
7. F_λ induces a family of pullback metrics (1) $g_{*\lambda\mathcal{M}}$ or distances $d_{*\lambda\mathcal{M}}$ on $\mathcal{M}$, respectively.
8. We optimize over this family of pullback distances/metrics in order to find the optimal such function/metric, according to some objective function.
9. This yields an optimal pullback metric $\hat{g}_*$ or distance function $\hat{d}_*$.
10. In the metric case, knowing the geodesics of $\mathcal{M}$ suffices to compute the geodesic distances on $\mathcal{M}$ based on $\hat{g}_*$.
11. The optimal distance function can finally be used to cluster or classify new "test" models/sequences.

[1] In the case considered here, a single coordinate chart actually spans the whole manifold.

2.4 Objective Functions: Classification Performance and Inverse Volume

When the data-set of models is *labeled*, we can exploit this information to determine the optimal metric/distance function. In particular, we can use cross-validation [7] to optimize its classification performance by dividing the overall sample into a number v of folds. The models belonging to $v-1$ folds are used as training sample, the remaining fold as testing sample, and the parameter of the pullback metric which optimizes the correct classification rate on fold v given the training set is selected. As the classification score is hard to describe analytically, we can extract a number of random samples from the parameter space and pick the maximal performance sample.

When the training set is *unlabeled*, manifold learning has to be based on purely geometrical considerations. Lebanon [21] has recently suggested in the context of document retrieval an approach that seeks to *maximize the inverse volume element* associated with a metric around the given training set of points [24]:

$$\mathcal{O}(D) = \prod_{k=1}^{N} \frac{(\det g(m_k))^{-\frac{1}{2}}}{\int_{\mathcal{M}} (\det g(m))^{-\frac{1}{2}}\, dm}, \tag{2}$$

where $g(m_k)$ denotes the Riemannian metric in the point m_k of the data-set D living on a Riemannian manifold $\mathcal{M}$. This amounts to finding a lower dimensional representation of the dataset, in a similar fashion to LLE [29] or Laplacian eigenmaps [3], where dimensionality reduction is often considered a factor in improving classification.

The computation of (2) requires that of the *Gramian* $\det g$. To find the expression of the Gramian associated with a pullback metric (1), we first need to choose a base of the space $T_m\mathcal{M}$ tangent to $\mathcal{M}$ in m. Let us denote by $\{\partial_i, i = 1, \dots, \dim \mathcal{M}\}$ the base of $T_m\mathcal{M}$. The push-forward of the vectors of this base yields a base for $T_{F(m)}\mathcal{M}$. By definition, the push-forward $F_{*\lambda}$ of a vector $v \in T_m\mathcal{M}$ under a an automorphism F_λ with parameter λ is given by [21]:

$$F_{*\lambda}(v) \doteq \frac{d}{dt} F_\lambda(m + t \cdot v)\bigg|_{t=0}, \qquad v \in T_m\mathcal{M}.$$

The automorphism F_λ induces a base for the space of vector fields on $\mathcal{M}$, $w_i \doteq \{F_{*\lambda}(\partial_i)\}$, for $i = 1, \dots, \dim \mathcal{M}$. We can rearrange these vectors as rows of a matrix:

$$J = [w_1; \dots; w_{\dim \mathcal{M}}].$$

The volume element of the pullback metric g_* in a point $m \in \mathcal{M}$ is the determinant of the Gramian [21]: $\det g_*(m) \doteq \det[g(F_{*\lambda}(\partial_i), F_{*\lambda}(\partial_j))]_{ij} = \det(J^T g J)$. If J is a square matrix (as in the rest of this chapter), we get simply:

$$\det g_*(m) = \det(J)^2 \cdot \det g(m). \tag{3}$$

After plugging (3) into (2), we obtain the function to minimize.

3 Pullback Metrics for Multidimensional Autoregressive Models

In virtue of their importance as a class of dynamical models, and their relative simplicity, we consider here the class of stable autoregressive discrete-time processes of order 2, $\mathcal{M} = \mathcal{AR}(2)$, in a stochastic setting in which the input signal is a Gaussian white noise with zero mean and unit variance.

3.1 *The Basis Manifold*

This set can be given a Riemannian manifold structure under Fisher metric. A natural parametrization uses as coordinates the nonunit coefficients (a_1, a_2) of the denominator of the transfer function:

$$h(z) = \frac{z^2}{z^2 + a_1 z + a_2} \tag{4}$$

which corresponds to the AR difference: $y(k) = -a_1 y(k-1) - a_2 y(k-2)$.

The basis manifold $\mathcal{M}$ and the associated Fisher metric *in the scalar case* have been studied in the context of control theory [25, 27]. We build here on these results to determine a coordinate chart and a product Fisher metric on the manifold $\mathcal{AR}(2, p)$ of *p-dimensional* AR models. We will then be able to design two different families of automorphisms on $\mathcal{AR}(2, p)$, and use the framework of Sect. 2 to determine there two families of pullback distance functions.

3.1.1 The Basis Manifold $\mathcal{AR}(2, 1)$ in the Scalar Case

Let us consider first the scalar case $p = 1$ of a single output channel. To impose the stability of the transfer function (4), the necessary conditions are $1 + a_1 + a_2 > 0$, $1 - a_1 + a_2 > 0$, and $1 - a_2 > 0$. The manifold of stable AR(2, 1) systems is then composed by a single connected component (see Fig. 3, left).

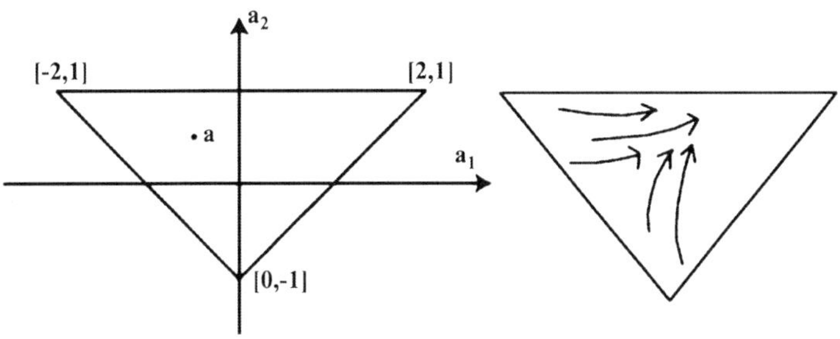

Fig. 3 The manifold of stable scalar autoregressive systems of order 2, $\mathcal{AR}(2, 1)$, with the nonunit coefficients of the denominator of $h(z)$ as parameters. It forms a simplex with vertices $[-2, 1], [2, 1], [0, -1]$. *Right*: effect of an automorphism of the form (10) on the $\mathcal{AR}(2, 1)$ simplex

The Fisher tensor is [25]:

$$g(a_1, a_2) = \frac{1}{(1 + a_1 + a_2)(1 - a_1 + a_2)(1 - a_2)} \begin{pmatrix} 1 + a_2 & a_1 \\ a_1 & 1 + a_2 \end{pmatrix} \tag{5}$$

with volume element

$$\det g_{\mathcal{AR}(2,1)}(a_1, a_2) = \frac{1}{(1 - a_2)^2(1 - a_1 + a_2)(1 + a_1 + a_2)}. \tag{6}$$

3.1.2 The Multidimensional Case

In the multidimensional case, an autoregressive model is composed of p separate channels, each characterized by a transfer function

$$h^i(z) = \frac{z^2}{z^2 + a_1^i z + a_2^i}$$

(assuming their independence). When using two coefficients a_1^i, a_2^i to describe each channel, each p-dimensional AR system has coordinates $\mathbf{a} = [a_1^i, a_2^i : i = 1, \dots, p]'$. $\mathcal{AR}(2, p)$ is therefore the *product manifold*

$$\mathcal{AR}(2, p) = \times_{i=1}^{p} \mathcal{AR}(2, 1), \tag{7}$$

that is, the Cartesian product of the manifolds associated with the individual channels. As a Cartesian product of a number of simplices (triangles), $\mathcal{AR}(2, p)$ turns out to be a *polytope* in $\mathbb{R}^{2p}$. Such polytope has in general $n_1 \times \cdots \times n_p$ vertices, the product of the number of vertices of the individual simplices. In our case, $\mathcal{AR}(2, p)$ is a polytope with 3^p vertices:

$$\mathcal{AR}(2, p) = Cl(\mathbf{v}_{i_1,\dots,i_p}, i_j = 1, \dots, 3\ \forall j = 1, \dots, p).$$

Each p-dimensional AR system $\mathbf{a}$ also possesses, therefore, a vector of simplicial coordinates *in the polytope* $\mathcal{AR}(2, p)$:

$$\mathbf{m} = [m_{i_1,\dots,i_p}, i_j = 1, \dots, 3\ \forall j = 1, \dots, p]' \tag{8}$$

such that

$$\mathbf{a} = \sum_{i_j=1, j=1,\dots,p}^{3} m_{i_1,\dots,i_p} \mathbf{v}_{i_1,\dots,i_p}.$$

3.1.3 Product Metric

On the Cartesian product $\mathcal{M}_1 \times \mathcal{M}_2$ of two Riemannian manifolds with metrics $g_p^{\mathcal{M}_1}$ and $g_q^{\mathcal{M}_2}$, respectively, one can define the *product metric* on $\mathcal{M}_1 \times \mathcal{M}_2$ as

$$g_{(p,q)}^{\mathcal{M}_1 \times \mathcal{M}_2} : T_{(p,q)}(\mathcal{M}_1 \times \mathcal{M}_2) \times T_{(p,q)}(\mathcal{M}_1 \times \mathcal{M}_2),$$

with

$$(u, v) \mapsto g_p^{\mathcal{M}_1}\big(T_{(p,q)}\pi_1(u), T_{(p,q)}\pi_1(v)\big) + g_q^{\mathcal{M}_2}\big(T_{(p,q)}\pi_2(u), T_{(p,q)}\pi_2(v)\big),$$

where $\pi_i : \mathcal{M}_1 \times \mathcal{M}_2 \to \mathcal{M}_i$ is the natural projection of a point of the Cartesian product onto one of the component manifolds. The definition can be extended to any finite number of manifolds. The product metric $g_{\mathcal{AR}(2,p)}$ is described by a $2p \times 2p$ block diagonal matrix, whose p 2×2 blocks are copies of the metric (5) valid in the scalar case:

$$g_{\mathcal{AR}(2,p)}(\mathbf{a}) = \mathrm{diag}\big(g_{\mathcal{AR}(2,1)}(a_1^i, a_2^i)\big).$$

Its volume element $\det g_{\mathcal{AR}(2,p)}$ is (given the expression (6) of the scalar volume element $\det g_{\mathcal{AR}(2,1)}$):

$$\begin{aligned} \det g_{\mathcal{AR}(2,p)}(\mathbf{a}) &= \prod_{i=1}^{p} \det g_{\mathcal{AR}(2,1)}(a_1^i, a_2^i) \\ &= \prod_{i=1}^{p} \frac{1}{(1-a_2^i)^2(1-a_1^i+a_2^i)(1+a_1^i+a_2^i)}. \end{aligned} \tag{9}$$

3.1.4 Geodesics

To compute the distance between two points of a Riemannian manifold (and therefore, in particular, between two dynamical models) the metric is not sufficient. It is necessary to compute (analytically or numerically) the shortest path connecting any such pair of points on the manifold (geodesic). All the geodesics of stable AR(2, 1) systems endowed with the Fisher metric (5) as a function of the Schur parameters $\gamma_1 = a_1/(1+a_2)$, $\gamma_2 = a_2$ have been analytically computed by Rijkeboer [27]:

$$4 \cdot (\ddot{\gamma}_1 + \ddot{\gamma}_2) + \frac{1}{1+(\gamma_2)^2} \cdot \dot{\gamma}_1\dot{\gamma}_2 + \frac{\gamma_2}{1-(\gamma_2)^2} \cdot (\dot{\gamma}_2)^2 - \frac{1}{1+\gamma_1} \cdot (\dot{\gamma}_1)^2 = 0.$$

In the general case AR(2, p), unfortunately, the manifold's geodesics are not analytically known. However, [26] shows the following.

Proposition 1 *The sub-manifolds of a product manifold are geodesic, that is, all geodesic paths on the individual sub-manifolds are geodesics of the product manifold too.*

In our case, as $\mathcal{AR}(2, p)$ is itself a product manifold (7), the (known) geodesics of the "scalar" manifold $\mathcal{AR}(2, 1)$ are also geodesics of $\mathcal{AR}(2, p)$. As an approximation, therefore, we can measure the geodesic distance between two generic p-dimensional autoregressive models by applying a generalization of Phytagoras' theorem $d(\mathbf{a}, \mathbf{a}') = \sqrt{\sum_p d_i(\mathbf{a}, \mathbf{a}')^2}$, where $d_i(\mathbf{a}, \mathbf{a}')$ is the distance of their projections on the ith sub-manifold.

3.2 An Automorphism for the Scalar Case

To build a parameterized family of Riemannian metrics for $\mathcal{AR}(2, p)$, it is necessary to choose a family of automorphisms of the manifold onto itself (Sect. 2). The more sophisticated is the set of automorphisms, the larger is the search space to optimize the metric on.

One possible choice for an automorphism of $\mathcal{AR}(2, 1)$ is suggested by the triangular form of the manifold, which has three vertices (see Fig. 3, left). Let $\mathbf{m} = [m_1, m_2, m_3]'$ collect the "simplicial" coordinates of a system $\mathbf{a} \in \mathcal{AR}(2, 1)$ in the manifold:

$$\mathbf{a} = [a_1, a_2]' = m_1[0, -1]' + m_2[2, 1]' + m_3[-2, 1]', \quad m_1 + m_2 + m_3 = 1.$$

A natural automorphism of a simplex onto itself is given by "stretching" the simplicial coordinates of its point by a set of weights $\lambda = [\lambda_1, \lambda_2, \lambda_3]'$ such that $\sum_j \lambda_j = 1, \lambda_j \geq 0$:

$$F_\lambda(\mathbf{m}) = F_\lambda\big([m_1, m_2, m_3]'\big) = \frac{[\lambda_1 m_1, \lambda_2 m_2, \lambda_3 m_3]'}{\lambda \cdot \mathbf{m}}, \tag{10}$$

where $\lambda \cdot \mathbf{m}$ denotes the scalar product of the two vectors λ, $\mathbf{m}$. The application (10) stretches the triangle towards the vertex with the largest λ_i (Fig. 3, right).

3.3 Product and Global Automorphisms for $\mathcal{AR}(2, p)$

A *product* automorphism for the whole manifold $\mathcal{AR}(2, p)$ of multidimensional, p channel autoregressive models can be obtained by using (10), designed for scalar systems, as a building block. If we denote by $\mathbf{m}^i = [m_1^i, m_2^i, m_3^i]'$ the simplicial coordinates of a system $\mathbf{a}$ in the ith sub-manifold, such a system will be identified by a vector $\mathbf{m} = [\mathbf{m}^i, i = 1, \ldots, p]'$ of $3p$ such coordinates.

The mapping

$$F_{\lambda^i, i=1,\ldots,p}(\mathbf{m}) = \big[F_{\lambda^i}\big(\mathbf{m}^i\big), i = 1, \ldots, p\big]' \tag{11}$$

with $3p$ parameters applies an automorphism (10) with parameter λ^i to the projection of $\mathbf{m}$ onto each sub-manifold.

In alternative, the global geometry of the product manifold $\mathcal{AR}(2, p)$ inspires a *global* automorphism which acts on the manifold as a whole, by multiplying its "polytopial" coordinates (8) by a set of convex weights

$$\mu = [\mu_{i_1,\ldots,i_p}, i_j = 1, \ldots, 3\ \forall j = 1, \ldots, p]'.$$

We obtain, up to normalization:

$$F_\mu(\mathbf{m}) \propto [\mu_{i_1,\ldots,i_p} \cdot m_{i_1,\ldots,i_p}, i_j = 1, \ldots, 3\ \forall j = 1, \ldots, p]'. \tag{12}$$

3.3.1 Volume Element for $\mathcal{AR}(2, p)$ Under Product Automorphism

Assume that the data-set of models is unlabeled. To select an optimal pullback metric for p-dimensional autoregressive models by volume minimization (2), we need to find the analytical expression of the determinant of the Gramian $\det g_{*\lambda}$ (3) as a function of the parameter vector λ. By plugging it into the expression for the inverse volume (2), we obtain the objective function to optimize. We consider here the relatively simpler product diffeomorphism (11).

Notice that, in the product simplicial coordinates $\mathbf{m} = [\mathbf{m}^i, i = 1, \ldots, p]'$ of a system of $\mathcal{AR}(2, p)$, the volume element of the product Fisher metric (9) reads as:

$$\det g_{\mathcal{AR}(2,p)}(\mathbf{m}) = \prod_{i=1}^{p} \frac{1}{(m_1^i)^2 m_2^i m_3^i}.$$

Theorem 1 *The volume element of the Fisher pullback metric on $\mathcal{AR}(2, p)$ induced by the automorphism* (11) *is*:

$$\det g_{*\lambda}(\lambda, \mathbf{m}) = \frac{1}{2^{2p}} \prod_{i=1}^{p} \frac{(\lambda_1^i \lambda_2^i \lambda_3^i)^2}{(\lambda^i \cdot \mathbf{m}^i)^6} \cdot \frac{1}{(m_1^i)^2 m_2^i m_3^i}. \tag{13}$$

Proof We need to compute the Gramian $\det g_{*\lambda}$ (3) of the pullback metric under the automorphism (11). Following the procedure of Sect. 2, we need to choose a basis of the tangent space

$$T\mathcal{AR}(2, p) = T\mathcal{AR}(2, 1) \oplus \cdots \oplus T\mathcal{AR}(2, 1)$$

of the product manifold (7). The size $2p$ vectors

$$\partial_1^i = [0, \ldots, 0, 1/2, 1/2, 0, \ldots, 0]', \qquad \partial_2^i = [0, \ldots, 0, -1/2, 1/2, 0, \ldots, 0]'$$

whose only nonzero entries are in positions $2i - 1$, $2i$, form such a basis. Let us express the product automorphism (11) in coordinates $\mathbf{a} = [a_1^1, a_2^1, \ldots, a_1^p, a_2^p]$. We have that, $\forall i = 1, \ldots, p$:

$$a_1^i = 2(m_2^i - m_3^i), \qquad a_2^i = m_2^i + m_3^i - m_1^i,$$
$$m_2^i = \frac{1 + a_1^i + a_2^i}{4}, \qquad m_3^i = \frac{1 - a_1^i + a_2^i}{4}, \qquad m_1^i = \frac{1 - a_2^i}{2}.$$

It follows that:

$$\begin{aligned} &F_{\lambda^1,\ldots,\lambda^p}(\mathbf{a})[2i-1, 2i] \\ &\quad = \frac{1}{\Delta_i}\big[2\lambda_2^i(1 + a_1^i + a_2^i) - 2\lambda_3(1 - a_1^i + a_2^i), \\ &\qquad\qquad \lambda_2(1 + a_1^i + a_2^i) + \lambda_3(1 - a_1^i + a_2^i) - 2\lambda_1(1 - a_2^i)\big]', \\ &F_{\lambda^1,\ldots,\lambda^p}(\mathbf{a})[h, l] = 0 \quad (h, l) \neq (2i-1, 2i), \end{aligned} \tag{14}$$

where $\Delta_i = 2\lambda_1^i(1 - a_2^i) + \lambda_2^i(1 + a_1^i + a_2^i) + \lambda_3^i(1 - a_1^i + a_2^i)$.

We seek for all channels $i = 1, \dots, p$ the push-forward tangent vectors

$$\mathbf{w}_1^i = \frac{d}{dt} F_{\{\lambda^j, j\}}(\mathbf{a} + t\partial_{\mathbf{1}}^{\mathbf{i}})\big|_{t=0}, \qquad \mathbf{w}_2^i = \frac{d}{dt} F_{\{\lambda^j, j\}}(\mathbf{a} + t\partial_{\mathbf{2}}^{\mathbf{i}})\big|_{t=0}.$$

We get:

$$\begin{aligned} \mathbf{w}_1^i[2i-1, 2i] = \big[&-2\lambda_1^i\lambda_3^i(3 - a_2^i + a_1^i) + 2\lambda_2^i(\lambda_1^i - 2\lambda_3^i)(1 + a_1^i + a_2^i), \\ &2\lambda_1^i\lambda_3^i(3 - a_2^i + a_1^i) + 2\lambda_1^i\lambda_2^i(1 + a_1^i + a_2^i)\big], \\ \mathbf{w}_1^i[h, l] = 0 \quad &(h, l) \neq (2i-1, 2i) \end{aligned} \tag{15}$$

and a similar expression for $\mathbf{w}_2^i$.

The matrix J formed by the stacked collection of the row vectors $\mathbf{w}_{1,2}^i$,

$$J = \big[\mathbf{w}_1^1; \mathbf{w}_2^1; \dots; \mathbf{w}_1^p; \mathbf{w}_2^p\big]$$

is clearly block diagonal. Its determinant is therefore the product of the determinants of the blocks:

$$\det J = \frac{1}{2^p} \prod_{i=1}^{p} \frac{\lambda_1^i \lambda_2^i \lambda_3^i}{(\lambda^i \cdot \mathbf{m}^i)^3}.$$

By plugging the expressions for $\det J$ and $\det g_{\mathcal{AR}(2,p)}$ into that (3) of the pullback volume element, we get (13). □

The function to maximize is finally obtained by plugging (13) in the general expression (2). The normalization factor $I(\lambda) = \int_{\mathcal{M}} (\det g_{*\lambda}(m))^{-\frac{1}{2}}\, dm$ can be approximated as:

$$I(\lambda) \simeq \sum_{k=1}^{N} \det g_{*\lambda}(\lambda, \mathbf{m}_k)^{-\frac{1}{2}}.$$

In the labeled case, instead, we find the optimal parameters λ by optimizing the classification performance on the available data-set by cross-validation.

4 Tests on Identity Recognition

To test the actual, empirical effect of our approach to manifold learning on the classification of dynamical models, we considered the problem of recognizing actions and identities from image sequences. We used the Mobo database [15], a collection of 600 image sequences of 25 people walking on a treadmill in four different variants (slow walk, fast walk, walk on a slope, walk carrying a ball), seen from 6 different viewpoints (Fig. 4). We selected all the sequences associated with the gaits "slow walk" and "walking on inclined slope", simulating this way the impact of nuisance factors actually present in gait identification, and making recognition more challenging.

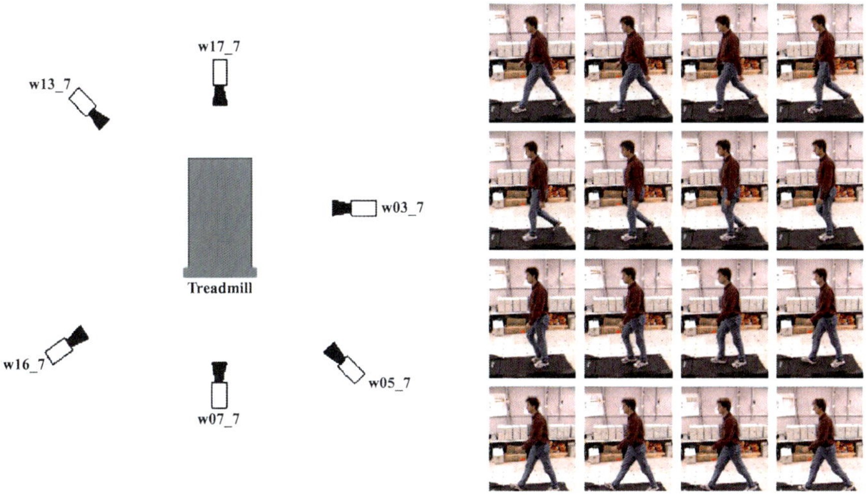

Fig. 4 *Left*: location and orientation of the six cameras in the Mobo experiment (the origin of the frame is roughly in the position of the walking person on the treadmill). *Right*: a sample image sequence from the Mobo database

4.1 Feature Representation

As the Mobo database comes with preprocessed silhouettes attached to each image, we decided to use silhouette based-features to represent images. In gait, ID images are usually preprocessed in order to extract the silhouettes of the walking person. However, this is by no means a limitation of the proposed approach. Indeed, more sophisticated 3D pose estimation methods could be used to run tests on the 3D setup [28]. We plan to run such tests in the near future.

We chose a simple but effective way of computing feature measurements for each frame. For each silhouette, we detected its center of mass, rescaled it to the corresponding bounding box, and projected its contours on to one or more lines passing through its barycenter (see Fig. 5). We favored this approach after testing a number of other different representations: the principal axes of the body-parts as they appear in the image [22], size functions [14], and a PCA-based representation of the contours. All turned out to be rather unstable.

4.2 Identification of a AR(2, p) Model for Each Sequence

According to the scheme of Fig. 2 each input sequence has to be represented by a dynamical model, in particular, an autoregressive model of order 2. Each component of the feature/observation vector, then, is associated with a different output channel of the AR(2, p) model (4). We used the Matlab routine $M = \text{ARX(DATA, ORDERS)}$ to identify by least-squares optimization the parameters a_1^i, a_2^i for each output

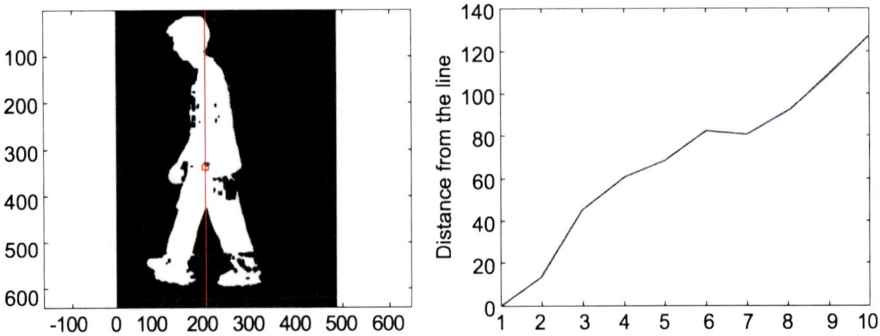

Fig. 5 Feature extraction. First, a number of lines passing through the center of mass of the silhouette are selected. Then for each such line the distance of the points on the contour of the silhouette from it is computed (here the segment is subdivided into 10 intervals). The collection of all such distance values for all lines eventually forms the feature vector representing the image

channel $i = 1, \ldots, p$. The parameters of the ARX model structure $A(q)y(t) = B(q)u(t - nk)e(t)$ are estimated using the classical least squares method. Both time and frequency-domain signals are supported. The only parameters are the orders of the different output channels: we invoked the routine as follows `seqmodel = arx(data,'na',2*eye(p,p));`, setting the order of each channel at 2.

For comparison, for each input sequence we also identified a hidden Markov model [13] by applying the Expectation Maximization [9] algorithm.

A *hidden Markov model* is a statistical model whose states $\{X_k\}$ form a *Markov chain*, and in which only a corrupted version $y_k \in \mathbb{R}^D$ of the state ("observation process") is observable. If we associate its n states with versors $e_i = [0, \ldots, 0, 1, 0, \ldots, 0]' \in \mathbb{R}^n$ [13], we can write the model[2] as

$$\begin{cases} X_{k+1} = AX_k + V_{k+1}, \\ y_{k+1} = CX_k + \mathrm{diag}(W_{k+1})\Sigma X_k. \end{cases} \tag{16}$$

Given a state $X_k = e_j$, the observations are assumed to have Gaussian distribution $p(y_{k+1}|X_k = e_j)$ centered around a vector $C_j = E[p(y_{k+1}|X_k = e_j)]$ which is the jth column of the matrix C. The parameters of a hidden Markov model (16) are therefore the *transition matrix* $A = [a_{ij}] = P(X_{k+1} = e_i|X_k = e_j)$, the matrix C collecting the means C_j of the state-output distributions $p(y_{k+1}|X_k = e_j)$, and the matrix Σ of their variances. Given a sequence of observations $\{y_1, \ldots, y_T\}$ they can be identified by means of the EM algorithm [9, 13].

Numerous distance functions between Markov models are offered in the literature. A classical pseudo-distance in the space of HMMs is derived from the Kullback–Leibler divergence [20] of two probability distributions. Arguably the

---

[2]Here $\{V_{k+1}\}$ is a sequence of martingale increments and $\{W_{k+1}\}$ is a sequence of i.i.d. Gaussian noises $\mathcal{N}(0, 1)$.

simplest possible choice is to pick the product metric obtained by applying the Frobenius norm to A and C matrices, respectively:

$$\|H_1 - H_2\| = \|A_1 - A_2\|_F + \|C_1 - C_2\|_F, \tag{17}$$

where $\|M - M'\|_F \doteq \sqrt{\mathrm{Tr}((M - M')(M - M')^T)}$, and $\mathrm{Tr}(M) = \sum_{ii} M[i,i]$ is the trace. The Frobenius norm is inexpensive to compute, and often produces surprisingly good classification results.

4.3 Performances of Optimal Pullback Metrics

To classify the test models, we adopted standard nearest neighbor classification: each testing sequence is attributed the label of the closest model in the training set. Note that *by no means* this is a limitation of the proposed approach: *any* advanced classification technique (Gaussian kernels, SVMs, etc.) can be used in cascade to our metric learning procedure. The classification performance was measured as the percentage of correctly classified sequences. For each run, we randomly selected a training and a testing set in the database.

Figure 6 plots the average classification performance (over 10 runs) of the following metrics on the space of autoregressive models of order 2 with p outputs: 1—product Fisher metric $g_{\mathcal{AR}(2,p)}$; 2—pullback Fisher metric induced by the product automorphism (11) optimizing the classification performance after cross-validation on the training set; 3—pullback Fisher metric induced by the *global* automorphism (12) for the same objective function; 4—pullback Fisher metric induced by (11) with optimal inverse volume; 5—Frobenius distance between HMMs.

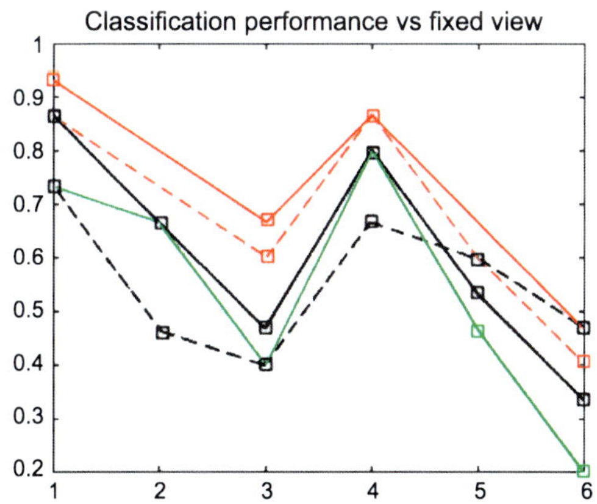

Fig. 6 In six separate experiments, the classification performance of the following metrics has been computed for image sequences coming from a single view, from 1 to 6. Fifteen identities, 200 samples extracted from the parameter space. Line styles: basis Fisher geodesic distance, *solid black*; Frobenius distance between HMMs (17), *dashed black*; optimal pullback Fisher under automorphism (11), *solid red*; optimal pullback Fisher under automorphism (12), *dashed red*; inverse volume optimal Fisher metric with automorphism (11), *solid green*

Fifteen identities are here considered, with the parameter space sampled 200 times to detect the optimal parameters. The optimal classification pullback Fisher metrics induced by the proposed automorphisms are clearly superior to the standard Fisher distance over all experiments. The improvement margin ranges from 5% up to even 20%.

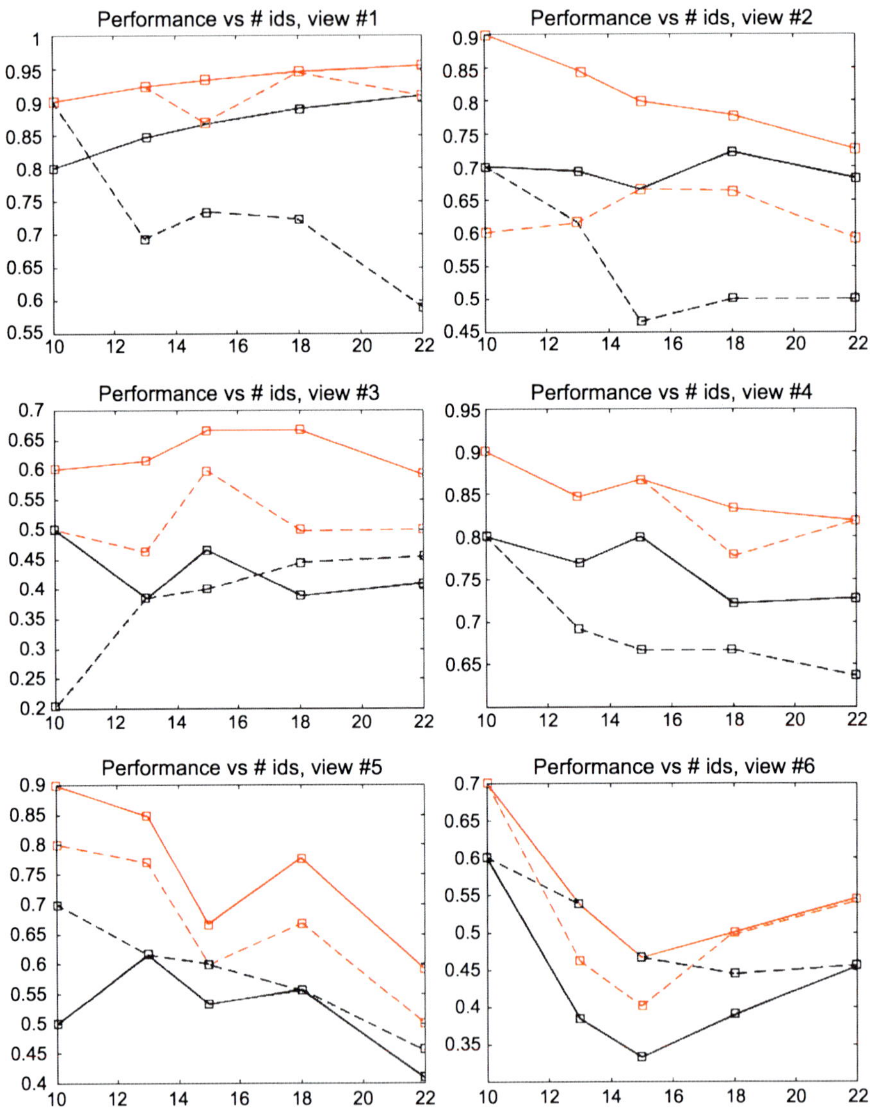

Fig. 7 Metrics' classification scores plotted versus the number of identities (from 10 to 22) in the testing set, for all viewpoints from 1 to 6. *Solid black*: Frobenius HMM distance. *Dashed black*: classical Fisher geodesic distance between AR(2, p) models. *Solid red*: optimal pullback Fisher geodesic distance induced by (11). *Dashed red*: optimal pullback Fisher induced by (12)

Figure 7 plots instead the classification score of the different competing metrics versus the number of identities considered in the experiments. We ran different tests involving a number of subjects ranging from 10 up to 22. Again, the performance of both pullback Fisher metrics obtained by maximizing classification score by n-fold validation (solid red and dashed red lines) is widely superior to that of the original Fisher distance (in solid black), or the naive Frobenius distance between HMMs (dashed black). The approach displays an interesting robustness to the expected decreasing performance as the problem grows more difficult, as optimal pullback classification rates are remarkably stable compared to those of classical metrics.

4.4 *Influence of Parameters*

It is natural to conjecture that, when optimizing the classification performance in the cross-validation procedure described in Sect. 2, a larger training set should lead to identify more effective automorphism parameters. Indeed, Fig. 8(left) shows the behavior of the considered metrics as a function of the size of the training set on which the optimal parameters are learnt. We can notice two facts here. First, as expected, optimization over larger training sets delivers better metrics, that is, better classification scores. Second, with its higher-dimensional parameter space, the global automorphism (12) of the $\mathcal{AR}(2, p)$ polytope generates more performing metrics, with a margin over the simpler, product automorphism ranging from 10% up to a remarkable 25%.

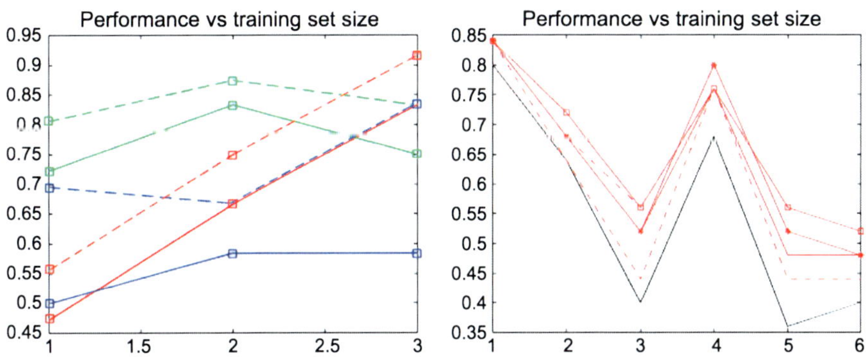

Fig. 8 *Left*: classification performance of the two optimal pullback Fisher metrics plotted versus increasing size of the training set over which the parameters of the optimal automorphism are learnt. Here, 12 identities, 200 parameter samples are chosen. Abscissa: 1–28 sequences in the training set 2–40 sequences 3–52 sequences. *Red*: experiment on view 1; *green*: view 4; *blue*: view 5. *Solid*: optimal metric induced by product automorphism (11); *dashed*: global automorphism (12). *Right*: classification performance of the optimal pullback Fisher metric induced by product automorphism, for experiments run on all viewpoints. All 25 identities are considered. Different classification scores are plotted for a number of samples varying from 10 (*red dotted line*) to 200 (*solid red with squares*). The score of the basis Fisher metric is plotted *in black* for comparison. When the optimal parameter is the result of a more extensive search, the classification performance is generally better

Finally, Fig. 8(right) illustrates how sampling more densely the parameter space when looking for the pullback metric that optimizes the n-fold classification score improves the performance of the resulting classifier. As an example, here the optimal pullback Fisher metric under product automorphism (11) is analyzed and compared with the baseline results obtained by using the basis Fisher geodesic distance between AR(2, p) models. As expected, the margin of improvement increases quite steadily as more samples are assessed in the parameter space. Here all 25 identities are considered, the margin ranging from a minimum of 5% for view 1 (for which the best silhouettes are available) to 15% for view 3, to a very substantial 23% for view 5, in which case the basis metric has the worst performance.

5 Perspectives and Conclusions

In this chapter, we proposed a differential-geometric framework for manifold learning given a data-set of linear dynamical models, based on optimizing over a family of pullback metrics induced by automorphisms. We adopted as basis metric tensor the classical Fisher information matrix, and showed tests on identity recognition that attest the improvement in classification performance one can gain from such a learnt metric. The method is fully general, and easily extendible to other classes of dynamical models or more sophisticated classifiers. For several classes of multi-dimensional linear systems both the Fisher metric and its geodesics can still be computed by means of an iterative numerical scheme [16, 25]. The extension to another popular class of stochastic model, hidden Markov models [13] requires an interesting analysis of its manifold structure and is already under way. Last but not least, the incorporation into the framework of objective functions that take into account a-priori knowledge on the training set, such as similarity relations [37], is highly desirable, and will be pursuit in the near future.

References

1. Amari, S.-I.: Differential Geometric Methods in Statistics. Springer, Berlin (1985)
2. Bar-Hillel, A., Hertz, T., Shental, N., Weinshall, D.: Learning distance functions using equivalence relations In: ICML03, pp. 11–18 (2003)
3. Belkin, M., Niyogi, P.: Semi-supervised learning on Riemannian manifolds. Mach. Learn. **56**, 209–239 (2004)
4. Bengio, Y., Paiement, J.-F., Vincent, P.: Out-of-sample extensions for LLE, Isomap, MDS, eigenmaps, and spectral clustering. Technical Report (2003)
5. Bilenko, M., Basu, S., Mooney, R.J.: Integrating constraints and metric learning in semi-supervised clustering. In: Proc. of ICML'04 (2004)
6. Bissacco, A., Chiuso, A., Soatto, S.: Classification and recognition of dynamical models: The role of phase, independent components, kernels and optimal transport. IEEE Trans. Pattern Anal. Mach. Intell. **29**(11), 1958–1972 (2007)
7. Burman, P.: A comparative study of ordinary cross-validation, v-fold cross-validation and the repeated learning-testing methods. Biometrika **76**(3), 503–514 (1989)

8. Chaudhry, R., Ravichandran, A., Hager, G., Vidal, R.: Histograms of oriented optical flow and Binet–Cauchy kernels on nonlinear dynamical systems for the recognition of human actions. In: Proc. of CVPR'09, pp. 1932–1939 (2009)
9. Dempster, A.P., Laird, N.M., Rubin, D.B.: Maximum likelihood from incomplete data via the EM algorithm. J. R. Stat. Soc. B **39**(1), 1–38 (1977)
10. Do, M.N.: Fast approximation of Kullback–Leibler distance for dependence trees and hidden Markov models. IEEE Signal Process. Lett. **10**(4), 115–118 (2003)
11. Doretto, G., Chiuso, A., Wu, Y.N., Soatto, S.: Dynamic textures. Int. J. Comput. Vis. **51**(2), 91–109 (2003)
12. Eick, C.F., Rouhana, A., Bagherjeiran, A., Vilalta, R.: Using clustering to learn distance functions for supervised similarity assessment. In: ICML and Data Mining (2005)
13. Elliot, R., Aggoun, L., Moore, J.: Hidden Markov Models: Estimation and Control. Springer, Berlin (1995)
14. Frosini, P.: Measuring shape by size functions. In: Proceedings of SPIE on Intelligent Robotic Systems, vol. 1607, pp. 122–133 (1991)
15. Gross, R., Shi, J.: The CMU motion of body (Mobo) database. Technical Report, CMU (2001)
16. Hanzon, B., Peeters, R.L.M.: Aspects of Fisher geometry for stochastic linear systems. In: Open Problems in Mathematical Systems and Control Theory, pp. 27–30 (2002)
17. Itoh, M., Shishido, Y.: Fisher information metric and poisson kernels. Differ. Geom. Appl. **26**(4), 347–356 (2008)
18. Kerckhove, M.: Computation of ridges via pullback metrics from scale space. In: Book Scale-Space Theories in Computer Vision. Lecture Notes in Computer Science, vol. 1682/1999, pp. 82–92. Springer, Berlin (1999)
19. Kim, T.K., Cipolla, R.: Canonical correlation analysis of video volume tensors for action categorization and detection. IEEE Trans. Pattern Anal. Mach. Intell. **31**(8), 1415–1428 (2009)
20. Kullback, S., Leibler, R.A.: On information and sufficiency. Ann. Math. Stat. **22**, 79–86 (1951)
21. Lebanon, G.: Metric learning for text documents. IEEE Trans. Pattern Anal. Mach. Intell. **28**(4), 497–508 (2006)
22. Lee, L., Grimson, W.: Gait analysis for recognition and classification. In: AFGR'02, pp. 155–162 (2002)
23. Martin, R.J.: A metric for ARMA processes. IEEE Trans. Signal Process. **48**(4), 1164–1170 (2000)
24. Murray, M.K., Rice, J.W.: Differential Geometry and Statistics. CRC Press, Boca Raton (1993)
25. Peeters, R.L.M., Hanzon, B.: On the Riemannian manifold structure of classes of linear systems (2003). Equadiff
26. Pitis, G.: On some submanifolds of a locally product manifold. Kodai Math. J. **9**(3), 327–333 (1986)
27. Rijkeboer, A.L.: Differential geometric models for time-varying coefficients of autoregressive processes. PhD Thesis, Tilburg University (1994)
28. Rogez, G., Rihan, J., Ramalingam, S., Orrite, C., Torr, P.H.S.: Randomized trees for human pose detection. In: CVPR'08 (2008)
29. Roweis, S., Saul, L.: Nonlinear dimensionality reduction by locally linear embedding. Science **290**(5500), 2323–2326 (2000)
30. Schultz, M., Joachims, T.: Learning a distance metric from relative comparisons. In: NIPS (2004)
31. Shental, N., Hertz, T., Weinshall, D., Pavel, M.: Adjustment learning and relevant component analysis. In: ECCV'02 (2002)
32. Smola, A.J., Vishwanathan, S.V.N.: Hilbert space embeddings in dynamical systems. In: Proc. of IFAC'03, pp. 760–767 (2003)
33. Sundaresan, A., Chowdhury, A.K.R., Chellappa, R.: A hidden Markov model based framework for recognition of humans from gait sequences. In: ICP'03, pp. II: 93–96 (2003)
34. Tsang, I.W., Kwok, J.T., Bay, C.W., Kong, H.: Distance metric learning with kernels. In: Proc. of the International Conference on Artificial Intelligence (2003)

35. Wang, Y., Mori, G.: Max-margin hidden conditional random fields for human action recognition. In: CVPR, pp. 872–879 (2009)
36. Xie, L., Ugrinovskii, A., Petersen, I.R.: Probabilistic distances between finite-state finite-alphabet hidden Markov models. In: Proc. of CDC'03, pp. 5347–5352 (2003)
37. Xing, E.P., Ng, A.Y., Jordan, M.I., Russel, S.: Distance metric learning with applications to clustering with side information. In: NIPS'03 (2003)
38. Yuan, J.S., Liu, Z.C., Wu, Y.: Discriminative subvolume search for efficient action detection. In: CVPR, pp. 2442–2449 (2009)
39. Zhang, Z.: Learning metrics via discriminant kernels and multidimensional scaling: toward expected Euclidean representation. In: ICML'03, Hong Kong (2003)

Part II
Tracking

Mixed-State Markov Models in Image Motion Analysis

**Tomás Crivelli, Patrick Bouthemy,
Bruno Cernuschi Frías, and Jian-feng Yao**

Abstract When analyzing motion observations extracted from image sequences, one notes that the histogram of the velocity magnitude at each pixel shows a large probability mass at zero velocity, while the rest of the motion values may be appropriately modeled with a continuous distribution. This suggests the introduction of mixed-state random variables that have probability mass concentrated in discrete states, while they have a probability density over a continuous range of values. In the first part of the chapter, we give a comprehensive description of the theory behind mixed-state statistical models, in particular the development of mixed-state Markov models that permits to take into account spatial and temporal interaction. The presentation generalizes the case of simultaneous modeling of continuous values and any type of discrete symbolic states. For the second part, we present the application of mixed-state models to motion texture analysis. Motion textures correspond to the instantaneous apparent motion maps extracted from dynamic textures. They depict mixed-state motion values with a discrete state at zero and a Gaussian distribution for the rest. Mixed-state Markov random fields and mixed-state Markov chains are defined and applied to motion texture recognition and tracking.

T. Crivelli (✉) · B. Cernuschi Frías
University of Buenos Aires, Buenos Aires, Argentina
e-mail: tomicrivelli@fibertel.com.ar

P. Bouthemy
INRIA, Rennes, France

B. Cernuschi Frías
CONICET, Buenos Aires, Argentina

J.-f. Yao
Unviersity of Rennes 1, Rennes, France

L. Wang et al. (eds.), *Machine Learning for Vision-Based Motion Analysis*,
Advances in Pattern Recognition,
DOI 10.1007/978-0-85729-057-1_4,

1 Introduction

Usually in statistics one is either interested in random variables that are discrete in nature, or that are continuous and hence they have a probability density function as is the case, for example, of the Gaussian random variables. In some situations, one may be interested in modeling a *mixed* variable that takes some discrete values in a set with non-zero probability mass, while for the other values not in the given set, the random variable may be modeled as a continuous distribution.

In particular, these models appear to be of interest in low level modeling of motion in computer vision [6, 14–16, 32]. When modeling the apparent motion in image sequences depicting textured motion patterns, that is, *motion textures*, as for example a video sequence of traffic, or a scene of a tree waving under the wind, we experimentally observe that such motion maps exhibit values of two types: a discrete component at zero (absence of motion) and continuous motion values. Both types are present in the motion maps forming a spatial configuration similar to intensity textures. Analog spatial properties as texture orientation and isotropy, repetitive local patterns and statistical interaction, are present and are characteristic of the dynamic content of the scene. Moreover, discrete and continuous values are not independent nor constitute two different processes. It is a single (motion) observation process that depicts what we call *mixed-state values*.

From motion detection to optical flow estimation [2, 7, 12], efforts have been devoted to extract reliable and representative motion quantities from a sequence of images. In the last years, there has been an increasing interest in retrieval, indexing, recognition, and classification of long sequences of video data for dynamic content analysis. In this context, motion information has been effectively used as a key feature for qualitative content characterization [18–20, 26, 32, 33].

We thus search for a modeling framework that can capture the statistical properties of the observed apparent motion, effectively integrating discrete and continuous values in a unified way. This motivates the introduction of mixed states random variables. That is, variables that have mass probability concentrated in discrete states, while they have a probability density for the rest. Moreover, these two types of variables may more deeply reflect two different classes of information: continuous real values (either in one-dimensional or multi-dimensional spaces) versus symbolic values (one or several symbols). These two classes should not be necessarily viewed as two exclusive states or as consecutive states (e.g., after a decision step).

Evaluating the distribution of mixed-state values, and accounting for local context, are of key importance in numerous image sequence analysis tasks (e.g., in motion modeling, detection, segmentation, estimation, recognition, or learning issues). Therefore, defining probabilistic mixed states models appears as an attractive objective.

The first main part of the current chapter is devoted to the development of the theory of mixed-state random fields. In dealing with mixed-state random variables, a measure theoretic approach is considered which is described here with necessary detail. Then, a full account is given for Markov models which includes mixed-state Markov random fields (MS-MRF) and mixed-state Markov chains (MS-MC). Fundamental issues as simulation, parameter estimation and inference in mixed-state random fields are treated by providing algorithms as extensions of classical methods.

In the second part of this chapter, we analyze two instances of mixed-state models for the characterization of motion textures. First, we consider the spatial modeling of motion textures with parametric MS-MRF's. Then we propose a motion texture recognition method based on statistical matching of mixed-state spatial models by means of the Kullback–Leibler divergence [13]. The advantage of our approach is that it considers an instantaneous motion map, instead of modeling the temporal evolution of image intensity along a video sequence as done in [9, 10, 17, 28, 41]. Consequently, a particular dynamic content can be learned and recognized in a frame-by-frame basis. This has important implications in motion texture detection, segmentation and tracking.

Second, we analyze the temporal modeling of motion textures with MS-MC. This implies modeling the interaction between motion random variables at time t and a temporal neighborhood in the previous motion map at $t-1$. We then discuss the advantages of the spatial model vs. the temporal model, particularly for motion texture modeling and recognition. Finally, we address the problem of motion texture tracking and we propose to use the MS-MC model as a content-characteristic feature for window matching.

1.1 Outline of the Chapter

The rest of this chapter is organized as follows. In Sect. 2, we review some previous work on mixed-state related approaches and dynamic texture characterization. In Sect. 3, the proposed motion measurements are defined leading to the definition of motion textures and the observation of their mixed-state statistical properties. This motivates the theoretical development of mixed-state Markov models in Sect. 4, accounting for mixed-state Markov random fields (MS-MRF) and mixed-state Markov chains (MS-MC). Estimation, sampling, and inference algorithms are provided in Sect. 5.

In Sect. 6, a MS-MRF is defined for the spatial modeling of motion textures and its application to motion texture classification. Then in Sect. 7, a causal MS-MC is discussed for the temporal modeling of motion textures. We compare both models for motion texture recognition in Sect. 8. Finally, a new motion texture tracking method is proposed in Sect. 9 based on the MS-MC model.

2 Related Work: Discrete-Continuous Models and Dynamic Textures

As said, this chapter is developed along two main parts: (i) a general theory of mixed-state random fields and (ii) the modeling, recognition and tracking of motion textures. In what follows, we briefly discuss different approaches related to both. Though in this chapter the first provides the theoretical basis for the second, they can be positioned in the state-of-the-art as two separate research streams.

2.1 Discrete-Continuous Approaches

Different approaches related to the concept of a random process that can take different types of values (either continuous or discrete) have been followed in the computer vision literature as an evidence of the necessity of considering discrete-continuous interaction.

In previous works on fuzzy pixels classifications as [36] and [35], the authors introduce a class of fuzzy MRF's or fuzzy Markov chains where each state variable, or classification variable, $x_i \in [0, 1]$ represents a classification rate. The fuzzy principle implies that the two hard classification states $x_i = 0$ or $x_i = 1$ have a positive probability while all the soft classification states, that is, $x_i \in (0, 1)$ follow a continuous distribution with some ad-hoc density function.

Also in computer vision decision problems, we can cite other models that exploit the interaction between symbolic and numeric values. Starting with the seminal paper of Geman and Geman [22] with the introduction of a *line process* for modeling edges between homogeneous image regions to be restored from different types of degradations. They proposed to introduce an unobservable process $\mathbf{L}$ for edge elements and regard the original image $\mathbf{I}$ as a marginal process from an extended joint field $\mathbf{X} = (\mathbf{I}, \mathbf{L})$. Then, they obtain a MAP solution for $\mathbf{X}$ solving for both the line process and the intensity process. Later in [5], for a similar problem, the line process is viewed as a way of rejecting the outlier measurements arisen from discontinuities between surfaces in the problem of reconstruction, giving an equivalence with robust estimators. In [39] the authors present a model for image segmentation that introduces a boundary model between regions in a more sophisticated way.

Here, we consider random variables that may take both discrete and continuous states and we present the theory of mixed-state random fields, generalizing previous approaches. We share a similar measure theoretic formulation with *fuzzy Markov random fields* (*FMRF*) and *fuzzy Markov chains* [35, 36] for defining mixed-state variables. However, the construction of the Gibbs potentials in [35, 36] is specific for the problem of image segmentation and the local conditional densities are obtained from this Gibbs distribution. On the other side, many situations are better modeled starting from the local characteristics that are of a mixed-state nature and thus it is not clear in this case which is the general form of the associated Gibbs distribution. Moreover, these authors define a mixed-state classification variable lying in [0, 1], which although it is absolutely natural for decision rules, it is restrictive for modeling arbitrary mixed-state data.

In [35, 36], a discrete-continuous random field is built upon classification states that are estimated exploiting the contextual modeling provided by nearest-neighbors

Gibbs distributions. In [22], a MAP approach exploits the line process for discontinuity preserving restoration and segmentation. However, discrete-continuous approaches are not only useful for doing inference of segmentation fields but also as generative models of mixed-state observations, which arise for example in motion analysis.

Discrete states also include symbolic information through abstract labels that (a) may not be numeric as in [35, 36], (b) may not appear as pairs of complementary states (e.g., $\{0, 1\}$) as in [22], (c) may have a very different physical meaning from the continuous states, which is different from the fuzzy-hard random fields where hard and soft values refer both to a classification rate. Though this case is not fully developed here, the general theory of mixed-state random fields remains valid. We refer the reader to [16] as an example of symbolic-numeric mixed-state fields.

2.2 Dynamic Texture Characterization

In the context of visual motion analysis, *motion textures* are closely related to *temporal* or *dynamic textures* [9, 10, 17, 28, 41], firstly introduced by Nelson and Polana [28]. Different from *activities* (walking, climbing, playing) and *events* (open a door, answer the phone), *temporal textures* show some type of stationarity or homogeneity, both in space and time. Mostly, they refer to dynamic video contents displayed by natural scene elements as rivers, sea-waves, smoke, moving foliage, fire, etc. They also encompass any dynamic visual information that, from the observer point of view, can be classified as a texture with motion (Fig. 1). For example, consider a walking person. This *activity* can be analyzed as attached to an articulated motion; however a group of persons or a crowd walking, observed from a wide angle may show a repetitive motion pattern, more adequate to be considered as a temporal texture.

A first distinction between different approaches, lies in the type of image features extracted from the image sequence. Soatto et al. [17] proposed the use of ARMA models directly on image intensities for dynamic texture synthesis. In [40], an improvement is proposed based on a control theory approach. In [37], linear dynamical systems (LDS) are also applied in combination with 2D translational models that permits to deal with a nonstatic camera and/or moving dynamic textures. Other intensity-based models can be found in [9, 10] for the simultaneous modeling of multiple dynamic texture regions.

Alternatively, there has been an increasing interest in the modeling of motion features extracted from dynamic textures instead of considering pixel-based intensity values [6, 14, 20, 21, 33]. Particularly, normal flow is an efficient and natural way of characterizing the local spatio-temporal dynamics of a dynamic texture [20]. A survey on dynamic texture characterization can be found in [11].

Much efforts on dynamic texture characterization have been devoted to the recognition and classification of these types of image sequences. Recent results have shown that methods based on motion features give the highest recognition and classification rates for dynamic textures depicting natural scenes [20, 27, 30]. They are

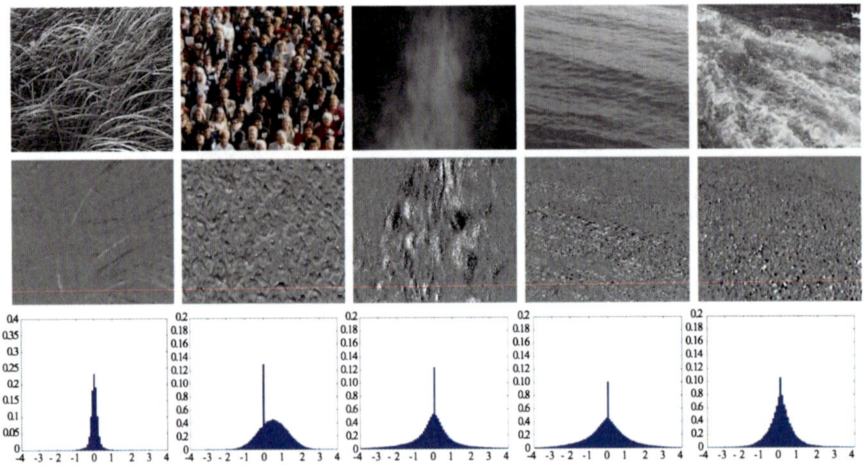

Fig. 1 *Top*: sample images from videos of dynamic textures (swaying trees, grass, crowd, steam, and water). *Middle*: scalar motion map based on normal flow computation and obtained between two consecutive frames of the sequence, which we call a motion texture. *Bottom*: histograms of motion values computed using (34) for each motion texture. Motion values display two components: a discrete value at zero and a continuous distribution for the rest

based on computing some motion statistics across the image sequence, and using them as class descriptors for the classification task. A different approach for dynamic texture recognition is proposed in [34], where a dissimilarity measure between linear dynamical models is utilized. Finally, the extension of the concept of dynamic textures to more complex scenes sets the necessity of considering more elaborated approaches for specific applications. See [41] as an interesting example.

The main difference of our approach with respect to existing dynamic texture characterization methods, is that we explicitly model motion information and that we analyze the instantaneous motion maps, instead of the temporal evolution of the image intensity. Given this, motion texture recognition and tracking can be naturally addressed in a frame-by-frame basis. In previous methods based on LDS, the need of processing several frames or even the whole image sequence for identifying the models, limits the application of such methods to complex problems as detection, segmentation and tracking.

3 The Mixed-State Nature of Motion Measurements

By definition, a motion texture is an instantaneous motion map obtained from a dynamic texture. Let $I_i(t)$ be a scalar function that represents the image intensity at image point $i \in S = \{1, \dots, N\}$ for time t where S denotes the image grid. A motion texture is extracted by computing scalar motion measurements between two consecutive images, say $I(t-1)$ and $I(t)$, for some given instant.

The Optical Flow Constraint [25] is a condition over intensity images from which velocity fields can be effectively estimated. It states that the intensity of a moving point remains constant along time, that is,

$$\frac{dI_i(t)}{dt} = \frac{\partial I_i(t)}{\partial t} + \nabla I_i(t) \cdot \upsilon_i(t) = 0, \tag{1}$$

where $\upsilon_i(t)$ is the velocity at location i and $\nabla I_i(t)$ the spatial intensity gradient. Though this condition allows to measure only the component of the velocity in the direction of the spatial intensity gradient, that is, *normal flow*, locally it gives valuable quantitative information about the spatio-temporal structure of the scene. From (1), one obtains the normal flow as

$$\upsilon_i^{(n)}(t) = -\frac{\frac{\partial I_i(t)}{\partial t}}{\| \nabla I_i(t) \|} \frac{\nabla I_i(t)}{\| \nabla I_i(t) \|}. \tag{2}$$

We introduce a weighted vectorial average of normal flow in order to smooth out noisy measurements and enforce reliability. The result is a smoothed measure of local motion and given by

$$\overline{\upsilon}_i(t) = \frac{\sum_{j \in W} \upsilon_j^{(n)}(t) \| \nabla I_j(t) \|^2}{\max(\sum_{j \in W} \| \nabla I_j(t) \|^2, \eta^2)}, \tag{3}$$

where η^2 is a regularization constant fixed to a small value, and W is a small window centered in location i. This average results in a local estimation of normal flow. The projection of this quantity over the intensity gradient direction gives rise to the following scalar motion observation:

$$v_i(t) = \overline{\upsilon}_i(t) \cdot \frac{\nabla I_i(t)}{\| \nabla I_i(t) \|}, \tag{4}$$

with $v_i(t) \in (-\infty, +\infty)$.

Experiments obtained from computing the proposed motion quantities for several different dynamic textures show that, if we observe the corresponding motion histograms (Fig. 1), the statistical distribution of the motion measurements has two elements: a discrete component at the null value $v_i = 0$, and a continuous distribution for the rest of the motion values.

The null value appears repeatedly in the motion maps, also following a textured binary pattern as well as it occurs for the rest of the motion values (Fig. 2). This is a typical structural characteristic of the motion measurements extracted from motion textures and not the result of a decision process as motion detection. As such, discrete and continuous values are not independently distributed in space, indeed displaying a mixed-state texture pattern. In other words, we want to model motion observations that display mixed-state values. This motivates the theoretical setting described in the following section.

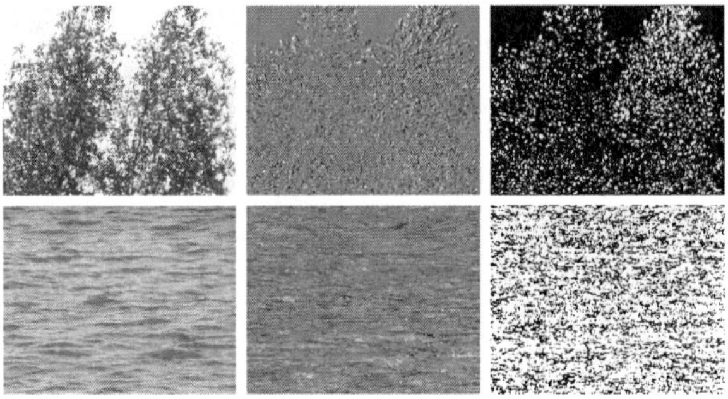

Fig. 2 (**a**) Sample images from motion textures. (**b**) The scalar motion values are spatially distributed forming a textured pattern. Here, we mapped the motion measurements to the *range of gray* [0, 255] where 128 corresponds to null motion. (**c**) The binary motion-no motion map also is distributed following a textured pattern. *White* represents a motion value different from zero

4 Mixed-State Markov Models

4.1 Mixed-State Random Variables

Mixed-state random variables are defined as to consider probability mass concentrated in discrete states, either symbolic labels or discrete numeric values in $\mathbb{R}^n$, and also continuous values described by a probability density function (p.d.f.). Formally:

Definition 1 (Mixed-state random variable) A mixed-state random variable is a function from a sample space Ω to a mixed-state space $\mathbb{M} = \mathbb{L} \cup \mathbb{R}^n$ where $\mathbb{L} = \{l_1, l_2, \dots\}$ is a countable set of symbolic labels and eventually at most a countable number of numeric values.

Associating a probability density function to a mixed-state random variable is not straightforward. A first direction would be to consider a generalized probability density function $p^m(x)$ in the distributional sense, where each discrete value is modeled by a Dirac delta "function" $\delta(x-l)$ at $x=l$, for each $l \in \mathbb{L}$, loosely in the form

$$p^m(x) = \rho \sum_{l \in \mathbb{L}} \pi_l \delta(x-l) + \rho^* p^c(x), \tag{5}$$

where $\rho\pi_l$ is the probability mass at $x=l$, $\rho \in [0,1]$, $\rho^* = 1-\rho$, $\sum_l \pi_l = 1$ and $p^c(x)$ is a continuous p.d.f. The corresponding distribution function has the form of the example given in Fig. 3(a). However, this formalization is only valid when all the discrete states are numeric values, that is, when $\delta(x-l)$ is indeed defined $\forall l \in \mathbb{L}$. In the case we want to define random variables that take symbolic discrete values, it is not possible to reside on the Dirac delta distribution for assigning them a positive probability mass. Hence, to correctly define these models in a general

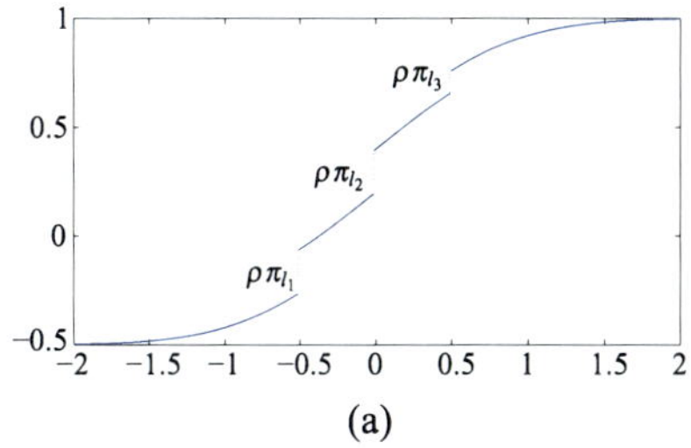

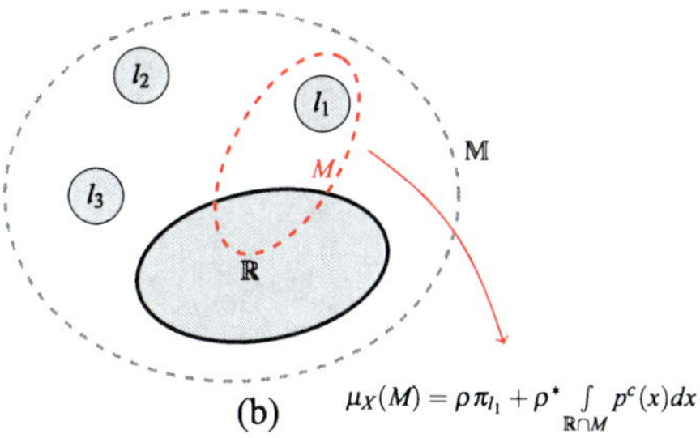

Fig. 3 (**a**) A probability distribution function for a mixed-state random variable with numeric discrete part. (**b**) A mixed-state space with symbolic discrete values. A probability measure is assigned to each subset (event) M of this space: a point mass for symbols and a Lebesgue measure for the continuous values contained in M

formulation a more appropriate direction would be to consider a measure theoretic approach, which is valid for any type of mixed-state random variables.

Since symbolic labels that may not have any algebraic structure are present, a distribution function cannot be defined to characterize the random variable. A possibility is to proceed directly to construct a probability measure μ_X, for the random variable X. Loosely, $\mu_X(M)$ is a function that assigns a probability in $[0, 1]$ to the sets $M \subset \mathbb{M}$. Note that for real random variables it is given by the distribution function so that $F_X(x) = \text{Prob}(X \leq x) = \mu_X(\{X \leq x\})$. We can thus define μ_X for a mixed-state random variable as a convex combination of measures

$$\mu_X(M) = \rho \sum_{l \in \mathbb{L}} \pi_l \mathbf{1}_M(l) + \rho^* \int_{M \setminus \mathbb{L}} p^c(x)\, dx, \tag{6}$$

where we define the characteristic function $\mathbf{1}_M(l) = 1$ if $l \in M$ and 0 otherwise, and also its complement, $\mathbf{1}^*_M(l) = 1 - \mathbf{1}_M(l)$. In other words, $\mu_X(M)$ accounts for the probability mass due to the discrete states and the continuous values in M (Fig. 3(b)).

A probability density function can not be defined as usually done for a real-valued random variable, by computing the derivative of the distribution function with respect to the Lebesgue measure (the length of an interval on the real line). We can instead construct a generalized probability density for X, defined from $\mu_X(M) = \int_M p^m(x) m(dx)$: a function that integrated across a mixed-state subset M gives the probability of $X \in M$. Such $p^m(x)$ exists thanks to the Radon–Nikodym theorem [38]. Here, we call $m(M)$ a reference measure which is defined as,

$$m(M) = \sum_{l \in \mathbb{L}} \mathbf{1}_M(l) + \int_{M \setminus \mathbb{L}} dx = m^d(M) + \lambda(M), \tag{7}$$

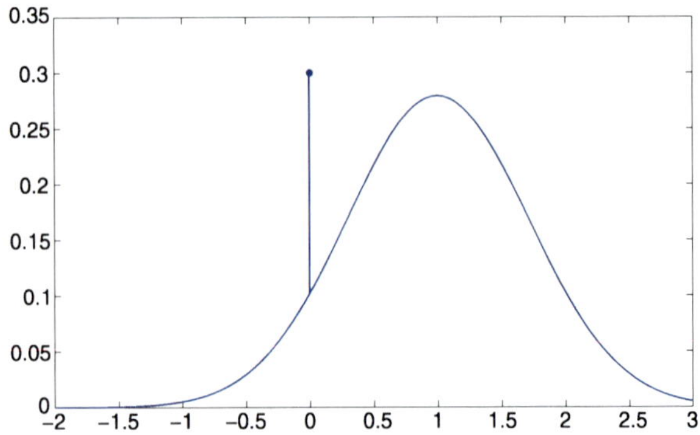

Fig. 4 A mixed-state p.d.f. with $\{0\}$ as the discrete value and Gaussian continuous part. Here, $\rho = P(x = 0) = 0.3$. This p.d.f. integrates to one with respect to the measure m

where $m^d(M)$ is a counting measure and $\lambda(M)$ is the Lebesgue measure. With this definition, we obtain

$$\mu_X(M) = \int_M p^m(x)m(dx) = \sum_{l\in\mathbb{L}} \mathbf{1}_M(l)p^m(l) + \int_{M\setminus\mathbb{L}} p^m(x)\,dx. \tag{8}$$

Interpret this equation as follows: the generalized density function p^m assigns a probability mass to each of the discrete values or symbols, and acts as a continuous density function for the continuous values. Combining (6) and (8), we get:

$$p^m(x) = \rho \sum_{l\in\mathbb{L}} \pi_l \mathbf{1}_l(x) + \rho^* \mathbf{1}^*_{\mathbb{L}}(x) p^c(x). \tag{9}$$

Equation (9) is the general shape of a mixed-state probability density which is valid for either numeric discrete values and/or symbolic abstract labels.

Example 1 (Mixed-state Gaussian random variable)
Consider the case of a random variable that is 0 with probability ρ or is distributed following a continuous Gaussian density with probability $1 - \rho$ (Fig. 4). Hence, $\mathbb{L} = \{0\}$ and

$$p^m(x) = \rho \mathbf{1}_0(x) + (1 - \rho)\mathbf{1}^*_0(x)\frac{1}{\sqrt{2\pi}\sigma} e^{-\frac{(x-c)^2}{2\sigma^2}}. \tag{10}$$

With this measure theoretic formulation, the discrete state need not be a numeric value as 0. We can define any abstract label l indicating for example the presence of an event of interest or some high-level state ('motion', 'occlusion', 'discontinuity', 'invalid value', 'true' etc.).

4.2 Mixed-State Markov Random Fields

A mixed-state random field is a collection of random variables $\mathbf{x} = \{x_i\}_{i\in S}$ on a lattice $S = \{1, \ldots, N\}$ of points or image locations, described by a joint generalized probability density function $p^m(\mathbf{x})$ with respect to the measure $\nu = m^N$ with m as in (7), and defined on a mixed-state space $\mathbb{M}$.

As the Hammersley–Clifford theorem states [3], Markov random fields with an everywhere positive density function, are equivalent to nearest neighbor Gibbs distributions. The joint p.d.f. of the random variables that compose the field has the form $p(\mathbf{x}) = \exp[Q(\mathbf{x})]/Z$, where $Q(\mathbf{x})$ is an energy function, and Z is called the partition function or normalizing factor of the distribution. The power of these models was primarily demonstrated in [3] with the introduction of the so-called *auto-models*, and its numerous applications.

Define $\mathbf{x}_A$ as the subset of random variables restricted to $A \subset S$, that is, $\mathbf{x}_A = \{x_i\}_{i\in A}$. Then the Markovian property yields $p(x_i \mid \mathbf{x}_{\{i\}^c}) = p(x_i \mid \mathbf{x}_{\mathcal{N}_i})$ where $\mathcal{N}_i \in S$ is a subset of S with $i \notin \mathcal{N}_i$, called the neighborhood of location i.

The Markovianity is as well expressed in the global form of the process. The energy $Q(\mathbf{x})$ can be expressed as a sum of potential functions, $Q(\mathbf{x}) = \sum_{\mathcal{C}\subset S} V_{\mathcal{C}}(\mathbf{x}_{\mathcal{C}})$, where the summation runs over those subsets $\mathcal{C}$ of S such that $V_{\mathcal{C}} \neq 0$, called *cliques* [3].

In order to construct a mixed-state Markov random field, we will assume that we have a single discrete (symbolic or numeric) value, that is, $\mathbb{L} = \{l\}$, avoiding to carry a cumbersome notation. We can extend (9) to a mixed-state conditional density by conditioning over the neighbors $\mathbf{x}_{\mathcal{N}_i}$, giving $p^m(x_i \mid \mathbf{x}_{\mathcal{N}_i})$, what motivates the following definition

Definition 2 (Mixed-state Markov random field MS-MRF) A *mixed-state Markov random field or MS-MRF* is a Markov random field for which the local conditional densities are:

$$p^m(x_i \mid \mathbf{x}_{\{i\}^c}) = p^m(x_i \mid \mathbf{x}_{\mathcal{N}_i}) = \rho_i(\mathbf{x}_{\mathcal{N}_i})\mathbf{1}_l(x_i) + \rho_i^*(\mathbf{x}_{\mathcal{N}_i})\mathbf{1}_l^*(x_i)p^c(x_i \mid \mathbf{x}_{\mathcal{N}_i}). \quad (11)$$

The symbol probability $\rho(\cdot) = \text{Prob}(x_i = l \mid \mathbf{x}_{\mathcal{N}_i})$ is now a function of the neighbors $\mathbf{x}_{\mathcal{N}_i}$. The definition corresponds to the property of Markovianity as for a classical MRF.

Next, we are interested in determining the joint probability density function for which the conditional ms-pdf take the form of (11). The equivalence between Gibbs distributions and MRF's is equally valid for probability densities obtained w.r.t. a mixed-state probability measure m, as the Hammersley–Clifford theorem [3] does not depend on the type of state space [8].

Effectively, one can define a mixed-state Gibbs distribution having the form $p^m(\mathbf{x}) = \exp -Q(\mathbf{x})/Z$, with $Q : \mathbb{M}^N \to \mathbb{R}$. The partition function Z is the normalizing factor obtained by computing the Lebesgue–Stieltjes integral $Z = \int_{\mathbb{M}} \exp -Q(\mathbf{x})\, dm$. The shape of the mixed-state potential functions $V_{\mathcal{C}}(\mathbf{x}_{\mathcal{C}})$ is discussed in what follows.

4.2.1 The Mixed-State Gibbs Distribution

At first, the form that the potentials may take is quite general and in fact for a given $Q(\mathbf{x}) = \sum_{\mathcal{C}} V_{\mathcal{C}}(\mathbf{x}_{\mathcal{C}})$ the conditional mixed-state density can always be obtained from:

$$\log \frac{p^m(x_i \mid \mathbf{x}_{\mathcal{N}_i})}{p^m(r \mid \mathbf{x}_{\mathcal{N}_i})} = \sum_{\mathcal{C} \subset S: i \in \mathcal{C}} V_{\mathcal{C}}(\mathbf{x}_{\mathcal{C}}), \tag{12}$$

where r is a reference value such that $V_{\mathcal{C}}(\mathbf{x}_{\mathcal{C}}) = 0$ if for any $i \in \mathcal{C}$, $x_i = r$. Such potentials are called canonical w.r.t. r, as in a classical MRF, which permits to obtain a unique expansion of the energy $Q(\mathbf{x})$. Finally, it can be written as

$$\begin{aligned} p^m(x_i \mid \mathbf{x}_{\mathcal{N}_i}) &= \mathbf{1}_l(x_i) p^m(l \mid \mathbf{x}_{\mathcal{N}_i}) + \mathbf{1}_l^*(x_i) p^m(x_i \mid \mathbf{x}_{\mathcal{N}_i}) \\ &= \mathbf{1}_l(x_i) \rho_i(\mathbf{x}_{\mathcal{N}_i}) + \mathbf{1}_l^*(x_i) \rho_i^*(\mathbf{x}_{\mathcal{N}_i}) \frac{p^m(x_i \mid \mathbf{x}_{\mathcal{N}_i})}{\rho_i^*(\mathbf{x}_{\mathcal{N}_i})}, \end{aligned} \tag{13}$$

where $p^m(l \mid \mathbf{x}_{\mathcal{N}_i}) = \rho_i(\mathbf{x}_{\mathcal{N}_i})$ and by calling $\frac{p^m(x_i | \mathbf{x}_{\mathcal{N}_i})}{\rho_i^*(\mathbf{x}_{\mathcal{N}_i})} = p^c(x_i \mid \mathbf{x}_{\mathcal{N}_i})$ for $x_i \neq l$, one obtains (11). However, we will see that imposing certain additional hypothesis on the conditional mixed-state p.d.f.'s and the shape of the mixed-state Gibbs distribution, more appealing conditions appear, which in turn provide a way of designing a mixed-state model. We refer the reader to [8] for a thorough theoretic discussion about the shape of the potentials for a mixed-state Markov random field.

4.2.2 Mixed-State Automodels

Most of the random fields used in image analysis and for describing physical systems, are encoded considering only the contributions from cliques containing no more than two sites, due to their simple form and low computational cost. In his seminal paper [3], Besag defined the class of Markov random fields named *auto-models*. He considered the particular scheme where the conditional probability densities that define the local characteristics of a Markov random field belong to a one-parameter exponential family of distributions, and the corresponding global Gibbs energy depends only on cliques that contain no more than two sites. With this setting, the expression for the parameter in the conditional density is given as an affine function of a sufficient statistic of the neighbors. This ideas can be generalized to the so-called *multi-parameter auto-models*, an extension given by Hardouin and Yao [24], and condensed in the following result.

Theorem 1 *For a Markov random field that satisfies the following conditions*:

(a) *The potential functions are at most of second order, that is,*

$$Q(\mathbf{x}) = \sum_{i \in S} V_i(x_i) + \sum_{\langle i,j \rangle \in S} V_{ij}(x_i, x_j).$$

(b) *The local conditional characteristics belong to the d-parameter exponential family*

$$\log p(x_i \mid \mathbf{x}_{\mathcal{N}_i}) = \boldsymbol{\Theta}_i^T(\mathbf{x}_{\mathcal{N}_i}) \mathbf{S}_i(x_i) + C_i(x_i) + D_i(\mathbf{x}_{\mathcal{N}_i}),$$

with $\mathbf{S}_i(x_i) \in \mathbb{R}^d$, $\boldsymbol{\Theta}_i(\mathbf{x}_{\mathcal{N}_i}) \in \mathbb{R}^d$, $C_i(x_i)$ *and* $D_i(\mathbf{x}_{\mathcal{N}_i}) \in \mathbb{R}$, *and the normalization conditions* $C_i(r) = 0$ *and* $\mathbf{S}_i(r) = \mathbf{0}$ *for some reference value* r.

(c) *The family of sufficient statistics* $\{\mathbf{S}_i(x_i)\}_{i \in S}$ *is regular in the sense that*

$$\forall i \in S, \quad \operatorname{Span}\{\mathbf{S}(x_i)\} = \mathbb{R}^d.$$

Then, the corresponding conditional parameter vector is given by,

$$\boldsymbol{\Theta}_i(\mathbf{x}_{\mathcal{N}_i}) = \boldsymbol{\alpha}_i + \sum_{j \in \mathcal{N}_i} \boldsymbol{\beta}_{ij} \mathbf{S}_j(x_j), \tag{14}$$

with $\boldsymbol{\beta}_{ij} \in \mathbb{R}^{d \times d}$ *and* $\boldsymbol{\alpha}_i = [\alpha_1 \ \dots \ \alpha_d]^T \in \mathbb{R}^d$. *Moreover, the canonical potential functions take the form,*

$$V_i(x_i) = \boldsymbol{\alpha}_i^T \cdot \mathbf{S}_i(x_i) + C_i(x_i), \tag{15}$$

$$V_{ij}(x_i, x_j) = \mathbf{S}_i(x_i)^T \boldsymbol{\beta}_{ij} \mathbf{S}_j(x_j). \tag{16}$$

Proof The proof is given in [24]. For $d = 1$, the original demonstration is due to Besag [3]. □

The first hypothesis is the assumption of pairwise interaction and is related to the structure of cliques which does not depend on the type of probability space. Thus, it is valid for any type of random field (i.e., discrete, continuous or mixed-state). Regarding the second assumption, we first have the following proposition.

Proposition 1 *Let* $p^c(x_i \mid \mathbf{x}_{\mathcal{N}_i})$ *be a continuous conditional p.d.f. such that it belongs to the d-parameter exponential family of distributions, that is,* $\log p^c(x_i \mid \mathbf{x}_{\mathcal{N}_i}) = \tilde{\boldsymbol{\Theta}}_i^T(\mathbf{x}_{\mathcal{N}_i})\tilde{\mathbf{S}}_i(x_i) + \tilde{C}_i(x) + \tilde{D}_i(\mathbf{x}_{\mathcal{N}_i})$. *For a discrete value* l, *the mixed-state probability density defined as*

$$p^m(x_i \mid \mathbf{x}_{\mathcal{N}_i}) = \rho_i(\mathbf{x}_{\mathcal{N}_i})\mathbf{1}_l(x_i) + \rho_i^*(\mathbf{x}_{\mathcal{N}_i})\mathbf{1}_l^*(x_i)p^c(x_i \mid \mathbf{x}_{\mathcal{N}_i}) \tag{17}$$

and w.r.t. $m(M) = \mathbf{1}_M(l) + \lambda(M)$ *with* $M \subset \{l\} \cup \mathbb{R}$ *and* $x \in \{l\} \cup \mathbb{R}$, *belongs to the* $(d+1)$*-parameter exponential family of distributions, that is,* $\log p^m(x_i \mid \mathbf{x}_{\mathcal{N}_i}) = \boldsymbol{\Theta}_i^T(\mathbf{x}_{\mathcal{N}_i})\mathbf{S}_i(x_i) + C_i(x_i) + D_i(\mathbf{x}_{\mathcal{N}_i})$, *and*

$$\begin{aligned}
\mathbf{S}_i(x_i) &= \left[\mathbf{1}_l^*(x_i), \mathbf{1}_l^*(x_i)\tilde{\mathbf{S}}_i(x_i)\right]^T, \\
\boldsymbol{\Theta}_i(\mathbf{x}_{\mathcal{N}_i}) &= \left[\log \frac{\rho_i^*(\mathbf{x}_{\mathcal{N}_i})}{\rho_i(\mathbf{x}_{\mathcal{N}_i})} + \tilde{D}_i(\mathbf{x}_{\mathcal{N}_i}), \tilde{\boldsymbol{\Theta}}_i^T(\mathbf{x}_{\mathcal{N}_i})\right]^T, \\
C_i(x_i) &= \mathbf{1}_l^*(x_i)\tilde{C}_i(x_i), \\
D_i(\mathbf{x}_{\mathcal{N}_i}) &= \log \rho_i(\mathbf{x}_{\mathcal{N}_i}).
\end{aligned} \tag{18}$$

Proof To simplify, write $\rho_i(\mathbf{x}_{\mathcal{N}_i}) \equiv \rho_i$ and $\mathbf{x}_{\mathcal{N}_i} = \cdot$. $p^m(x_i \mid \cdot)$ is a sum of excluding terms, allowing us to express the logarithm of the left-hand side of (17) as the sum of the logarithms of the two terms on the right-hand side. Some calculations and rearrangements yield:

$$\log p^m(x_i \mid \cdot) = \log \rho_i + \mathbf{1}_l^*(x_i) \log \left[p^c(x_i \mid \cdot) \frac{\rho_i^*}{\rho_i} \right], \tag{19}$$

and replacing $p^c(x_i \mid \cdot)$,

$$\begin{aligned}
\log p^m(x_i \mid \cdot) &= \log \rho_i + \mathbf{1}_l^*(x_i) \big[\tilde{\boldsymbol{\Theta}}_i^T(\cdot) \tilde{\mathbf{S}}_i(x_i) + \tilde{C}_i(x_i) + \tilde{D}_i(\cdot) \big] + \mathbf{1}_l^*(x_i) \log \frac{\rho_i^*}{\rho_i} \\
&= \left[\log \frac{\rho_i^*}{\rho_i} + \tilde{D}_i(\cdot), \tilde{\boldsymbol{\Theta}}_i^T(\cdot) \right] \cdot \left[\begin{matrix} \mathbf{1}_l^*(x_i) \\ \mathbf{1}_l^*(x_i) \tilde{\mathbf{S}}_i(x_i) \end{matrix} \right] \\
&\quad + \mathbf{1}_l^*(x_i) \tilde{C}_i(x_i) + \log \rho_i,
\end{aligned} \tag{20}$$

which is the desired result. Finally, it is straightforward to check that $\int_{\{l\} \cup \mathbb{M}} p^m(x_i \mid \cdot) m(dx_i) = 1$ given that $\int_{\mathbb{M}} p^c(x_i \mid \cdot)\, dx_i = 1$. □

To pin things down, let us consider the following example.

Example 2 (Gaussian mixed-state Markov random field with $l = \{0\}$)
Assume that the continuous component of the mixed-state model, that is, $p^c(x_i \mid \mathbf{x}_{\mathcal{N}_i})$, follows a Gaussian law with mean $m_i(\mathbf{x}_{\mathcal{N}_i})$ and variance $\sigma_i(\mathbf{x}_{\mathcal{N}_i})$ (the reader should not confuse the mean $m_i(\mathbf{x}_{\mathcal{N}_i})$ with the measure m). The discrete state is set to $l = \{0\}$ with conditional probability of occurrence $\rho_i(\mathbf{x}_{\mathcal{N}_i})$. The local characteristics are then expressed as:

$$p^m(x_i \mid \mathbf{x}_{\mathcal{N}_i}) = \rho_i(\mathbf{x}_{\mathcal{N}_i}) \mathbf{1}_0(x_i) + \rho_i^*(\mathbf{x}_{\mathcal{N}_i}) \mathbf{1}_0^*(x_i) \frac{1}{\sqrt{2\pi}\sigma_i(\mathbf{x}_{\mathcal{N}_i})} e^{-\frac{(x_i - m_i(\mathbf{x}_{\mathcal{N}_i}))^2}{2\sigma_i^2(\mathbf{x}_{\mathcal{N}_i})}}. \tag{21}$$

For simplicity, we abbreviate $m_i \equiv m_i(\mathbf{x}_{\mathcal{N}_i})$, $\sigma_i \equiv \sigma_i(\mathbf{x}_{\mathcal{N}_i})$, $\rho_i \equiv \rho_i(\mathbf{x}_{\mathcal{N}_i})$, and in exponential form:

$$\begin{aligned}
\boldsymbol{\Theta}_i^T(\mathbf{x}_{\mathcal{N}_i}) &= [\theta_{1,i}, \theta_{2,i}, \theta_{3,i}] = \left[-\frac{m_i^2}{2\sigma_i^2} + \log \frac{1}{\sigma_i \sqrt{2\pi}} + \log \frac{\rho_i^*}{\rho_i}, \frac{1}{2\sigma_i^2}, \frac{m_i}{2\sigma_i^2} \right], \\
\mathbf{S}_i^T(x_i) &= \big[\mathbf{1}_l^*(x_i), -x_i^2, x_i \big], \\
C_i(x_i) &= 0, \\
D_i(\mathbf{x}_{\mathcal{N}_i}) &= \log \rho_i.
\end{aligned} \tag{22}$$

Corresponding to a 3-parameter distribution. Note that in this case, where $l = 0$ is a numeric value, $\mathbf{1}_l^*(x_i) x_i = x_i$. The parametrization of the conditional distribution in terms of $\boldsymbol{\Theta}_i$ allows us to express the dependence of a point on its neighbors through (14). Moreover, the parameters of the original parametrization, ρ_i, m_i, σ_i

are also functions of the neighborhood and can be obtained easily from the first line of (22), yielding:

$$\rho_i = \frac{(\sigma_i\sqrt{2\pi})^{-1}}{(\sigma_i\sqrt{2\pi})^{-1} + e^{\theta_{1,i} + \frac{m_i^2}{2\sigma_i^2}}}, \quad \sigma_i^2 = \frac{1}{2\theta_{2,i}}, \quad m_i = \frac{\theta_{3,i}}{2\theta_{2,i}}. \tag{23}$$

Finally, the set of parameters that define the mixed-state automodel with Gaussian conditional densities are the matrices $\boldsymbol{\beta}_{ij} \in \mathbb{R}^{3\times 3}$ and the vector $\boldsymbol{\alpha}_i \in \mathbb{R}^3$. The design of this a model is thus restricted to the choice of these parameters to assure that the joint mixed-state Gibbs distribution is well-defined.

4.3 Causal Mixed-State Markov Models

So far, we have considered the modeling of spatial interaction through the development of mixed-state Markov random fields. However, it is also interesting to consider the case of causal Markovian dependency which leads to the definition of mixed-state Markov chains (MS-MC). Indeed, if one is to study the temporal evolution of a sequence of mixed-state fields as one may observe from a video sequence, it is crucial not only to study the spatial distribution at a given instant, but also statistical interaction in the temporal dimension.

In the general case, we consider a chain of mixed-state random fields. At each instant t, we have a spatial arrangement $\mathbf{x}_t = \{x_{i,t}\}_{i\in S}$ on a lattice S and the Markovian first order dependency is associated to a sequence $\{\mathbf{x}_0, \mathbf{x}_1, \ldots, \mathbf{x}_T\}$ of such fields giving

$$p^m(\mathbf{x}_0, \mathbf{x}_1, \ldots, \mathbf{x}_T) = p^m(\mathbf{x}_0) \prod_{t=1}^{T} p^m(\mathbf{x}_t \mid \mathbf{x}_{t-1}). \tag{24}$$

The evolution of the mixed-state Markov chain is then governed by the transition kernel

$$P(\mathbf{x}_t, \mathbf{x}_{t-1}) = p^m(\mathbf{x}_t \mid \mathbf{x}_{t-1})\nu(d\mathbf{x}_t), \tag{25}$$

where $\nu = m^N$ is a mixed-state measure as before. Thus, the transition kernel behaves as a positive probability for the discrete states and as a continuous transition kernel for the rest. Apart from the construction of the mixed-state random variables developed in Sect. 4.1, the definition of MS-MC is equivalent to that of classical Markov chains.

At this point, we can introduce both a model for spatial and temporal interaction. Spatial interaction can be considered defining a mixed-state Markov random field on $\mathbf{x}_t$ for the specification of $p^m(\mathbf{x}_t \mid \mathbf{x}_{t-1})$ and considering $\mathbf{x}_{t-1}$ as an observed process in modeling the temporal evolution.

5 Sampling, Estimation and Inference in MS-MRF

We now consider three fundamental aspects in the modeling and application of mixed-state Markov random fields: sampling for simulation, parameter estimation for model learning, and inference of mixed-state variables.

5.1 Sampling

Given a MS-MRF with distribution $p^m(\mathbf{x})$ the Gibbs sampler algorithm [22] permits to generate a sequence of configurations $\{\mathbf{x}_0, \mathbf{x}_1, \ldots, \mathbf{x}_\tau\}$ such that for any starting configuration η and each configuration ω:

$$\lim_{\tau \to \infty} p^m_{\mathbf{x}_\tau}(\mathbf{x}_\tau = \omega \mid \mathbf{x}_0 = \eta) = p^m(\omega) \tag{26}$$

that is, the distribution of $\mathbf{x}_\tau$ converges to the given distribution $p^m(\mathbf{x})$ regardless of $\mathbf{x}_0$. The process $\{\mathbf{x}_0, \mathbf{x}_1, \ldots, \mathbf{x}_\tau\}$ behaves as a Markov chain with $p^m(\mathbf{x})$ as equilibrium distribution. This is achieved by applying the following strategy. At each epoch t or step of the algorithm, only one site undergoes a possible change, so that $\mathbf{x}_{t-1}$ and $\mathbf{x}_t$ can differ in at most one coordinate. For that site, say i, a sample s is drawn from the local characteristic $p^m(x_i \mid \mathbf{x}_{t-1,\mathcal{N}_i})$. The new configuration $\mathbf{x}_t$ has $x_{i,t} = s$ and $x_{j,t} = x_{j,t-1}$ for $j \neq i$. The only assumption is that every site is visited an infinite number of times, which is necessary to ensure convergence.

The classical Gibbs sampler can be extended for mixed-state random variables with any number of discrete (possibly symbolic) states, for sampling a mixed-state random field on the state space $\mathbb{M} = \{l\} \cup \mathbb{R}$. The procedure is as follows:

- *Step* 1: *Specify an arbitrary initial configuration* $\mathbf{x}_0$. A possibility is to draw, for each site, a discrete state from the set $k+1$ elements $\{l_1, l_2, \ldots, l_k, \varsigma\}$ each with uniform probability $1/(k+1)$, with ς indicating the continuous state. Then if ς occurs, draw a continuous value from some adequate continuous distribution (uniform, Gaussian, etc.). The other possibility, which somehow helps convergence, is to sample each site assuming independence in the Gibbs distribution, that is, only considering the singleton potentials.
- *Step* 2: *Visit each site and sample the local conditional mixed-state p.d.f.* Sites are to be visited in random order or applying a raster scan. For drawing a value from the mixed-state conditional density, one first proceeds by computing the conditional probability $\rho_i(\mathbf{x}_{\mathcal{N}_i})$ using the current values of the neighbors. Then the site variable is set to l with this probability. If it is not the case, a continuous numeric value is drawn from $p^c(x_i \mid \mathbf{x}_{\mathcal{N}_i})$.
- *Step* 3: *Repeat* 2 *a fixed number of iterations.* The Gibbs sampler converges to the target distribution as the number of iterations become large. Normally a good result is achieved by assuring that each site is visited 100–300 times.

5.2 Parameter Estimation

For a mixed-state automodel (Sect. 4.2.2), the parameters are the set of matrices $\boldsymbol{\beta}_{ij}$ and the vectors $\boldsymbol{\alpha}_i$ for each $i \in S$. In general, the model parameters is a vector of real-valued coefficients $\boldsymbol{\phi}$.

In order to estimate $\boldsymbol{\phi}$, considering the log-likelihood function $L(\boldsymbol{\phi}) = \log p^m(\mathbf{x}; \boldsymbol{\phi}) = Q(\mathbf{x}; \boldsymbol{\phi}) - \log Z(\boldsymbol{\phi})$ becomes intractable in writing the log-partition function $\log Z(\boldsymbol{\phi})$ despite the expression of the Gibbs energy $Q(\mathbf{x}; \boldsymbol{\phi})$ may show a more or less simple dependence with the parameters. The problem of implementing a maximum-likelihood estimator therefore rests upon the evaluation of the partition function, the treatment of which becomes unmanageable, both numerically and analytically, specially when the number of possible configurations is big.

A tractable alternative was given by Besag [3] who proposed to write the so-called *pseudo-likelihood function* which in logarithmic form yields

$$L_{\mathrm{PL}}(\boldsymbol{\phi}) = \sum_{i \in S} \log p^m(x_i \mid \mathbf{x}_{\mathcal{N}_i}, \boldsymbol{\phi}), \tag{27}$$

where the need to evaluate Z is avoided. The Maximum-pseudo-likelihood estimates can then be obtained by solving

$$\nabla_{\boldsymbol{\phi}} L_{\mathrm{PL}}(\boldsymbol{\phi}) = 0. \tag{28}$$

Though $L_{\mathrm{PL}}(\boldsymbol{\phi})$ is not a true likelihood function as the product of the conditional densities does not necessarily result in a true p.d.f., this conceptually intuitive approach results in a consistent estimator (it converges with probability one) of the true parameters [23], which gives a mathematical justification of its widely accepted use.

When the conditional mixed-state p.d.f.'s belong to an exponential family of distributions as in the multiparameter auto-model, we have

$$p^m(x_i \mid \mathbf{x}_{\mathcal{N}_i}, \boldsymbol{\phi}) = \frac{e^{-\boldsymbol{\phi}^T \cdot \mathbf{H}_i(x_i, \mathbf{x}_{\mathcal{N}_i})}}{Z_i(\boldsymbol{\phi}, \mathbf{x}_{\mathcal{N}_i})}, \tag{29}$$

where we can identify $-\boldsymbol{\phi}^T \cdot \mathbf{H}_i(x_i, \mathbf{x}_{\mathcal{N}_i}) = \boldsymbol{\Theta}_i^T(\mathbf{x}_{\mathcal{N}_i}) \cdot \mathbf{S}_i(x_i)$ given the affine form of $\boldsymbol{\Theta}_i^T(\mathbf{x}_{\mathcal{N}_i})$ (14), and $Z_i(\boldsymbol{\phi}, \mathbf{x}_{\mathcal{N}_i}) \equiv Z_i$ is the normalization factor of the conditional mixed-state density. Applying the logarithm and differentiating twice with respect to the parameter vector $\boldsymbol{\phi}$,

$$\begin{aligned} \frac{\partial^2}{\partial \boldsymbol{\phi}^2} \log p^m(x_i \mid \mathbf{x}_{\mathcal{N}_i}, \boldsymbol{\phi}) &= -\frac{\partial^2}{\partial \boldsymbol{\phi}^2} \log Z_i = \frac{1}{Z_i^2}\left(\frac{\partial Z_i}{\partial \boldsymbol{\phi}}\right)^T \left(\frac{\partial Z_i}{\partial \boldsymbol{\phi}}\right) - \frac{1}{Z_i}\frac{\partial^2 Z_i}{\partial \boldsymbol{\phi}^2} \\ &= -E\big[\mathbf{H}_i(x_i, \mathbf{x}_{\mathcal{N}_i})\mathbf{H}_i^T(x_i, \mathbf{x}_{\mathcal{N}_i})\big] \\ &\quad + E\big[\mathbf{H}_i(x_i, \mathbf{x}_{\mathcal{N}_i})\big]E\big[\mathbf{H}_i^T(x_i, \mathbf{x}_{\mathcal{N}_i})\big] \\ &= -\mathrm{Cov}\big(\mathbf{H}_i(x_i, \mathbf{x}_{\mathcal{N}_i})\big) \le 0, \end{aligned} \tag{30}$$

where $\mathrm{Cov}(\mathbf{H}_i(x_i, \mathbf{x}_{\mathcal{N}_i}))$ is the covariance matrix which is positive definite, $\frac{\partial Z_i}{\partial \boldsymbol{\phi}}$ is a row-vector (i.e., the gradient) and the second equality results from differentiating $Z_i = \int \exp\{-\boldsymbol{\phi}^T \cdot \mathbf{H}_i(x_i, \mathbf{x}_{\mathcal{N}_i})\}m(dx)$. The pseudo-likelihood (as well as the classical likelihood) is then a concave function of the parameters, which assures the unicity of the associated estimator. It also assures the convergence of gradient-based algorithms for iterative optimization.

5.3 *Inference of Mixed-State Random Fields*

When applying mixed-state random fields within a Maximum-a-posteriori formulation, as for example in [16], the MAP energy to be maximized depends on discrete and continuous states. Thus, a method is to be defined for performing inference of mixed-state values. Given a set of observations $\mathbf{y}$, the maximum-a-posteriori estimate maximizes the posterior distribution $p^m(\mathbf{x} \mid \mathbf{y})$.

We specifically consider the ICM method [4] because it is valid regardless of the nature of the space of values. The formulation of the algorithm can be understood from the following identity:

$$p^m(\mathbf{x} \mid \mathbf{y}) = p^m(x_i \mid \mathbf{x}_{S\setminus\{i\}}, \mathbf{y}) p^m(\mathbf{x}_{S\setminus\{i\}} \mid \mathbf{y}). \tag{31}$$

By choosing the value at site i that maximizes the conditional probability $p^m(x_i \mid \mathbf{x}_{S\setminus\{i\}}, \mathbf{y}) = p^m(x_i \mid \mathbf{x}_{\mathcal{N}_i}, \mathbf{y})$, it results that $p^m(\mathbf{x} \mid \mathbf{y})$ increases [23]. The difficulty of maximizing the joint probability of an MRF is avoided, by applying an iterative and greedy procedure exploiting relation (31), and visiting each site infinitely often.

The ICM algorithm for a mixed-state random field is as follows:

- *Step* 1: *Choose an initial estimate* $\mathbf{x}_0$ *conveniently.* One of the complications of the ICM method is that the final solution depends very much on the initial estimate $\mathbf{x}_0$. A possible choice is to assume independence and set $\mathbf{x}_0 = \arg\max_{\mathbf{x}} \prod_{i \in S} p^m(x_i \mid \mathbf{y})$ so as to base the first estimate only on the observations and discarding higher-order interaction potentials.
- *Step* 2: *Visit each site and maximize the local conditional p.d.f.* Sites are to be visited in random order or applying a raster scan. For each site i, obtain the maximizing value $\hat{x}_i$ of $p^m(x_i \mid \mathbf{x}_{\mathcal{N}_i}, \mathbf{y})$ given the current state of the neighbors $\mathbf{x}_{\mathcal{N}_i}$. Then set $x_i = \hat{x}_i$. For a mixed-state p.d.f. (11), the maximization step implies applying the following procedure:

  (i) compute $\rho_i(\mathbf{x}_{\mathcal{N}_i}, \mathbf{y}) = p^m(x_i = l \mid \mathbf{x}_{\mathcal{N}_i}, \mathbf{y})$
  (ii) maximize $p^c(x_i \mid \mathbf{x}_{\mathcal{N}_i}, \mathbf{y})$ w.r.t. the continuous values to obtain x_i^c
  (iii) if $\rho_i(\mathbf{x}_{\mathcal{N}_i}, \mathbf{y}) > \rho_i^*(\mathbf{x}_{\mathcal{N}_i}, \mathbf{y}) p^c(x_i = x_i^c \mid \mathbf{x}_{\mathcal{N}_i}, \mathbf{y})$, set $\hat{x}_i = l$, else $\hat{x}_i = x_i^c$

- *Step* 3: *Repeat* 2 *until convergence.* Convergence is difficult to determine, and usually a fixed number of iterations is applied, which is set empirically, but often in the order of 10–20 passes by each site.

6 Characterizing Motion Textures with MS-MRF

In Sect. 3, we have analyzed the statistical properties of motion textures and we have seen that motion observations are obtained depicting a spatial arrangement of mixed-state values. A first approach consists in modeling the instantaneous motion maps associated to dynamic textures, as a mixed-state spatial random field. In general terms, a MS-MRF could be defined by a different set of parameters for each location of the image [see (14)]. This would give rise to a motion texture model with a number of parameters proportional to the image size. Unfortunately, such high-dimensional representation is unfeasible in practice and does not constitute a compact description of motion textures. Moreover, an increasing number of frames would be necessary for the estimation process. This is against a formulation oriented to efficient content recognition and retrieval. However, the framework could deal with spatially non-stationary motion textures.

We will assume that the extracted motion fields can be considered as realizations of an homogeneous MS-MRF spatial model. Indeed, the visual information attached to a dynamic texture is mostly displayed from spatially homogeneous motion regions, and moreover, mostly associated to statistically homogeneous textured intensity patterns.

The motion histograms in Fig. 1 suggest that the continuous values can be appropriately modeled with a Gaussian distribution. Thus, in what follows we define an instance of the Gaussian mixed-state auto-model described through (21), (22), and (23) for motion texture modeling.

6.1 Defining the Set of Parameters

Regarding the neighborhood structure, we define $\mathcal{N}_i = \{i_E, i_W, i_N, i_S, i_{NW}, i_{SE}, i_{NE}, i_{SW}\}$ as the set of the 8-nearest neighbors for location i, where for example, i_E is the east neighbor of i in the image grid, i_{NW} the north-west neighbor, etc. Moreover, a necessary condition in order to define an homogeneous and, in fact, stationary spatial process, is that the parameters related to symmetric neighbors (E-W, N-S, NW-SE, NE-SW) must be the same.

It is desirable that the conditional mean of the continuous values for a site, depends linearly on the neighbors, in order to effectively obtain a Gaussian texture for the continuous values. With this assumption, the model is able to extract the main properties of the field, also keeping a reduced number of parameters to be estimated. The parameters are then chosen to be

$$\boldsymbol{\beta}_{ij} = \begin{pmatrix} d_{ij} & 0 & 0 \\ 0 & 0 & 0 \\ 0 & 0 & h_{ij} \end{pmatrix}, \qquad \boldsymbol{\alpha} = \begin{bmatrix} a & b & c \end{bmatrix}^T, \tag{32}$$

so that with this choice and from (14) and (23), we obtain

$$\rho_i = \frac{(\sigma_i\sqrt{2\pi})^{-1}}{(\sigma_i\sqrt{2\pi})^{-1} + e^{a+\sum_{j\in\mathcal{N}_i} d_{ij}\mathbf{1}_0(x_j)+\frac{m_i^2}{2\sigma_i^2}}}, \quad m_i = \frac{c}{2b} + \sum_{j\in\mathcal{N}_i} \frac{h_{i,j}}{2b} x_j, \ \sigma_i = \frac{1}{2b}. \tag{33}$$

Note that the homogeneity of the field was applied to set $\boldsymbol{\alpha}_i = \boldsymbol{\alpha}$ for the first order potentials and as already said, this also implies the symmetry of the parameters for second order potentials. Thus, for the 8-point neighborhood, we have 4 interacting directions: vertical (V), horizontal (H), diagonal (D), and anti-diagonal (AD). Then, $\boldsymbol{\beta}_{ij} = \boldsymbol{\beta}_k$ with $k \in \{H, V, D, AD\}$ and an homogeneous Gaussian mixed-state model is defined by the 11 parameters $\boldsymbol{\phi} = \{a, b, c, d_H, h_H, d_V, h_V, d_D, h_D, d_{AD}, h_{AD}\}$.

Another aspect, related to spatial interaction, is considered in the definition of the model. The type of motion textures that we want to study show some local motion smoothness, mostly associated to *cooperative* schemes. Then, this condition is explicitly imposed in the model, resulting in a constraint on the parameters. Formally, in a mixed-state cooperative model, the conditional mean of the continuous component for a site has to be an increasing function of its neighbors. See [24] for further comments. Following equation (33), this implies that $h_{ij} \geq 0$. Finally, we can write the full expression of the global mixed-state Gibbs energy to obtain:

$$Q(\mathbf{x}) = \sum_i a\mathbf{1}_0^*(x_i) - bx_i^2 + cx_i + \sum_{\langle i,j\rangle: j\in\mathcal{N}_i} h_{ij}x_ix_j + d_{ij}\mathbf{1}_0^*(x_i)\mathbf{1}_0^*(x_j). \tag{34}$$

In order to check that the Gibbs density defined by the energy function in equation (34) is integrable, a sufficient and necessary condition for the proposed homogeneous cooperative MS-MRF to converge is $b > \sum_j \frac{h_{ij}}{2}$. The proof is given in [14].

6.2 Recognition of Motion Textures

One of the key aspects of a model oriented to dynamic content recognition, as the one discussed here, is the ability to define a way of computing some similarity measure between models, in order to embed it in a decision-theoretic-based application. In this context, the Kullback–Leibler (KL) divergence is a well-known distance (more precisely, a pseudo-distance) between statistical models [13]. This quantity can be computed between general Gibbs distributions and thus MS-MRF's, and will allow us to find a strategy for classifying motion textures.

Recall the expression of the KL divergence from a density $p_1(\mathbf{x})$ to $p_2(\mathbf{x})$ [13]:

$$\mathrm{KL}(p_1 \| p_2) = \int_\Omega p_1(\mathbf{x}) \log \frac{p_1(\mathbf{x})}{p_2(\mathbf{x})} m(d\mathbf{x}), \tag{35}$$

where $m(d\mathbf{x})$ is a reference measure, possibly a mixed-state measure. Given that $\mathrm{KL}(p_1 \| p_2)$ is not symmetric one usually considers the symmetrized KL divergence as $d_{\mathrm{KL}}(p_1, p_2) = \frac{1}{2}[\mathrm{KL}(p_1(\mathbf{x}) \| p_2(\mathbf{x})) + \mathrm{KL}(p_2(\mathbf{x}) \| p_1(\mathbf{x}))]$. Now, if $p_1(\mathbf{x})$ and $p_2(\mathbf{x})$ are Markov random fields, then $\log \frac{p_1(\mathbf{x})}{p_2(\mathbf{x})} = -\Delta Q(\mathbf{x}) + \log \frac{Z_2}{Z_1}$, where $\Delta Q(\mathbf{x}) = Q_2(\mathbf{x}) - Q_1(\mathbf{x})$, and

$$d_{\mathrm{KL}}\big(p_1(\mathbf{x}), p_2(\mathbf{x})\big) = \frac{1}{2}\big(E_{p_2}\big[\Delta Q(\mathbf{x})\big] - E_{p_1}\big[\Delta Q(\mathbf{x})\big]\big). \tag{36}$$

We observe from this general equation, that we do not need to have knowledge of the partition functions of the Gibbs distributions which simplifies enormously the handling of this expression. Now, let $p_1(\mathbf{x})$ and $p_2(\mathbf{x})$ be two Gaussian MS-MRF. Then,

$$E_{p_k}\big[\Delta Q(\mathbf{x})\big] = \sum_i \Delta\boldsymbol{\alpha} E_{p_k}\big[\mathbf{S}(x_i)\big] + \sum_{\langle i,j\rangle} E_{p_k}\big[\mathbf{S}(x_i)\Delta\boldsymbol{\beta}_{ij}\mathbf{S}(x_j)\big], \tag{37}$$

where $\Delta\boldsymbol{\alpha} = \boldsymbol{\alpha}^{(2)} - \boldsymbol{\alpha}^{(1)}$ and $\Delta\boldsymbol{\beta}_{ij} = \boldsymbol{\beta}_{ij}^{(2)} - \boldsymbol{\beta}_{ij}^{(1)}$. Finally, for the motion texture model

$$\begin{aligned} E_{p_k}\big[\Delta Q(\mathbf{x})\big] &= \sum_i \Delta a E_{p_k}\big[\mathbf{1}_0^*(x_i)\big] - \Delta b E_{p_k}\big[x_i^2\big] + \Delta c E_{p_k}[x_i] \\ &\quad + \sum_{\substack{\langle i,j\rangle \\ j\in\mathcal{N}_i}} \Delta h_{ij} E_{p_k}[x_i x_j] + \Delta d_{ij} E_{p_k}\big[\mathbf{1}_0^*(x_i)\mathbf{1}_0^*(x_j)\big]. \end{aligned} \tag{38}$$

The involved expectations w.r.t. each p_k in (38) can be computed easily from the observed motion textures given that we have a spatial homogeneous model and they are equal for each site of the motion field.

6.2.1 Application to Motion Texture Classification

The following experiment is intended to validate the recognition performance of the mixed-state motion texture models. Let us consider 10 different motion texture classes (Fig. 5): Steam, Straw, Traffic, Water, Candles, Shower, Flags, Water-Rocks, Waves, Fountain. A total of 30 different sequences were considered from the DynTex dynamic texture database [31], and for each one, 5 pairs of consecutive images were selected at frames 1, 20, 40, 60, 80, for a total of 150 samples. All sequences were initially composed by gray scale images with a resolution of 720×576 pixels, given at a rate of 25 frames per second. In order to reduce computation time, the original images were filtered and subsampled to a resolution of 180×144 pixels. The motion maps were computed for this images of reduced resolution by applying (4). Then, each motion texture class parameter set is learned from a single pair of images picked from only one of the sequences belonging to each type of motion texture.

Table 1 Parameters learned for the 10 motion texture classes (C: center, H: horizontal, V: vertical, D: diagonal, AD: anti-diagonal)

	a	b	c	d_H	h_H	d_V	h_V	d_D	h_D	d_{AD}	h_{AD}
Steam	−2.72	0.274	0.012	0.441	0.047	0.604	0.082	0.249	0.064	0.323	0.077
Straw	−1.93	4.07	−0.065	0.61	0.00	0.79	1.13	0.07	0	0.41	2.91
Traffic	−4.96	4.5	−0.05	1.87	4.27	1.95	0.177	0.195	0	0.562	0
Water	−2.29	8.18	0.031	1.23	8.1	0.927	0	0.168	0	0.017	0
Candles	−4.98	17	−0.647	2.47	8.89	2.55	5.82	0.482	2.11	0.066	0
Shower	−2.09	2.03	−0.005	0.515	0	0.832	1.18	0.0649	0	0.392	0.837
Flags	−4.91	0.434	−0.006	1.07	0.018	2.08	0.322	0.219	0.047	0.292	0.021
Water-rocks	−2.88	2.68	0.01	1	1.53	0.979	1.1	0.228	0.314	0	0.153
Waves	−2.36	2.2	−0.016	1.08	1.61	0.693	0.439	0.151	0	0.153	0.135
Fountain	−3.26	3.97	0.005	1.11	0.639	1.11	3.29	0.331	0	0.465	0

Table 2 Motion texture class confusion matrix. Each row indicates how the samples for a class were classified

	Steam	Straw	Traffic	Water	Candles	Shower	Flags	Water-rocks	Waves	Fountain
Steam	**100%**	–	–	–	–	–	–	–	–	–
Straw	–	**93.3%**	–	–	–	6.7%	–	–	–	–
Traffic	–	–	**86.7%**	–	–	–	–	13.3%	–	–
Water	–	–	–	**100%**	–	–	–	–	–	–
Candles	–	–	13.3%	–	**73.4%**	–	–	13.3%	–	–
Shower	20%	–	–	–	–	**80%**	–	–	–	–
Flags	–	–	–	–	–	–	**100%**	–	–	–
Water-rocks	–	–	–	–	–	–	–	**93.3%**	–	6.7%
Waves	–	–	–	6.7%	–	–	–	13.3%	**80%**	–
Fountain	–	–	–	–	–	–	–	–	–	**100%**

The MS-MRF model parameters for each of the 10 training samples are learned by applying the pseudo-likelihood maximization criterion discussed in Sect. 5.2 (Table 1). We use a gradient descent technique for the optimization as the derivatives of $L_{\mathrm{PL}}(\boldsymbol{\phi})$ w.r.t. $\boldsymbol{\phi}$ are known in closed form. For the testing stage, we estimate $\boldsymbol{\phi}$ for each test sample and compute the Kullback–Leibler distance (36) with each learned parameter vector.

In Table 2, we show the confusion matrix for the 10 motion texture classes. A correct recognition is considered when both, the model estimated for the test sample and the closest reference model, belong to the same class. An overall classification rate of 90.7% was achieved. With our non-optimized implementation in MATLAB running on a laptop with an Intel Core 2 Duo 1.83 GHz processor and 1 GB RAM, the process of recognizing a motion texture takes 3.2 seconds in average. The bottleneck of the process is the estimation of the parameters for the test sample.

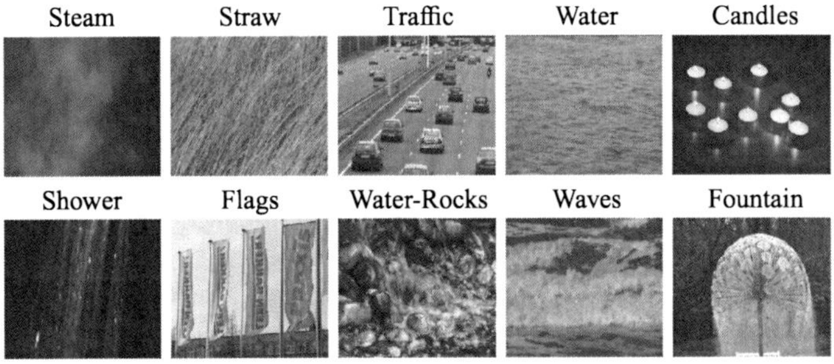

Fig. 5 Sample images from the 10 motion texture classes used for the recognition experiments. For some of the samples the dynamic texture of interest does not occupy the whole image, violating the assumption of spatial homogeneity. However the model is sufficiently robust to allow a correct classification

As for the confusion matrix, let us note that it is likely that waves are classified as water or water-rocks as they correspond to close dynamic contents, straw is confused with shower, they have a close vertical orientation, and candles can be classified as traffic, as both classes show a motion pattern consisting of isolated blobs. The non-symmetry of the confusion matrix is associated to the nature of the tested data set, where for some classes, the tested sequences have a closer resemblance to the training sample, while for others, there are notorious variations, that may lead to misclassifications. In 4 out of 10 classes the recognition rate is as high as 100% and except for the Candles class we always get rates above 80%. Note also in Fig. 5 that for some of the samples the dynamic texture of interest does not occupy the whole image, violating the assumption of spatial homogeneity. However, the model is sufficiently robust to allow a correct classification.

Reported experiments for dynamic texture recognition using a model-based approach as the one presented in [34], have shown a classification rate of 89.5%, on a data set composed by dynamic texture classes that are similar to the ones used here, as plants, smoke, waterfalls, ocean waves, etc. Their method is based on computing a subspace distance between the linear models learned for each class, that describe the evolution of the image intensity over time. Although the effectiveness of this approach is similar to ours, the method proposed here has a big advantage, that is, we only need two consecutive frames to estimate and recognize the mixed-state models, while in [34] subsequences of 75 frames are used. This is a consequence of modeling the spatial structure of motion rather than the time evolution of the image.

Motion-based methods for dynamic texture classification [20, 30] have shown an improved performance with recognition rates of over 95% using invariant flow statistics. Although they are the most accurate reported results for addressing this problem, they do not provide a general characterization of dynamic textures, as model-based approaches do. Consequently, more complex scenarios with combined problems, as simultaneous detection, segmentation, and recognition of these type of sequences are not directly addressable. Our framework provides an unified statistical representation suitable of being applicable to other complex problems, as well.

6.2.2 Temporal Consistency

The previous experiments on motion texture classification have shown that training a motion texture class from a single pair of images is sufficient in order to achieve a high recognition rate. However, a question may remain about the validity of the spatial model to represent these dynamic contents, given that in fact they have a much longer temporal extent. So in what follows, we show that for modeling temporal stationary motion textures, the spatial model estimated from only two consecutive frames is representative of the rest of the sequence, that is, the MS-MRF model parameters are consistent over time.

We take two sequences, each one composed by two different motion textures (Fig. 6): Ocean–Steam[1] (a circular region associated to steam superimposed to a sequence of ocean waves) and Trees (two different kinds of trees moved by the wind, and in different planes with respect to the camera). We estimate the set of parameters over manually fixed regions (Figs. 6(a) and 6(d)) corresponding to each of the

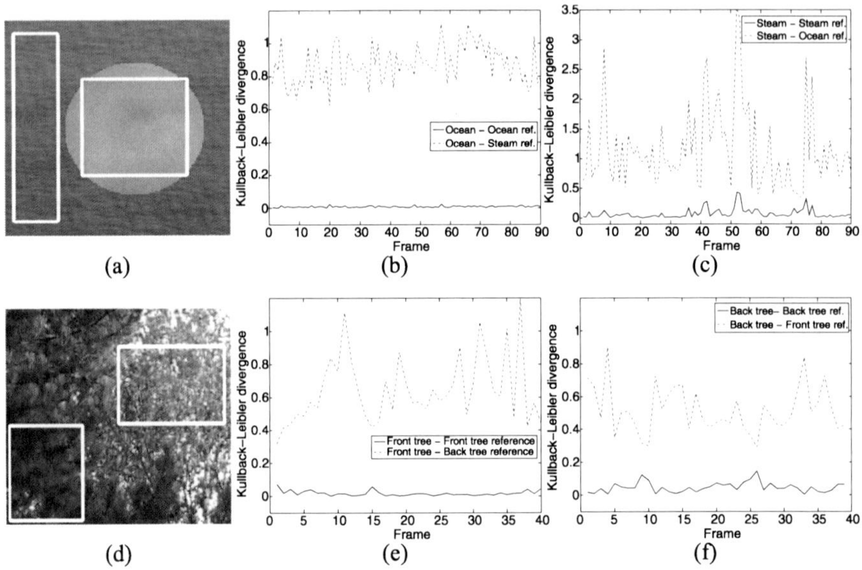

Fig. 6 Temporal consistency of the motion texture model parameters. (**a**) Ocean–Steam sequence. The *white squares* indicate manually fixed regions along which the parameters for each class were estimated. (**b**) Kullback–Leibler distance between the parameters for the Ocean class at each time instant and the reference model obtained from the first pair of frames for Ocean (*solid line*) and Steam (*dashed line*). (**c**) KL distance between Steam and the reference model for Steam (*solid line*) and Ocean (*dashed line*). (**d**) Trees sequence, where we have a Front tree and a Back tree. (**e**) KL distance Front tree–Front tree reference (*solid line*) and Front tree–Back tree reference (*dashed line*). (**f**) KL distance Back tree–Back tree reference (*solid line*) and Back tree–Front tree reference (*dashed line*)

[1]Copyright (c) 2003, UCLA Vision Lab. Thanks to Daniel Cremers and Stefano Soatto for providing this sequence.

motion texture classes. From the first pair of frames, we obtain a set of representative reference parameters for each class. Finally, we compute the Kullback–Leibler divergence between the reference models obtained at $t = 1$, and the parameters estimated for the rest of the sequence within each region (for each $t > 1$).

In Fig. 6(b), we plot the KL distance between the MS-MRF model estimated for the Ocean motion texture at each instant, and the reference MS-MRF models for Ocean (solid line) and Steam (dashed line). We see that the model for Ocean is very close (in term of KL) to the reference Ocean model extracted from the first pair of frames. At the same time, the distance between Ocean and Steam is large along the whole sequence. Figure 6(c) is the complementary experiment for the Steam motion texture and the same behavior is confirmed for the Trees sequence, where we have two motion textures: a Front tree and a Back tree (Fig. 6(e)).

In summary, the motion texture model parameters obtained for a given instant are indeed consistent in time in the sense that they are representative of the whole sequence in the case of temporal stationary motion textures.

7 Mixed-State Causal Modeling of Motion Textures

So far, we have studied dynamic textures through the modeling and characterization of the instantaneous apparent (and scalar) motion maps, that is, motion textures. As a spatial arrangement of mixed-state values, we have developed purely spatial and noncausal mixed-state Markov models. However, as pointed out before, the dynamic video content associated to a dynamic texture usually extents over some time interval. If the process is stationary within this interval, it is interesting to analyze a model able to integrate this temporal information in a compact representation.

We analyze now a motion texture model based on the mixed-state Markov chain (MS-MC) defined in Sect. 4.3. We consider a Markov chain of random fields $\mathbf{x} = \{\mathbf{x}_t\}_{t:0,\dots,T}$, as in (24), where $\mathbf{x}_t = \{x_i(t)\}_{i \in S}$ represents a motion texture field computed at time t. As before, each (motion) random variable $x_i(t)$ is assumed to be a mixed-state random variable in $\mathbb{M} = \{0\} \cup \mathbb{R}$.

Here we study a purely causal temporal model, for which a first assumption is to consider spatial conditional independence within a motion texture for time t. Consequently, given the previous instant, and assuming a local dependency on a neighborhood $\mathcal{N}_{i,t-1}$ of 'past' random variables at time $t-1$ (Fig. 7), we consider the following transition densities

$$p^m(\mathbf{x}_t \mid \mathbf{x}_{t-1}) = \prod_{i \in S} p^m(x_{i,t} \mid \mathbf{x}_{\mathcal{N}_{i,t-1}}), \tag{39}$$

where $x_{i,t} = x_i(t)$. In our case, we will assume that the temporal neighborhood is a 9-point set which includes the previous (at $t-1$) center, diagonal, anti-diagonal, horizontal and vertical motion variables for a point at time t as depicted in Fig. 7. Note that conditional independence does not imply independence and actually for a given t and $i, j \in S$, $x_{i,t}$ and $x_{j,t}$ are spatially correlated.

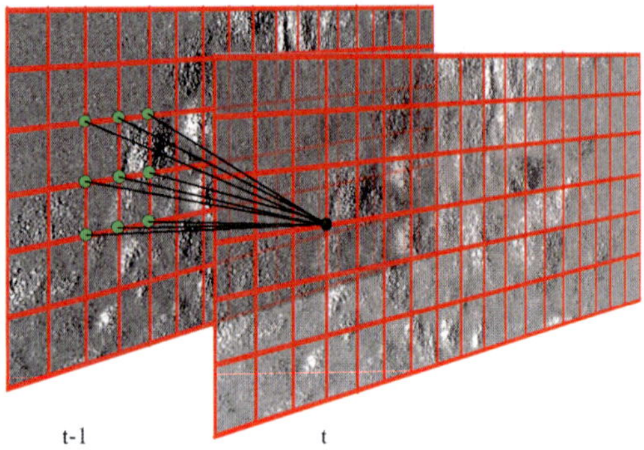

Fig. 7 Temporal neighborhood structure for the mixed-state Markov chain. At a given time instant t the motion values within $\mathbf{x}_t$ are considered conditionally independent given $\mathbf{x}_{t-1}$

From a physical point of view, the assumption of causal interaction (instead of spatial interaction) is still valid as one can consider that the motion textures between consecutive time instants are identically distributed.

Let us consider a Gaussian MS-MC for which

$$p^m(x_{i,t} \mid \mathbf{x}_{\mathcal{N}_{i,t-1}}) = \rho_{i,t}\mathbf{1}_0(x_{i,t}) + \rho^*_{i,t}\mathbf{1}^*_0(x_{i,t})\frac{1}{\sqrt{2\pi}\sigma_{i,t}}e^{-\frac{(x_{i,t}-m_{i,t})^2}{2\sigma^2_{i,t}}}. \tag{40}$$

Analogously to the spatial motion texture model, we take the mean $m_{i,t}$ as a weighted average of its past neighbors,

$$m_{i,t} = c + \sum_{j\in\mathcal{N}_{i,t-1}} h_j x_{j,t-1}, \tag{41}$$

and $\sigma^2_{i,t} = \sigma^2$ is a constant for every point. The expression for $\rho_{i,t}$ may be chosen arbitrarily for a Markov chain, as there are not consistency conditions to be fulfilled as it occurs in a MS-MRF. We can then take a similar expression as for the spatial model

$$\rho_{i,t} = \left[1 + \sqrt{2\pi}\sigma e^{a+\sum_{\mathcal{N}_{i,t-1}} d_{ij}\mathbf{1}^*_0(x_{j,t-1})+\frac{m^2_{i,t}}{2\sigma^2}}\right]^{-1}, \tag{42}$$

where we see that there is a simultaneous coupling between the continuous and discrete states of the past neighbors. This shape of $\rho_{i,t}$ takes into account a cooperative interaction between neighbors: more non-null neighboring values diminishes $\rho_{i,t}$ as well as a larger motion value does, through m_i.

Finally, the MS-MC motion texture model is defined by 13 parameters (considering the five interacting directions described before), which are

$$\boldsymbol{\phi} = \{a, \sigma^2, c, d_C, h_C, d_H, h_H, d_V, h_V, d_D, h_D, d_{AD}, h_{AD}\}.$$

Note that we are assuming spatial homogeneity as for every point $x_{i,t}$ the same set of parameters applies.

7.1 Learning MS-MC Motion Texture Models

A maximum likelihood formulation can be exploited to estimate the MS-MC model from the likelihood function

$$p^m(\mathbf{x}_0)\prod_{t=1}^{T} p^m(\mathbf{x}_t \mid \mathbf{x}_{t-1};\boldsymbol{\phi}). \tag{43}$$

In general, one can take any number of consecutive frames ($T > 1$) to obtain a sequence of motion textures and estimate the parameters. However, considering an homogeneous model permits to eventually consider a pair of motion textures and obtain the parameters by maximizing the likelihood $p^m(\mathbf{x}_t \mid \mathbf{x}_{t-1};\boldsymbol{\phi}) = \prod_{i\in S} p^m(x_{i,t} \mid \mathbf{x}_{\mathcal{N}_{i,t-1}};\boldsymbol{\phi})$. Refer to Sect. 8 for a discussion on the advantages and disadvantages of considering a larger temporal window.

In any case, a gradient descent algorithm is used as the expressions for the gradient of the cost function $\log p^m(\mathbf{x}_t \mid \mathbf{x}_{t-1};\boldsymbol{\phi})$ w.r.t. the parameters can be obtained in closed form.

7.2 Model Matching

In the same way as for the spatial model, a measure of similarity between MS-MC motion texture models is given by the Kullback–Leibler divergence. One can use in this case the conditional Kullback–Leibler divergence [13] between the mixed-state conditional distributions $p(\mathbf{x}_t \mid \mathbf{x}_{t-1})$ in (39). As a matter of fact, it is more formally correct to compute KL for the joint distribution $p^m(\mathbf{x}_t, \mathbf{x}_{t-1})$, but this involves knowing $p^m(\mathbf{x}_{t-1})$ which, as said before, is indeed a correlated spatial field. A joint spatio-temporal model would then be necessary, for example by combining MS-MRF's and MS-MC's in a more complex setting.

Equation (40) can be more conveniently expressed in the form $p^m(x_{i,t} \mid \mathbf{x}_{\mathcal{N}_{i,t-1}}) = \exp Q_{i,t}(x_{i,t}, \mathbf{x}_{\mathcal{N}_{i,t-1}})/Z_i(\boldsymbol{\phi})$ where $Z_i(\boldsymbol{\phi})$ is a normalizing factor and:

$$\begin{aligned} Q_{i,t}(x_{i,t}, \mathbf{x}_{\mathcal{N}_{i,t-1}}) = {} & -\frac{x_{i,t}^2}{2\sigma^2} + c\frac{x_{i,t}}{\sigma^2} + \sum_{j\in\mathcal{N}_{i,t-1}} \frac{h_j}{\sigma^2} x_{i,t} x_{j,t-1} + a\mathbf{1}_0^*(x_{i,t}) \\ & + \sum_{j\in\mathcal{N}_{i,t-1}} d_{ij}\mathbf{1}_0^*(x_{i,t})\mathbf{1}_0^*(x_{j,t-1}) + \log \rho_{i,t}, \end{aligned} \tag{44}$$

where we distinguish the continuous and discrete terms. The conditional KL divergence from the density $p_1 = p^m(\mathbf{x}_t \mid \mathbf{x}_{t-1}, \boldsymbol{\phi}_1)$ to the density $p_2 = p^m(\mathbf{x}_t \mid \mathbf{x}_{t-1}, \boldsymbol{\phi}_2)$ is defined as

$$\mathrm{KL}(p_1 \| p_2) = E_{p_1}\left[\log \frac{p_1}{p_2}\right] = \int_{\Omega} \log \frac{p^m(\mathbf{x}_t \mid \mathbf{x}_{t-1}, \boldsymbol{\phi}_1)}{p^m(\mathbf{x}_t \mid \mathbf{x}_{t-1}, \boldsymbol{\phi}_2)} p^m(\mathbf{x}_t, \mathbf{x}_{t-1} \mid \boldsymbol{\phi}_1)\, dm, \tag{45}$$

which is independent of the measure m. Define as before the symmetrized KL divergence as $d_{\mathrm{KL}}(p_1, p_2) = \frac{1}{2}[\mathrm{KL}(p_1 \| p_2) + \mathrm{KL}(p_2 \| p_1)]$ and analogously to equation (36):

$$d_{\mathrm{KL}}(p_1, p_2) = \frac{1}{2} \sum_i E_{p_2}\big[\Delta Q_{i,t}(x_{i,t}, \mathbf{x}_{\mathcal{N}_{i,t-1}})\big] - E_{p_1}\big[\Delta Q_{i,t}(x_{i,t}, \mathbf{x}_{\mathcal{N}_{i,t-1}})\big]. \quad (46)$$

Observing the expression of $Q_{i,t}^{(k)}(\cdot)$ in (44), we note that $d_{\mathrm{KL}}(p_1, p_2)$ can be computed by estimating the following expectations w.r.t. each model $\boldsymbol{\phi}_k$:

$$\begin{aligned} &E_{p_k}\big[\mathbf{1}_0^*(x_{i,t})\big] = P_{p_k}[x_{i,t} \neq 0], \qquad E_{p_k}\big[\mathbf{1}_0^*(x_{i,t})\mathbf{1}_0^*(x_{j,t-1})\big], \\ &E_{p_k}[x_{i,t}], \qquad E_{p_k}\big[x_{i,t}^2\big], \qquad E_{p_k}[x_{i,t}x_{j,t-1}], \qquad E_{p_k}\big[\log \rho_{i,t}^{(1)}/\rho_{i,t}^{(2)}\big], \end{aligned} \quad (47)$$

where $\rho_{i,t}^{(k)}$ is as in (42), using the corresponding parameters $\boldsymbol{\phi}_k$. As we assume that we have a spatially homogeneous model, the latter expectations involved in equation (47) are equal for each site of the motion field, and thus, they can be efficiently estimated by simple averaging from the observed data and using the estimated parameters.

8 Mixed-State Markov Chain vs. Mixed-State Markov Random Field Motion Texture Models

Two different mixed-state models for motion texture characterization have been considered so far: a spatial noncausal MS-MRF and a purely causal MS-MC. In order to design an effective and efficient representation of motion textures, it is important to analyze when one method should be preferred over the other:

- Besides having a particular spatial extension, a motion texture extents in the temporal dimension as well. The dynamic content is then present during some time interval. If the spatial extent is sufficiently large, the MS-MRF spatial model can be applied given that the model is learned using enough image points (Fig. 8(b)). If the spatial support of the motion texture is small, the parameters of the MS-MRF cannot be estimated accurately. In contrast, the MS-MC causal model can be applied by assuming an adequately temporal window of sufficient length T in (43), so that it involves $T \times N$ points (Fig. 8(a)), where N is the spatial size.
- As for classical Markov random field models, the problem of partition function calculation of the mixed- state Gibbs distribution can be problematic, as for example when trying to apply a maximum likelihood estimator as discussed in Sect. 5.2. For the causal model, this problem is not present as the joint distribution factorizes in the product of the conditional densities.
- The spatial MS-MRF motion texture model permits to model an instantaneous motion map without the need to observe a temporal window. If one deals with a non-stationary dynamic content (Fig. 8(c)), as for example a motion texture

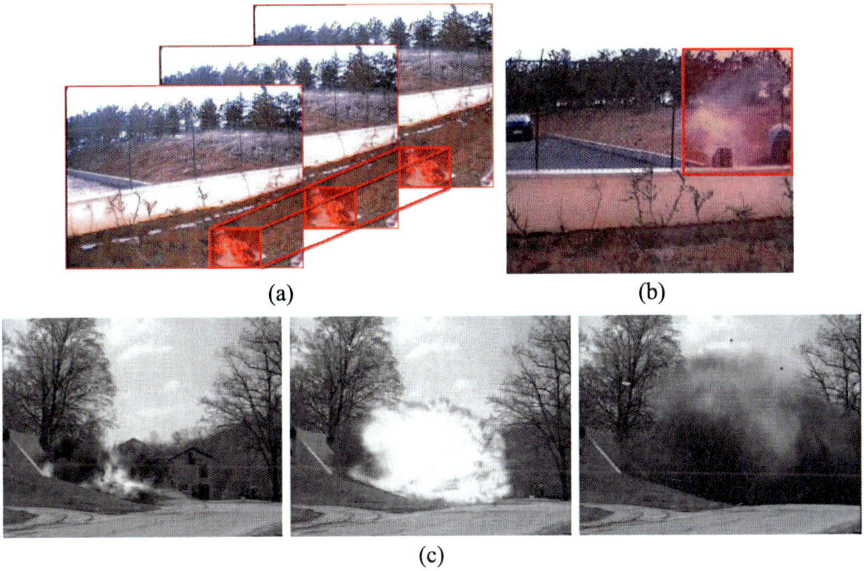

Fig. 8 (a) A motion texture of fire whose spatial extension is small. In this case, a mixed-state Markov chain model is more appropriate as it incorporates several frames within a time window. (b) A motion texture of smoke with a sufficiently large spatial support for considering a spatial model (MS-MRF) of the instantaneous motion map. (c) Explosion: a non-stationary motion texture. Here, a spatial model for each time instant is able to capture the instantaneous statistical properties of the dynamic content which varies along the sequence. A temporal model assumes that within a time window the motion texture is stationary, which is false in this case

whose statistical properties vary over time or whose spatial support is rapidly changing, the purely spatial model is suitable, while assuming a stationary and homogeneous MS-MC over a time window would lead to a wrong assumption.

- The computation time for estimating the MS-MC grows with T, as it considers $N \times T$ points. For $T = 1$ the MS-MRF and MS-MC models are similar in terms of speed given the parameters are estimated from N data points. However, the MS-MC involves 13 parameters, two more than the MS-MRF, so its computational cost is slightly bigger.

In order to analyze the performance of each model depending on the size of the image region occupied by the motion texture, and to be able to determine when one model has to be preferred over the other, we have performed the following experiment. First, we took the ten motion texture classes displayed in Fig. 5 and used in Sect. 6.2 for motion texture classification with MS-MRF. For each processed image of the video sequence we considered a square region of variable size centered at the middle location (Fig. 9(a)). For each class, a MS-MRF model was learned over that region and from a single pair of images picked from only one of the sequences belonging to each type of motion texture. Also a MS-MC model was learned for each class, by taking a temporal window of 5 frames and over the same region size. The temporal window was taken starting from the same frame used for learning the MS-

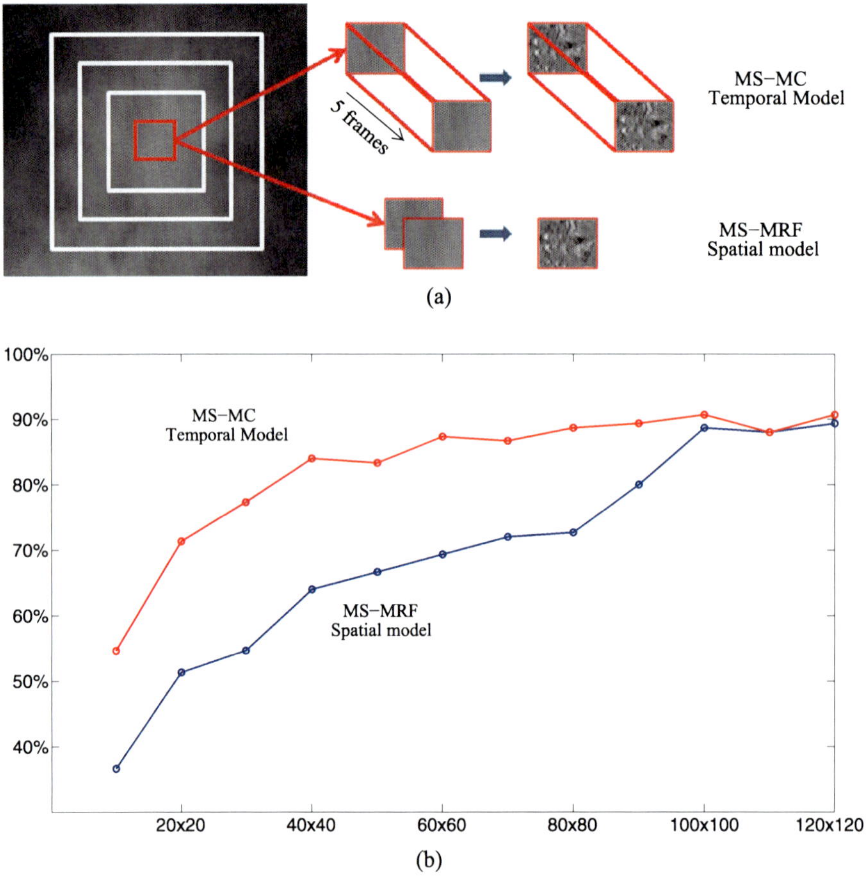

Fig. 9 Comparison between the spatial mixed-state Markov random field motion texture model and the mixed-state Markov chain motion texture model. (**a**) Different region sizes were tested for assessing the recognition performance of each model. For the MS-MC, a temporal window of 5 frames is considered and a temporal sequence of motion textures are extracted. For the MS-MRF, two consecutive frames are used to obtain a single motion texture at each time instant. (**b**) Classification rate for the ten class dataset and for different region sizes

MRF. As in Sect. 6.2, we then took several test samples at different instants for each test sequence (frame pairs for the MS-MRF model and 5-frame temporal windows for the MS-MC model). Finally, we classified each test sample by assigning it the closest class in the sense of the Kullback–Leibler divergence (36) for the MS-MRF motion texture model and the conditional Kullback–Leibler divergence (46) for the MS-MC motion texture model.

We computed the overall classification rate for different region sizes: from 10×10 to 100×100 with increments of 10 pixels. In Fig. 9(a), we plot the recognition performance of each method as a function of the region size. For large regions, the behavior of both models is similar as they consider a sufficient number of motion texture points for reliably estimating the model parameters. As the region size decreases, for the MS-MRF model the classification rate clearly drops faster compared with the MS-MC. Even for small regions the MS-MC achieves acceptable rates of for example 70% at a size of 20×20, while the MS-MRF shows a poor performance. The cost of a better recognition ability of the temporal model is of course an increased computational burden as it has to process, for each test sample in this case, a set of 4 motion textures obtained from 5 consecutive frames.

9 Motion Texture Tracking

In the dynamic texture literature, efforts have been devoted mainly to modeling, classification and segmentation of dynamic textures. However, the particular problem of dynamic texture tracking is still an open issue of great relevance. Critical vision-based surveillance applications such as detecting fire, monitoring pollutants or tracking an agitated crowd of people need for a compact representation of this type of dynamic phenomena. Here, we exploit the temporal model of motion textures and the matching strategy described in Sect. 7.2, as a powerful representation for tracking motion textures over time.

The objective is to track and follow a motion texture as for example the fire flame that moves towards left in Fig. 10. For this purpose, we exploit the proposed mixed-state Markov chain model and the model matching strategy.

We first consider that the initial position of the motion texture is given or set manually by defining a starting window W_o of a given size centered at initial location o. Then the maps of motion measurements (4) $\mathbf{x}_t$ and $\mathbf{x}_{t+1}$ are obtained for that window (for $t = 0$) using three consecutive frames and a MS-MC motion texture model is estimated. In this case, we take $T = 1$. Considering a larger temporal window would be incorrect given that the motion texture is moving. We thus obtain a reference model $\boldsymbol{\phi}_{\text{ref}}$ that characterizes the dynamic content we want to track. Let q_t be the position of the window at time t. Then we estimate q_t by applying the following rule:

$$\hat{q}_t = \arg\min_{q_t \in \Lambda_{q_{t-1}}} d_{\text{KL}}(W_o, W_{q_t}), \tag{48}$$

where we abuse of notation denoting $d_{\text{KL}}(W_o, W_{q_t})$ the KL divergence (46) between the reference motion texture model estimated in W_o and that estimated in W_{q_t}. A diamond search algorithm [42] is applied over a search area $\Lambda_{q_{t-1}}$ for obtaining the $\hat{q}_t$ that minimizes d_{KL}. For each possible location q_t tested by the diamond search algorithm, we estimate the motion texture parameters for the window W_{q_t} and compute $d_{\text{KL}}(W_o, W_{q_t})$. For that we need to extract the expectations defined in (47) from W_o and the candidate W_{q_t}. Note that except for $E_{p_k}[\log \rho_{i,t}^{(1)} / \rho_{i,t}^{(2)}]$ in (47), all the remaining expectations need to be computed only once for the reference model, at the initialization step.

We then apply a simple Kalman filter [1] to the estimate $\hat{q}_t$ in order to reduce the measurement noise in the estimated paths. Here, we have considered a constant

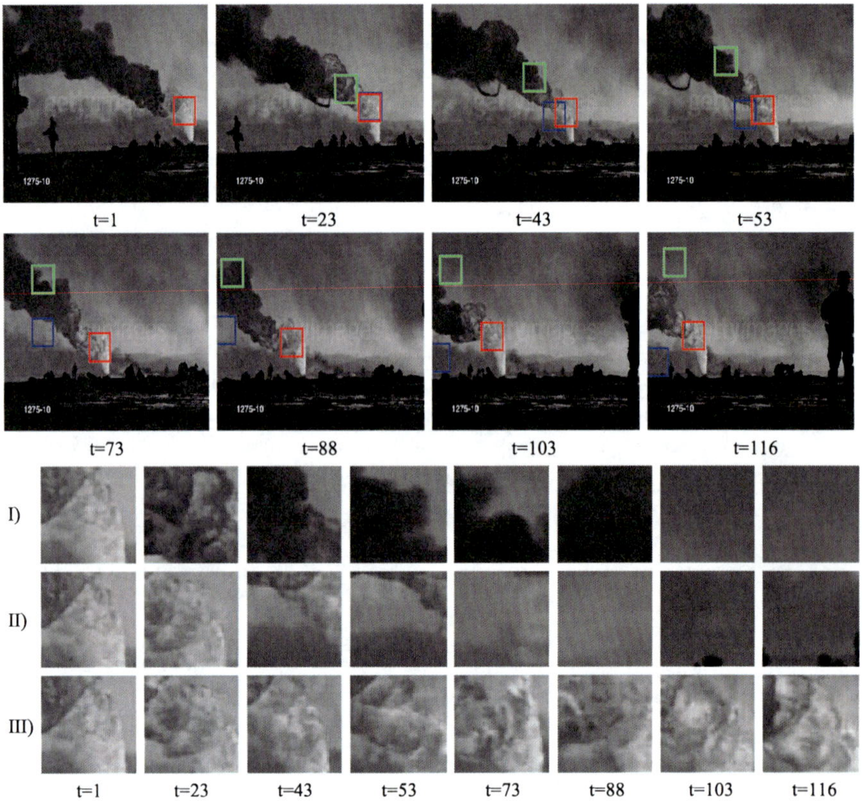

Fig. 10 Results of tracking the fire flame with methods **I** (*green*), **II** (*blue*) and our method **III** (*red*) for frames 1, 23, 43, 53, 73, 88, 103 and 116. (**I**–**II**–**III**) Content of the tracked window for each frame and methods **I** (SSD), **II** (Histogram) and **III** (MS-MC)

velocity state model. Finally, once a filtered position $\tilde{q}_t$ is obtained, the window is moved to the new position and the process starts all over.

More sophisticated tracking approaches can be applied as well, also exploiting the motion features introduced here, which for example involve more complex filtering techniques [29]. We now report comparative experimental results of motion texture tracking.

9.1 Experimental Results

We have considered three different methods for window matching:

I. Pixel-wise intensity matching by minimizing the Sum of Squared Differences (SSD)

II. Intensity histogram matching by minimization of a Bhattacharyya-coefficient-based distance

$$d(\gamma_1, \gamma_2) = \left(1 - \sum_{1}^{N} \sqrt{\gamma_1(n)\gamma_2(n)}\right)^{\frac{1}{2}}, \qquad (49)$$

where $\gamma_k(n)$ $k = 1, 2$ are the N-bin intensity histograms to match, as in [29]

III. MS-MC method: mixed-state Markov chain motion texture model matching by minimizing the Kullback–Leibler divergence (46)

Method I tends to be more suitable for rigid motion where the tracked objects keep a constant geometry while Method II is much more robust to pose variations and moderate deformations.

The diamond search matching strategy [42] at each time instant was performed for all the three methods over a maximum displacement of 15 pixels in both vertical and horizontal directions.

In Fig. 10, we display the results for the sequence 'Fire Flame' for different frames where the motion is given by a panning camera. We see that the motion texture is far from being localized in space and its extent is wider than the size of the tracked window, which was set to 50×50 pixels. The Method I (green square in Fig. 10) is not able to track the flame and tends to match the ascending smoke until it breaks down. Method II (blue square) is more coherent with the expected trajectory but it does not keep the target correctly located at every frame. The mixed-state causal model in Method III (red square), performs very well, specially considering that there is another motion texture of smoke that could interfere the matching process. In Figs. 10(I), 10(II) and 10(III), we display the content of the tracked window for the different methods. We can see that our algorithm always keeps the target on the motion texture of fire. In Fig. 11(a), we display the estimated paths for each method.

We observe the results for the 'Antorch' sequence in Fig. 12. Again it corresponds to fire, but note that it is very different than before. Moreover, the flame is partially occluded in several frames. The rapid variations of the motion texture (in size and

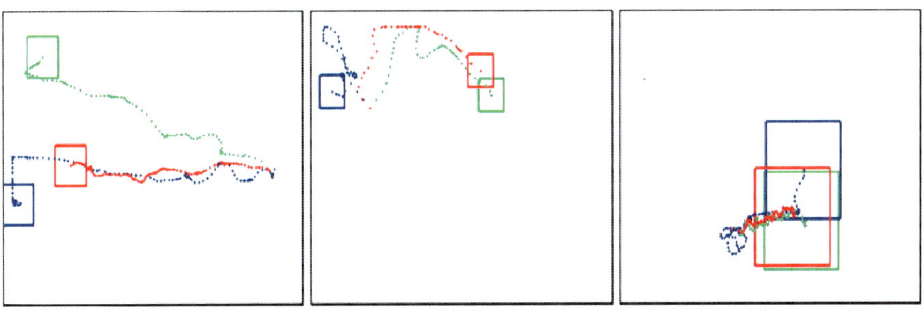

(a) (b) (c)

Fig. 11 Estimated tracks for the sequences (**a**) fire–flame, (**b**) antorch and (**c**) crowd, and for methods SSD (*green*), histogram (*blue*) and our method (*red*)

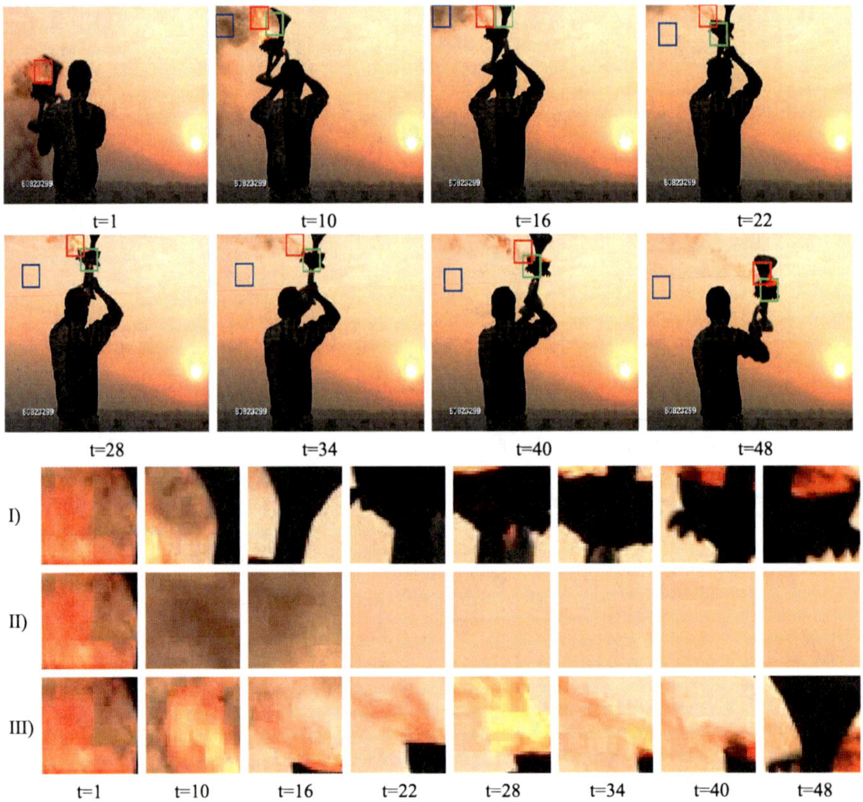

Fig. 12 Results of tracking fire (antorch) with Methods **I** (*green*), **II** (*blue*) and our Method **III** (*red*) for frames 1, 10, 16, 22, 28, 34, 40 and 48. (**I**–**II**–**III**) Content of the tracked window for each frame and Methods **I** (SSD), **II** (Histogram) and **III** (MS-MC). Note the large occlusion in the last frame

location) may produce some perturbation on the smoothness of the estimated trajectory, compared to what one would expect. Method II (blue square in Fig. 12(a) and blue track in Fig. 11(b)) fails completely, loosing the target just after the start of the sequence. Method I (green square) performs better but still does not keep the target correctly located. With the mixed-state causal model (red square), the flame is tracked satisfactory even in the presence of great variations of shape and intensity as seen in Fig. 12(III). Note also the large occlusions in the last depicted frame.

Finally, we have processed a challenging sequence (Fig. 13) of a motion texture that corresponds to a crowd of people crossing a street. Human motion, viewed from a long distance, can be considered as a repetitive motion pattern. Note that this motion texture is more sparse than the previous ones, showing many null motion values between persons. This is explicitly modeled within the mixed-state framework and exploited as a particular characteristic of the dynamic content. The different individuals enter and exit the texture, making very difficult to track the group in a compact way. However, Method III (red square in Figs. 13 and 11(c) performs satisfactory,

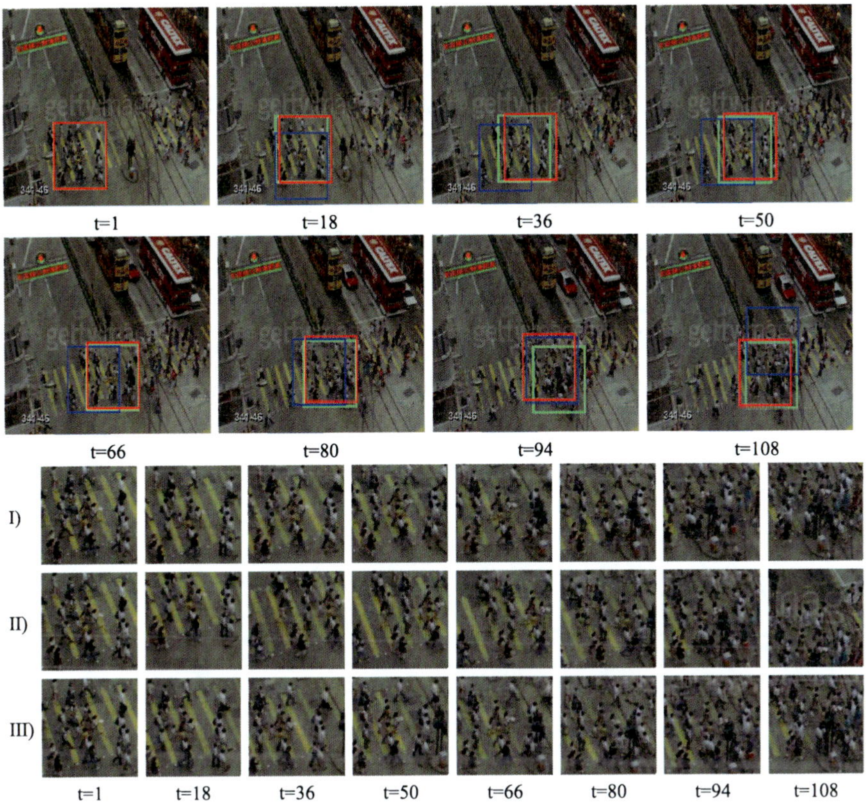

Fig. 13 Results of tracking a crowd with Methods **I** (*green*), **II** (*blue*) and our Method **III** (*red*) for frames 1, 18, 36, 50, 66, 80, 94, 108. (**I–II–III**) Content of the tracked window for each frame and Methods **I** (SSD), **II** (Histogram) and **III** (MS-MC)

even when the tracked group merges with the one that goes in the opposite direction. The estimated trajectory shows some expected variations due to the complexity of the scene, but it is globally correct. Method I (green square) also performs satisfactory but it shows some deviations from the expected trajectory, specially in the last depicted frame. Method II (blue square) behaves erratically, specially at the beginning, probably due to the fact that it is invariant to the spatial distribution of intensity and thus the two approaching persons from behind are incorrectly included in the tracked window, despite their distance from the rest.

As for the computation time, the Methods I (SSD) and II (Histogram) run at a rate of near 5 frames per second with our non-optimized MATLAB implementation and for windows of size 50×50 pixels (fire-flame sequence). The MS-MC method (III) consumes on average 6 seconds per frame in this case. As before, the bottleneck is the estimation of the MS-MC model for each searched window. For the moment, our method is not able to run in real time. We are currently investigating a way of computing directly the KL divergence from the motion data without estimating the

Table 3 Parameters estimated in the tracked window for the three tested sequences at four different time instants. t_0 corresponds to the MS-MC reference model learned in the initialization step (C: center, H: horizontal, V: vertical, D: diagonal, AD: anti-diagonal)

		$(2\sigma^2)^{-1}$	α	c	d_C	h_C	d_H	h_H	d_V	h_V	d_D	h_D	d_{AD}	h_{AD}
Fire flame	$t_0 = 1$	0.21	−3.28	−0.55	0.58	−0.12	0.63	−0.01	0.29	−0.07	0.34	0.02	0.24	0.02
	$t_1 = 23$	0.20	−1.65	−0.03	0.29	−0.06	0.31	0.02	0.16	−0.08	0.19	0.04	0.22	0.04
	$t_2 = 53$	0.17	−1.65	−0.23	0.44	−0.03	0.36	0.08	0.19	−0.08	0.07	0.06	0.19	0.07
	$t_3 = 116$	0.24	−1.41	0.11	0.48	−0.03	0.41	−0.01	0.18	−0.01	0.12	0.01	0.13	0.06
Antorch	$t_0 = 1$	0.01	−2.12	1.36	0.92	−0.01	1.18	−0.01	0.46	0.00	0.31	−0.03	−0.19	−0.03
	$t_1 = 10$	0.10	−2.23	1.61	0.71	−0.07	0.63	−0.04	0.21	0.01	0.16	−0.01	−0.12	−0.01
	$t_2 = 28$	0.06	−2.16	0.47	0.49	−0.01	0.55	−0.03	0.35	0.01	0.05	−0.04	−0.14	−0.03
	$t_3 = 48$	0.01	−0.22	−9.12	−0.32	0.32	−0.48	0.23	−0.01	0.08	−0.33	0.70	−0.36	0.20
Crowd	$t_0 = 1$	0.87	−3.44	−0.11	0.08	0.06	0.41	0.06	0.22	0.01	0.61	0.08	0.61	0.03
	$t_1 = 36$	1.85	−2.30	−0.03	0.30	0.07	0.46	0.15	0.35	0.09	0.52	0.14	0.42	0.08
	$t_2 = 66$	1.63	−2.04	−0.02	0.29	0.03	0.50	0.21	0.42	0.07	0.45	0.14	0.34	0.10
	$t_3 = 96$	1.96	−1.65	0.03	0.32	0.03	0.37	0.18	0.21	0.10	0.40	0.15	0.31	0.08

parameters. This will reduce considerably the computation time, while still exploiting the proposed modeling framework.

In Table 3, we display the MS-MC parameters estimated for the tracked window at four different time instants. The values should not be compared in the sense of the Euclidean distance, but by means of the KL divergence. Nevertheless, one can observe that for each tested sequence the parameters show coherent values for different instants. Note the coherency in the sign and the order of magnitude. For the antorch sequence, the occlusion observed in Fig. 12(III) at t_3 is the cause of a noticeable difference in the parameter values w.r.t. the reference model.

10 Conclusions

The mixed-state framework opens a novel statistical scenario for the simultaneous modeling of continuous and discrete (numeric or symbolic) states. The approach permits to deal, in a joint and coupled way, with values of different statistical nature. Standard Markov models have been extended to account for mixed-state values by means of a measure theoretic formulation.

In particular, the approach permits to model motion observations arising from dynamic or motion textures, with a numeric discrete state, that is, the zero value. The main contribution of our method is that we have designed a model of the instantaneous motion maps, which allows to learn, recognize, track, and detect motion textures in a frame-by-frame basis. The mixed-state Markov model has shown to be a powerful nonlinear representation for describing complex dynamic content with only a few parameters. The results presented here for tracking and recognition of this type of phenomena show that the proposed parametric representation can be effectively used as class-characteristic features.

Furthermore, the presentation of the model provided here permits also to define non-numeric discrete values, as abstract labels that may arise in the context of decision problems. This possibility is particularly interesting when solving simultaneous decision–estimation problems where two tasks as estimation (related to continuous values) and decision (related to symbolic states) are to be addressed jointly [16]. The implications of considering these types of mixed-state models are important in computer vision, where high-level information, represented by abstract labels, can be introduced in an optimal way. Sample applications include introduction of symbolic states for: borders (e.g., estimating discontinuous optical flow fields), detection of regions of interest (defined abstractly) and structural change detection (e.g., in remote sensing).

References

1. Anderson, B., Moore, J.: Optimal filtering. Englewood Cliffs, Prentice-Hall (1979)
2. Barron, J.L., Fleet, D.J., Beauchemin, S.S.: Performance of optical flow techniques. Int. J. Comput. Vis. **12**(1), 43–77 (1994)

3. Besag, J.: Spatial interaction and the statistical analysis of lattice systems. J. R. Stat. Soc., Ser. B **36**, 192–236 (1974)
4. Besag, J.: On the statistical analysis of dirty pictures. J. R. Stat. Soc., Ser. B **48**(3), 259–302 (1986)
5. Black, M., Rangarajan, A.: On the unification of line processes, outlier rejection, and robust statistics with applications in early vision. Int. J. Comput. Vis. **19**(1), 57–91 (1996)
6. Bouthemy, P., Hardouin, C., Piriou, G., Yao, J.F.: Mixed-state auto-models and motion texture modeling. J. Math. Imaging Vis. **25**(3), 387–402 (2006)
7. Bruhn, A., Weickert, J., Feddern, C., Kohlberger, T., Schnorr, C.: Variational optical flow computation in real time. IEEE Trans. Image Process. **14**(5), 608–615 (2005)
8. Cernuschi-Frias, B.: Mixed-states Markov random fields with symbolic labels and multidimensional real values. Rapport de Recherche, INRIA No 6255 (2007). http://hal.inria.fr/docs/00/16/59/37/PDF/RR-6255.pdf
9. Chan, A., Vasconcelos, N.: Modeling, clustering, and segmenting video with mixtures of dynamic textures. IEEE Trans. Pattern Anal. Mach. Intell. **30**(5), 909–926 (2008)
10. Chan, A.B., Vasconcelos, N.: Layered dynamic textures. IEEE Trans. Pattern Anal. Mach. Intell. **31**(10), 1862–1879 (2009). http://doi.ieeecomputersociety.org/10.1109/TPAMI.2009.110
11. Chetverikov, D., Peteri, R.: A brief survey of dynamic texture description and recognition. In: Proc. 4th International Conference on Computer Recognition Systems, CORES'05, Advances in Soft Computing, pp. 17–26. Springer, Berlin (2005)
12. Corpetti, T., Memin, E., Perez, P.: Dense estimation of fluid flows. IEEE Trans. Pattern Anal. Mach. Intell. **24**(3), 365–380 (2002)
13. Cover, T., Thomas, J.: Elements of Information Theory. Wiley, New York (1991)
14. Crivelli, T., Cernuschi, B., Bouthemy, P., Yao, J.: Segmentation of motion textures using mixed-state Markov random fields. In: Proc. of SPIE, vol. 6315, p. 63150J. SPIE, Bellingham (2006)
15. Crivelli, T., Cernuschi-Frias, B., Bouthemy, P., Yao, J.F.: Mixed-state Markov random fields for motion texture modeling and segmentation. In: Proc. IEEE International Conference on Image Processing ICIP'06, Atlanta, USA, pp. 1857–1860 (2006)
16. Crivelli, T., Piriou, G., Bouthemy, P., Cernuschi-Frías, B., Yao, J.F.: Simultaneous motion detection and background reconstruction with a mixed-state conditional Markov random field. In: ECCV '08: Proc. of the 10th European Conference on Computer Vision, Marseille, France, pp. 113–126 (2008)
17. Doretto, G., Chiuso, A., Wu, Y., Soatto, S.: Dynamic textures. Int. J. Comput. Vis. **51**(2), 91–109 (2003)
18. Fablet, R., Bouthemy, P.: Motion recognition using non-parametric image motion models estimated from temporal and multiscale co-ocurrence statistics. IEEE Trans. Pattern Anal. Mach. Intell. **25**(12), 1619–1624 (2003)
19. Fablet, R., Bouthemy, P., Perez, P.: Non-parametric motion characterization using causal probabilistic models for video indexing and retrieval. IEEE Trans. Image Process. **11**(4), 393–407 (2002)
20. Fazekas, S., Chetverikov, D.: Normal versus complete flow in dynamic texture recognition: a comparative study. In: Texture 2005: 4th International Workshop on Texture Analysis and Synthesis, ICCV'05, Beijing, pp. 37–42 (2005)
21. Fazekas, S., Amiaz, T., Chetverikov, D., Kiryati, N.: Dynamic texture detection based on motion analysis. Int. J. Comput. Vis. **82**(1), 48–63 (2009)
22. Geman, S., Geman, D.: Stochastic relaxation, Gibbs distributions, and the Bayesian restoration of images. IEEE Trans. Pattern Anal. Mach. Intell. **6**, 721–741 (1984)
23. Guyon, X.: Random Fields on a Network: Modeling, Statistics and Applications. Springer, New York (1995)
24. Hardouin, C., Yao, J.F.: Spatial modelling for mixed-state observations. Electron. J. Stat. **2**, 213–233 (2008)
25. Horn, B., Schunck, B.: Determining optical flow. Artif. Intell. **17**(1–3), 185–203 (1981)
26. Hsieh, J., Yu, S., Chen, Y.: Motion-based video retrieval by trajectory matching. IEEE Trans. Circuits Syst. Video Technol. **16**(3), 396–409 (2006)

27. Lu, Z., Xie, W., Pei, J., Huang, J.: Dynamic texture recognition by spatio-temporal multiresolution histograms. In: IEEE Workshop on Motion and Video Computing, WACV/MOTION, pp. 241–246 (2005)
28. Nelson, R.C., Polana, R.: Qualitative recognition of motion using temporal texture. CVGIP, Image Underst. **56**(1), 78–89 (1992). doi:10.1016/1049-9660(92)90087-J
29. Pérez, P., Hue, C., Vermaak, J., Gangnet, M.: Color-based probabilistic tracking. In: ECCV '02: Proc. of the 7th European Conference on Computer Vision-Part I, pp. 661–675. Springer, London (2002)
30. Peteri, R., Chetverikov, D.: Dynamic texture recognition using normal flow and texture regularity. In: Proc. of IbPRIA, Estoril, pp. 223–230 (2005)
31. Peteri, R., Huiskes, M., Fazekas, S.: Dyntex: a comprehensive database of dynamic textures. http://www.cwi.nl/projects/dyntex/index.html
32. Piriou, G., Bouthemy, P., Yao, J.F.: Recognition of dynamic video contents with global probabilistic models of visual motion. IEEE Trans. Image Process. **15**(11), 3417–3430 (2006)
33. Rahman, A., Murshed, M.: Real-time temporal texture characterisation using block based motion co-occurrence statistics. In: Proc. of the 11th IEEE International Conference on Image Processing, ICIP'04, pp. 1593–1596 (2004)
34. Saisan, P., Doretto, G., Wu, Y., Soatto, S.: Dynamic texture recognition. In: Proc. of the IEEE Conference on Computer Vision and Pattern Recognition, CVPR'01, Hawaii, pp. 58–63 (2001)
35. Salzenstein, F., Collet, C.: Fuzzy Markov random fields versus chains for multispectral image segmentation. IEEE Trans. Pattern Anal. Mach. Intell. **28**(11), 1753–1767 (2006)
36. Salzenstein, F., Pieczynski, W.: Parameter estimation in hidden fuzzy Markov random fields and image segmentation. Graph. Models Image Process. **59**(4), 205–220 (1997)
37. Vidal, R., Ravichandran, A.: Optical flow estimation and segmentation of multiple moving dynamic textures. In: Proc. of CVPR'05, San Diego, vol. 2, pp. 516–521 (2005)
38. Wheeden, R., Zygmund, A.: Measure and Integral: An Introduction to Real Analysis. Dekker, New York (1977)
39. Wu, J., Chung, A.C.S.: A segmentation model using compound Markov random fields based on a boundary model. IEEE Trans. Image Process. **16**(1), 241–252 (2007). 10.1109/TIP.2006.884933
40. Yuan, L., Wen, F., Liu, C., Shum, H.: Synthesizing dynamic textures with closed-loop linear dynamic systems. In: Proc. of the 8th European Conference on Computer Vision, ECCV'04, Prague. Lecture Notes in Computer Science, vol. 3022, pp. 603–616. Springer, Berlin (2004)
41. Zhao, G., Pietikainen, M.: Dynamic texture recognition using local binary patterns with an application to facial expressions. IEEE Trans. Pattern Anal. Mach. Intell. **29**(6), 915–927 (2007)
42. Zhu, S., Ma, K.K.: A new diamond search algorithm for fast block-matching motion estimation. IEEE Trans. Image Process. **9**(2), 287–290 (2000)

Learning to Detect Event Sequences in Surveillance Streams at Very Low Frame Rate

Paolo Lombardi and Cristina Versino

Abstract Some camera surveillance systems are designed to be autonomous—both from the energy and storage points of view. Autonomy allows operation in environments where wiring cameras for power and data transmission is not feasible. In these contexts, for cameras to work unattended over long periods of time requires choosing a low frame rate to match the speed of the process to be supervised while minimizing energy and storage usage. The result of surveillance is a large stream of images acquired sparsely over time with limited visual continuity from one frame to the other. Reviewing these images to detect events of interest requires techniques that do not assume traceability of objects by visual similarity. When the process surveyed shows recurrent patterns of events, as it is often the case for industrial settings, other possibilities open up. Since images are time-stamped, techniques which use temporal data can help detecting events. This contribution presents an image review tool that combines a scene change detector (SCD) with a temporal filter. The temporal filter learns to recognize relevant SCD events by their time distribution on the image stream. Learning is supported by image annotations provided by end-users during past reviews. The concept is tested on a benchmark of real surveillance images stemming from a nuclear safeguards context. Experimental results show that the combined SCD-temporal filter significantly reduces the workload necessary to detect safeguards-relevant events in large image streams.

1 Introduction

In surveillance, one or more cameras look over a scene. The image stream is transmitted to monitors for online inspection or to storage for *offline review*. Cameras are

---

P. Lombardi · C. Versino (✉)
European Commission, Joint Research Centre, Nuclear Security Unit, Via Fermi 2749, 21027 Ispra, Italy
e-mail: cristina.versino@jrc.ec.europa.eu

P. Lombardi
e-mail: paolo.lombardi.vision@gmail.com

L. Wang et al. (eds.), *Machine Learning for Vision-Based Motion Analysis*, Advances in Pattern Recognition,
DOI 10.1007/978-0-85729-057-1_5,

positioned to view large areas to minimize the number of independent data streams to be analyzed, while guaranteeing that a human reviewer will see events of interest online or at a later time.

Some surveillance systems (for example [5] for process auditing in nuclear safeguards) are designed under the constraint that the camera operates as an *autonomous system*—both from the *energy* and *storage* points of view. The image *frame rate* reflects the speed of the process to be surveyed so that (i) energy consumption is limited (ii) images acquired fit on the camera's storage until downloaded for review.

Choosing a *low* frame rate allows cameras to work unattended for long periods in facilities where cameras cannot be wired for power or data transmission.

Low data rates are necessary in several other contexts. For instance, battery-powered devices for ubiquitous computing need to switch between data acquisition, processing, communication and the 'idle' state to prolongue operational lifetime. This issue is a central for emerging monitoring technology such as wireless sensor networks [2].

The low frate rate approach to surveillance requires *review techniques* that can deal with data acquired *sparsely over time*. In the case of surveillance images, sparness translates to *very limited continuity in visual content across consecutive frames*. Most state-of-the-art review techniques based on traceability of objects by visual continuity become non applicable in these contexts.

When the process to be surveyed shows *recurrent patterns of events over time*, other possibilities open up. This is the case for process monitoring in industrial settings where key events repeat themselves, occur in typical sequences with quite regular durations. Since surveillance *images are time-stamped*, this allows for review techniques, which, besides analyzing the image content also use *temporal data* to detect events of interest. This is the theme of the present chapter.

We discuss machine learning (ML) for the review of surveillance images from an applicative standpoint, the test bed being video surveillance for nuclear safeguards. Sections 2, 3 and 4 are introductory. We first establish the difference between uninformed and ML-informed image filters, describe surveillance for nuclear safeguards, and explain the significance of using ML in this context. Sections 5 and 6 cover specific filters. Section 5 presents an uninformed filtering technique (a scene change detector), today's default filter for safeguards reviews. Section 6 focuses on the core contribution of this chapter, namely a ML filter for the detection of events based on temporal data. This temporal filter extends the uninformed filter described in Sect. 5. Section 7 presents tests on the use of these filters on real safeguards image streams, and provides a comparison with alternative informed techniques that search events by image content. Section 8 wraps up and concludes.

2 Approaches for Image Reviews

The task of inspecting streams of images can require the application of different techniques dependent on the goal of surveillance. For example, in security when the police is looking for a person wearing known clothes, a human supervisor would screen images for people wearing colors matching the description. The same approach can be used by an automatic system: a *filter* may be designed to select images containing blobs of matching colors corresponding to a human shape.

This is called *informed search*: the supervisor (human or filter) checks for the presence of data known to correspond to a meaningful event. The approach can be used when the supervisor has expectations on the target of the search. The more precise the information, the better the search.

When the supervisor has only general clues on the type of events, *uninformed search* techniques must be employed. In the example above, if the police are not searching any specific person, attention may be focused to violations of prohibited areas or counter-crowd movements. These events are not always meaningful, but they can be related to events relevant to security.

Informed and uninformed searches apply also to *tools for image reviews, i.e., IT tools designed to assist a human supervisor in scanning large image streams to detect relevant events*. Informed search techniques consist in matching a *model* of a significant event with the image data. They require the image review software to incorporate such a model. They are typical of top–down approaches that start from knowledge and expectations and verify these on the data. By contrast, uninformed search techniques *detect image features that show a correlation with significant events*. Uninformed search goes bottom–up, starting from evidence of stimuli to build knowledge.

In the literature, uninformed techniques are also referred to as *novelty detection* techniques [18, 19]. Novelty detection is the identification of signals different from a reference. Fundamental components of these techniques are: (i) the *reference signal* to which data are compared, and (ii) the *conditional test* that triggers the identification of 'novel' data. In video surveillance, a common form of uniformed techniques is *background maintenance for motion detection* [20]. Here machine learning is proposed to bootstrap and maintain the background model. Methods include statistical models [30], kernel-based classifiers [1], Wiener filters [35], Kalman filter [21], and mean shift [14]. Section 5 illustrates a state-of-the-art filter used in nuclear safeguards and security applications, which relies on simple form of background model.

Informed techniques for video analysis are numerous. Those pertinent to our discussion fall in two categories: (i) *behavior analysis* and (ii) *video annotation*.

Learning for video surveillance has been focused around behavior analysis, where the goal is to learn patterns of 'normal' behavior to find 'anomalous' situations. The concept is that trajectories of *tracked objects* are clustered and labeled by machine learning to add semantics. Methods include polynomial fitting, multi-resolution quantization, hidden Markov models, kernel methods, neural networks, and k-means [23]. The currently accepted reference model for these systems has been introduced in [17]. Here, machine learning builds a graph representation of an image in which nodes represent points of interest (e.g., entry/exit points, stop points) and arcs represent activity paths (e.g., motion, change of activity). Groupings of points of interest and activity paths are then labeled (e.g., 'car enters', 'car moves' and car stops' may be labeled as 'parking'). Behavior-analysis assumes the availability of *motion tracking data* extracted with uninformed techniques [12, 39].

turn requires visual correlation (i.e., similar appearance) or temporal cor- (i.e., produced by a high frame rate) of *objects in consecutive images*. Let us defer a deeper discussion on this to Sect. 6. *Here, we underline that research on behavior analysis is focused on real-time, video-rate streams. Reviewing image streams stored at a far lower frame rate (e.g., one image per minute or less) cannot be treated with the same methods.*

Video annotation is rarely used for surveillance. However the filter discussed in Sect. 6 is highly related to this topic. A learning-based annotation system takes a video segmented into short units and extracts low-level *features* from each unit to describe its *content*. Given classes of events of interest, some units are manually *annotated* to be 'positive' or 'negative' examples of events and used to infer the annotation model [7]. Examples of learning-based video annotation include rule-based systems [6], multi-cue statistical learning [13, 25], support vector machines [7], and graph matching [38]. We are interested in cases where video segmentation techniques are not applicable across all classes of events due to weak visual signatures for some classes. In these cases, learning algorithms can be extended to use *noncontent related information*, such as temporal data or other meta-data where stronger features exist. When using temporal data, the *interactivity* of the review algorithm must be enhanced by embedding it in the image search work flow in a clever way.

In later sections, we discuss the application of uninformed and informed filters using machine learning similar to those developed for video annotation. The field of use is the analysis of surveillance images stemming from the *real* application setting of nuclear safeguards.

3 Surveillance for Nuclear Safeguards

Nuclear safeguards verify that a State's nuclear material is not diverted to build weapons or explosive devices.

Camera surveillance in nuclear facilities helps to attain safeguards at a reasonable cost without interfering with a facility's operations (Fig. 1). To this goal, the International Atomic Energy Agency (IAEA) maintains about 800 cameras in 170 safeguarded sites worldwide [28]. Surveillance image streams are reviewed by safeguards inspectors one-by-one.

The surveillance of nuclear plants poses several challenges to state-of-the-art image filters. *First*, the field of view covers all the locations where important processing takes place (Fig. 1, second row). These locations are many meters apart. The visual appearance of *flasks of nuclear material*, the objects of interest, significantly changes during the process. *Second*, the image acquisition rate is very low—one frame every several minutes. This frame rate is designed to match the speed of the process to be supervised. It guarantees that all images acquired are stored on the camera's storage during months of unattended operation. *Third*, flasks of nuclear material are visible only in a small fraction of images. Typically, each camera acquires several thousands of images (10 to 100 thousands) before these are

Fig. 1 Inspectors setting-up surveillance cameras in a nuclear plant (*first row*). A camera's view (*second row*) (© D. Calma/IAEA)

reviewed by inspectors. Of these, less than 1% is expected to relate to safeguards-relevant events. The remaining either present no change between consecutive frames or contain events not involving flask movements (e.g., moving cranes, trolleys, illumination changes). *Fourth*, nuclear flasks are *indistinguishable*, i.e., there is no sign (feature, color, label, etc.) visible from surveillance images to distinguish one flask from the others.

Inspectors eliminate the no-change images by applying an uninformed scene change detection filter [10] (Sect. 5). This operation reduces the image set to about 10% of the original size. The latter set is reviewed by inspectors frame-by-frame. Safeguards-relevant images are annotated to produce a *review report* (Fig. 2), i.e., a list of time-stamped, chronologically ordered images labeled by the classes of events recognized by the inspectors.

4 Filtering Surveillance Streams by Combining Uninformed and Informed Search Strategies

We have developed a set of tools based on machine learning, collectively named 'Safeguards Review Station' (SRS), to assist nuclear inspectors in the review of

Date	Time	Scene Nr.	Event annotation
2006/11/07	09:38:08	Scene #4314	Flask visible over hatch
2006/11/07	09:52:08	Scene #4316	Flask visible in decontamination area
2006/11/07	13:29:08	Scene #4347	Flask visible over hatch
2006/11/07	13:43:08	Scene #4349	Flask visible in decontamination area
2006/11/07	15:00:08	Scene #4360	Flask visible over pond
2006/11/08	08:30:08	Scene #4510	Flask visible over pond
2006/11/08	08:44:08	Scene #4512	Flask visible in decontamination area
2006/11/08	15:16:08	Scene #4568	Flask visible over pond
2006/11/09	09:14:08	Scene #4722	Flask visible over pond
2006/11/09	10:17:08	Scene #4731	Flask visible in decontamination area
2006/11/10	08:06:08	Scene #4918	Flask visible over hatch
2006/11/10	12:11:08	Scene #4953	Flask visible over hatch
2006/11/13	10:11:08	Scene #5553	Flask visible over hatch
2006/11/13	10:25:08	Scene #5555	Flask visible in decontamination area
2006/11/13	14:30:08	Scene #5590	Flask visible over hatch
2006/11/13	14:37:08	Scene #5591	Flask visible in decontamination area

Fig. 2 Report resulting from an image review

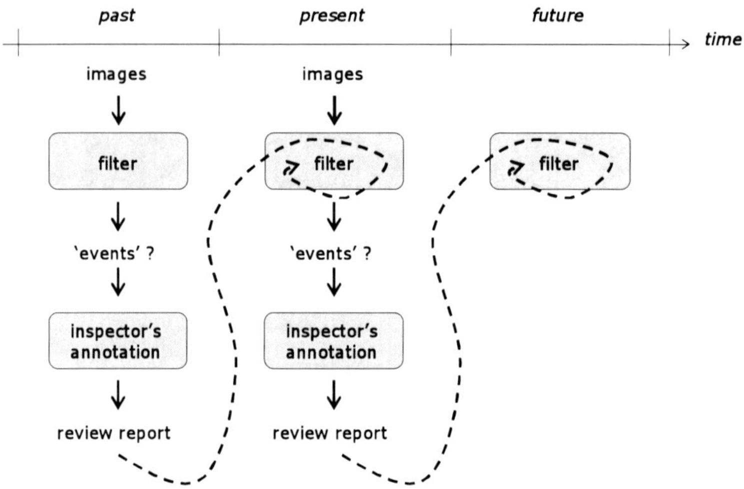

Fig. 3 The state-of-the-art review flow (*solid arrows*) and the augmented flow in the Safeguards Review Station (*dashed arrows*)

images. The approach is that *SRS tools learn to improve their detection performance by using information available from past reviews and from the on-line compilation of a review report.*

Figure 3 shows the review flow in the SRS. Image data is generated over time continuously, but it is divided in batches to ensure timeliness of reviews. Given a batch of images, a filter is applied to extract *events*. This step is largely automatic and relies on scene change detection (SCD). The *main role of the inspector is in the annotation of relevant SCD events for the review report*. When a new batch of images is to be reviewed, the same process is *repeated unchanged*, i.e., the SCD filter *does not learn from past review results*. By contrast, the SRS 'reconnects' the sequence of reviews *given that the stream of images over time is generated by the*

Table 1 Plant properties that can be derived from review reports

Property ID	Property description
P1	Classes of typical events taking place in the plant as per the events' annotations (e.g., 'flask over hatch', 'flask in decontamination area', 'flask over pond').
P2	Examples of the visual appearance of safeguards-relevant events as seen from a specific camera (by retrieving the corresponding image files).
P3	Sequence of the events (e.g., a 'hatch' event is followed by a 'decontamination' event, then by a 'pond' event).
P4	Duration of events (by computing the time interval between events).

same nuclear process. By this point of view, *we design filters that learn from past reviews*.

The departure point for the SRS is the archive of review reports for a given plant and surveillance camera. Table 1 lists plant- and camera-specific information derived from these reports. Besides highlighting *classes of safeguards-relevant events* annotated by inspectors (P1), reports associate classes of events with their *visual appearance* by the corresponding image files (P2). The annotated *sequence of events* provides information on the stages of the processing of flasks of nuclear material within the plant (P3). The events' time-stamps give an indication on the *duration* of each stage of the processing (P4). On these properties, we build *informed filters* to detect safeguards-relevant with less false positives with respect to SCD.

The SRS includes three filters, each used in cascade to SCD. The SRS filters are based on: (i) image retrieval (IR), (ii) decision trees (DT) and (iii) Markov models (MM). Filters IR and DT rely on the *visual appearance of events* (Properties P1 and P2 in Table 1). By contrast MM, performs a 'meta-classification' of the *sequence of events extracted* by SCD, and it is trained on statistics about P1, P2 and P4. MM is described in detail in Sect. 6, while the SCD filter is covered in the next section. IR and DT [37] are not covered in this chapter, but experimental results on the use of these filters are provided in Sect. 7 as term of comparison to the performance of MM.

5 Searching Events by Scene Change Detection

Uninformed event detection arises from monitoring signals computed on the image stream over time. An event is declared when a signal alters its value with respect to a condition. The conditional test for the event can be the breaking of a threshold value, or a complex frequency component analysis or case-based reasoning [1, 18, 30], to name a few possibilities. The parameters for the conditional check *do not relate specifically to the event type being searched*. They are set a priori so as to detect a broader family of events that *includes* the target events.

As an example of uninformed technique, we illustrate the scene change detection (SCD) filter in use in official nuclear inspection software [10] (Fig. 4).

The technique is a *two-frame differencing* based on the average intensity value of pixels inside one or more area of interest (AOI). Before starting the algorithm,

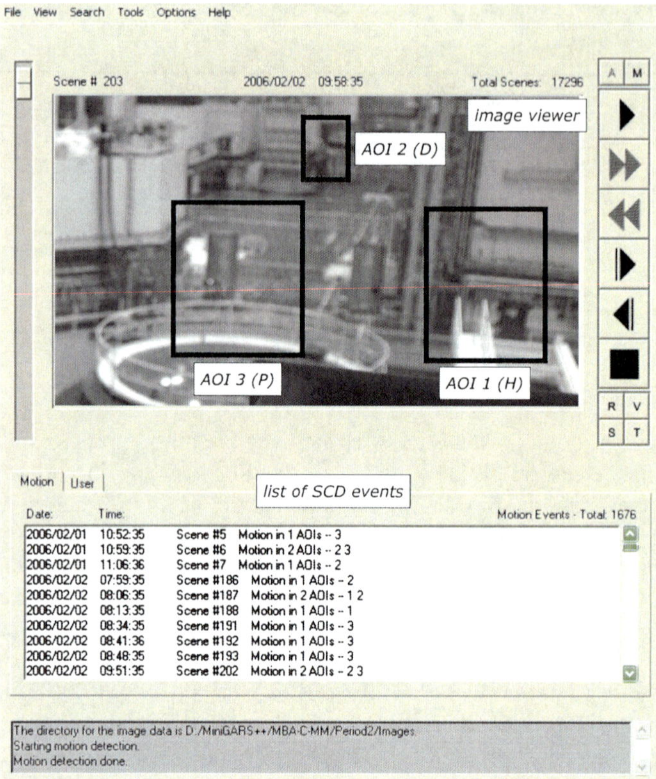

Fig. 4 Mock-up interface of the image review tool used in nuclear safeguards. It shows three areas of interest (AOIs) defined by the user and the list of events (1676 events, in this case) generated by running SCD over these AOIs on a stream of surveillance images (17296 images, in this example)

the user draws AOIs on a reference image (Fig. 4) around locations of interest. For an image at a given time, SCD computes the average intensity of pixels belonging to an AOI. This is compared to the value computed on the previous image to derive a measure of relative change. If this breaks a threshold, a SCD event is marked for that image and that AOI. The same computation is run for every AOI, possibly with different thresholds. As a result, for each image the SCD filter detects a number of events up to the number of AOIs. By altering the thresholds, the user increases or decreases the number of events detected by SCD. Figure 4 shows the list of SCD events as it appears on the interface of the review tool. Inspectors browse the list of SCD events one-by-one and annotate the relevant images by the event class.

There are three locations of interest in a nuclear plant assigned to AOIs:

1. the *hatch* ('*H*', AOI nr. 1)
2. the *decontamination area* ('*D*', AOI nr. 2)
3. the *pond* ('*P*', AOI nr. 3)

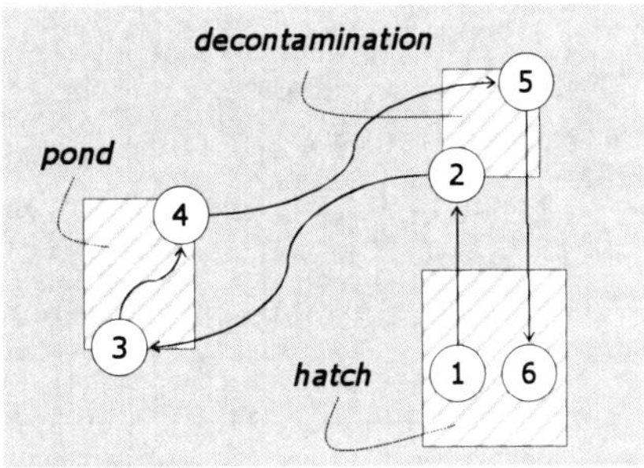

Fig. 5 Schematic flow of the movements of a flask of nuclear material within a plant

In a normal process a flask of nuclear fuel enters the hatch and reaches the decontamination area, whence it is moved to the pond (Fig. 5). From the pond, the flask moves back to decontamination and exits through the hatch. In SCD, each AOI is assigned to a label, a convenient label set being $\{h, d, p\}$. Images for which a SCD event has been detected are associated with one or more of these labels. For instance, if image t is labeled $[h, p]$, this means that the AOIs of hatch and pond changed from $t - 1$ to t.

Due to the loose correlation between target events and the SCD signal, uninformed filters like SCD are employed in the first stage of the image stream analysis to eliminate clear no-change images. The parameters for the conditional test must be set to attain a true positive rate of 100%. The down-side is that the number of false positives is high.

In our studies of image surveillance for nuclear safeguards, most parts of the image sequences contain little or no activity. Sequences of about 20,000 images must be reviewed offline by nuclear inspectors operating under time pressure. The use of SCD reduces the amount of images to be inspected on average by 90%, from 20,000 to 2,000. However, reviewing 2,000 images one-by-one is still a hard job.

Inspectors are interested movements of nuclear material. Typically, out of 2,000 SCD Images, only 0–30 are safeguards-relevant. Thus, the rate of false positives is close to 99%. It is then reasonable to follow SCD with informed techniques that use a model of safeguards-relevant events to highlight SCD images of interest.

Why do inspectors employ this uninformed SCD? The answer has to do mainly with the degree of *programmability* offered by uninformed techniques. Nuclear inspectors, like most users of image review software for surveillance, are not expert of image processing or automatic pattern recognition, but understand event detection by setting a threshold. This *understanding* and the total *degree of control* given by changing one or two simple parameters give the inspector confidence of mastering the tool.

6 Searching Events by Sequence and Time Attributes

Informed event detection can be employed when the review tool has an internal *model* of expected event sequences. To build such a model, image streams must exhibit at least one of the following properties:

Property 1. Traceable objects.
Property 2. Recurrent sequencing of events.
Property 3. Recurrent timing of events.

Property 1 has to do with *continuity of information*, whereas Properties 2 and 3 have to do with *recurrence of information*.

Property 1, i.e., *traceable objects*, corresponds to situations where object movements can be tracked. Objects can be tracked when they have a video signature with *high self-correlation* and *low cross-correlation with other objects or the background*. In turn, this requires either highly characterized visual features (e.g., color, motion, shape, texture, etc.)—for visual self-correlation—or a high acquisition *frame rate*—for temporal self-correlation. In most research benchmarks, both these elements are present. When objects are traceable, event detection is based on the verification (or absence) of expected motion patterns by tracked objects. Property 1 is the assumption underlying gesture recognition and most surveillance and behavior analysis systems (e.g., tracking people and animals) [1, 21, 23, 30, 35, 36, 39], or video analysis for compression [31].

In a scenario with low acquisition frame rate and low visual characterization of objects, an informed event-detection system must resort upon Property 2 or Property 3.

Property 2, i.e., *recurrent sequencing of events*, is verified when events happen following a regular sequence repeated over time. For instance, in video-based human-machine interaction a user's hands or face movements are checked for model gestures that correspond to commands [22]. Regular patterns can be observed in surveillance applications in constrained scenarios [3]. Also process monitoring in a manufacturing plant can benefit from model-based temporal analysis. Manufacturing processes follow standard paths, and deviations can indicate errors and anticipate defects in the final product.

Property 3, i.e., *recurrent timing of events*, consists in event sequences characterized by regular *durations* and/or *intervals* between events. An example is the monitoring of abandoned objects in metro stations: if an image region is occupied by a foreground object for more than the expected 'regular' time, an 'event' is detected [9].

In the specific scenario of nuclear safeguards surveillance illustrated in Sect. 3 (in short: varying appearance of nuclear flasks across different image regions, a low frame-rate, rare events, and indistinguishable flasks), it is clear that Property 1 does not hold for what regards detecting events of nuclear flask movements. *Hence, in this particular scenario, most well-known techniques cannot be applied.* However, the process a flask undergoes in a nuclear plant is *structured* and *recurrent*. Therefore, tracking can be performed by modeling the *regularities of the process*, instead of those of the flask.

In this section, we describe an informed filter to perform event detection based purely on Properties 2 and 3. More specifically, the filter is implemented by a hidden semi-Markov model (HSMM) used in semi-automatic mode.

To better understand the contribution of event *sequencing and timing*, we isolate the filtering based on temporal properties from the filtering based on image content.

We do this by using SCD (Sect. 5) to transform the sequence of images into a *sequence of symbols* corresponding to the active AOIs and the associated image *time stamps*, in all similar to Fig. 2 with the only difference of having AOI identifiers instead of annotations (Fig. 2, rightmost column).

6.1 Modeling Nuclear Flask Processing with a HSMM

Hidden semi-Markov models (HSMMs) are hidden Markov models where each state has an explicit duration model to represent sojourn times in nonabsorbing states [15, 24, 27].

A semi Markov model is made of an embedded first-order Markov chain X, and of discrete distributions of sojourn times S. The embedded chain is described by (T, χ_o), where χ_o is the initial state distribution and T is the transition matrix, such that $T_{ij} = P(X_{t+1} = j | X_t = i)$. For a semi-Markov model, $T_{ii} = 0,\ \forall i$. The sojourn time distributions are a set of discrete distributions depending only on the current state, $S = \{S_i,\ \forall i\}$. The model is hidden when the relation between the state and the observation is probabilistic. The emission distributions for every state are summarized in the emission matrix E, $E_{is} = P(O_t = y | X_t = i)$, y being an emitted symbol (Fig. 6). A HSMM is thus defined by the set (T, E, χ_o, S).

The reasons for using an indirect observation model (*hidden*) in our scenario are the following:

1. Flasks have no distinctive identifier and even a trained supervisor cannot discern one flask from another. Hence, when more than one flask are being processed inside the plant and one of them is moved, there is a probabilistic relation between the motion event and the flask that generated it, and not deterministic.
2. The low frame rate confuses the perception of motion direction: deciding whether a flask is entering or exiting a location requires more reasoning than simply inspecting the visual content of the previous and present frames.

The reason for using explicit duration models is that processing stages have typical durations dictated by the process itself. The duration of a processing step depends of the step's nature, and it is similar for all flasks. In such a scenario, duration models can give a significant contribution in helping understand which flask is being moved, when more than one is undergoing the same process step (typically, the first flask that moves in a step will be the first to be moved out).

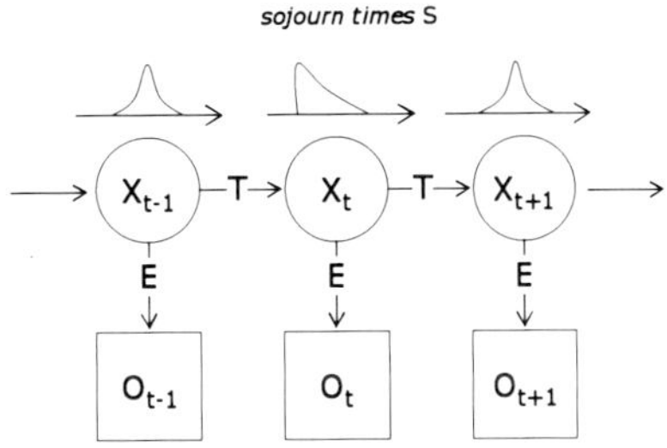

Fig. 6 A hidden semi-Markov model represents events with generic durations

6.1.1 State Space

Let us elaborate on the number of states needed for the Markov chain. A processing system can be described by three parameters (F, K, N):

- F is the maximum number of flasks that can be processed in parallel.
- K is the number of available cranes to move the flasks around ($K \leq F$).
- N is the number of processing stages that make up the nuclear process. In the example of Fig. 5, $N = 6$.

In realistic plants, $1 \leq F \leq 3$, $K = 1$, and $3 \leq N \leq 10$.

For building the state space, let us first define the *real states* of the plant as F-arrays of labels indicating the progression stage of each flask in its individual process. This space includes all possible permutations (with repetition) of flask positions. The *real states* of a plant with $F = 3$ and $N = 6$ are $[1\ 1\ 1], [1\ 1\ 2], [1\ 2\ 2], \ldots, [6\ 6\ 6]$. State [2 3 5] means that one flask is in stage 2, a second is in stage 3, and a third is in stage 5. We alter the number of real states for two reasons:

1. As flasks are *indistinguishable*, we can reduce the state space to the number of *combinations with repetitions*. For instance, [1 2 2], [2 1 2], and [2 2 1] are equivalent.
2. Emissions *depend on the transition*: by design, we associate the emission of a symbol to the event of a flask *entering* a state. So the starting state in a transition determines which symbol is emitted when the landing state is reached. For instance, consider a plant with $F = 2$. Its real state [2 3] can be entered from $[u\ 3]$, $u \neq 2$, thus triggering an emission linked to a flask entering stage 2. Or it can be entered from $[2\ v]$, $v \neq 3$, thus triggering an emission typical of a flask entering stage 3. To allow for multiple emission distributions, each *real state* is designed to be represented by F *virtual states*.

Given these premises, the size of the state space for a plant (F, K, N) is M:

$$M = F\frac{(N+F-1)!}{F!(N-1)!} = \frac{(N+F_1)!}{(F-1)!(N-1)!}. \tag{1}$$

6.1.2 Transition Matrix

The transition matrix is $M \times M$. Even if M is independent of the number of cranes K, K constrains the transition probabilities so that T_{ij} is 0 if two states i and j differ for more than K single-process states. If $K = 1$, only one process at a time can change state, hence for instance the transition $[1\ 1] \rightarrow [2\ 2]$ has probability 0. All virtual states referring to the same real state have the same transition probabilities towards other real states (equal rows in T).

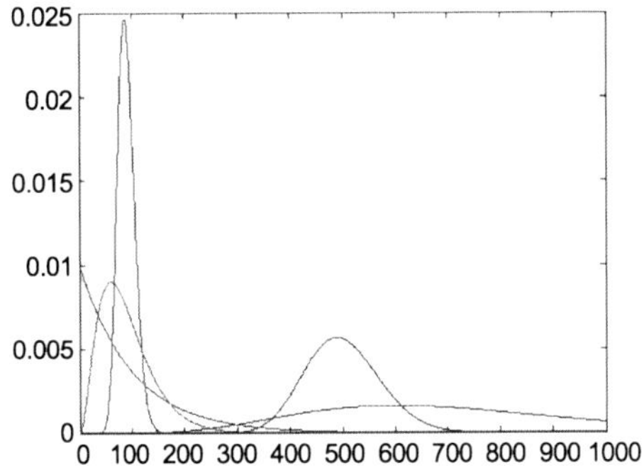

Fig. 7 Different shapes of a Gamma distribution obtained by changing it's a and b parameters

6.1.3 Sojourn Times

We refer the distributions of *sojourn times* S to single-flask processes, so a real state has F associated distributions. We have found this approach to produce more accurate predictions than assigning a single distribution to each real state. We use parametric distributions of Gamma shape (2) for single-flask stage durations. Gamma distributions can flexibly assume a shape ranging from an exponential to a bell Gaussian-like shape (Fig. 7). For example, when $a = 1$, Gamma is an exponential distribution; when $b = 2$, it is the chi-square distribution; the Gamma distribution closely approximates a Gaussian for large a, with mean $\approx a^*b$

$$Y = \left(b^a \Gamma(a)\right)^{-1} x^{a-1} e^{-x/b}. \tag{2}$$

The only exception to the use of the Gamma distribution is for the state corresponding to the empty plant (i.e., no flask under processing). In this case, we fetch the next SCD event over the hatch area, because the start of a new processing cycle is *independent* from previous cycles.

6.1.4 Emissions

Because the analysis is performed on the labels outputted by scene change detection (SCD), we define the *emission alphabet* of size A as the set of SCD labels associated to the areas of interest drawn by inspectors. In the example used throughout this chapter the emission alphabet is $\{H, D, P\}$, and $A = 3$. The matrix E is of size $M \times A$. For the use as filter, the emission probabilities must be set to 1 for the symbol expected for that transition, and 0 otherwise. This will constrain the model to admit only the correct symbols for each state and to filter out the false alarms.

A trivial example of T for a plant with $F = 2$, $K = 1$, and $N = 2$ is given in (3). Note that, with $N = 2$, the only possible events are either a flask going from stage 1 to stage 2 or vice versa (T_{ii} being null by definition)

$$T = \begin{pmatrix} 0 & 0 & 0 & 1 & 0 & 0 \\ 0 & 0 & 0 & 1 & 0 & 0 \\ 0.5 & 0 & 0 & 0 & 0 & 0.5 \\ 0.5 & 0 & 0 & 0 & 0 & 0.5 \\ 0 & 0 & 1 & 0 & 0 & 0 \\ 0 & 0 & 1 & 0 & 0 & 0 \end{pmatrix} \begin{matrix} \text{Real states} \\ [1\ 1] \\ [1\ 1] \\ [1\ 2] \\ [1\ 2] \\ [2\ 2] \\ [2\ 2] \end{matrix} \quad E = \begin{pmatrix} 1 & 0 \\ 1 & 0 \\ 1 & 0 \\ 0 & 1 \\ 0 & 1 \\ 0 & 1 \end{pmatrix}. \tag{3}$$

6.2 Training the HSMM

Some parameters of the HSMM are plant-specific and known:

1. The plant's related parameters F (nr. of flasks), K (nr. of cranes), and N (nr. of flask processing steps) of Sect. 6.1.
2. The expected sequence of processing steps of an *individual* flask, i.e. the schematics shown in Fig. 5. These steps determine the single-flask emission matrix. For the example in Fig. 5, the emission matrix is given in (4).
3. The initial state χ_o is the last state reported by the analysis of the previous image stream in the most recent inspector's report available

$$\begin{matrix} \text{Symbols} & h & d & p \end{matrix}$$
$$E = \begin{pmatrix} 1 & 0 & 0 \\ 0 & 1 & 0 \\ 0 & 0 & 1 \\ 0 & 0 & 1 \\ 0 & 1 & 0 \\ 1 & 0 & 0 \end{pmatrix}. \tag{4}$$

Moving from the single-flask emission matrix to the full-process matrix for multiple flasks requires replicating the correct emission based on the description of the corresponding real state, as in (3).

Other parameters, i.e., the transition matrix and duration models, must be trained on real examples of flask processing sequences. The training set is provided by past review annotations for the current plant. These reports are reliable in that they are double-checked for precision. They contain all the necessary information to train both the symbolic and duration models, i.e., annotated events and relative time stamps (Fig. 2).

An algorithm for HSMM training is described in [11]. For practical reasons, we pursue approximate likelihood maximization by separating the training of the transition matrix and the duration models. The transition matrix is trained with the Baum–Welch algorithm [27] using the emission matrix for the full process. The training set is the sequence of states annotated on inspectors' reports. An example

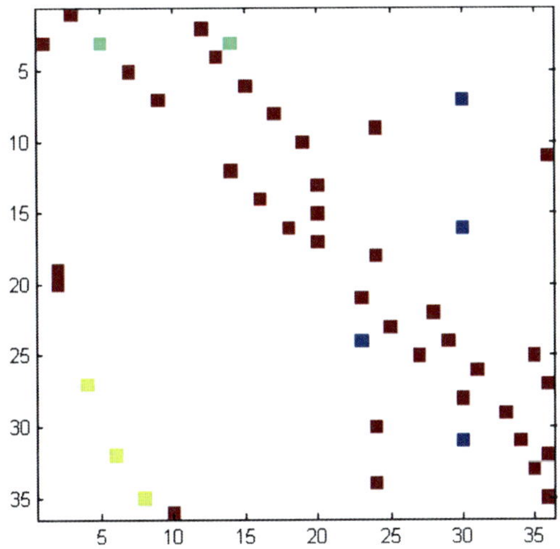

Fig. 8 A trained transition matrix for a 2-flask processing plant with flask processing as shown in Fig. 5. A *white square* indicates a null probability of transition; *dark red* means a probability ≥ 0.5; *shades of light blue*, *green* and *yellow* indicate intermediate values

result for a 2-flask plant with one crane is shown in Fig. 8. The duration models are parametric distribution functions of Gamma (Sect. 6.1). We model the duration of single-flask steps, as it was shown to be more consistent than the duration of real (global) states. To determine a and b for each state we use maximum-likelihood fitting of duration data extracted from review reports of single-flask processes.

In the course of a review, the model is used as *is*, i.e., without online adaptation. After completing a review on a batch of images, inspectors double-check the resulting list of events to approve it. The new consolidated report is added to the tail of the previous reports to form a continuous history of reviews, and the HSMM model is retrained on this extended history. The new model is then tested in simulation[1] to assess its performance and compare it to the performance of the previous model. It is retained only if the detection performance improves. Although nuclear flasks' processes tend to be stationary over time, this update procedure allows the model to eventually adapt to process changes.

6.3 The MM Image Review Tool

We use the HSMM built as in Sects. 6.1 and 6.2 to filter images during a review, by integrating it in a semi-automated software tool called the MM review tool (MM in short). The graphical user interface is the one shown in Fig. 4.

The general idea of the MM review tool is that, given the *history of annotations* produced by the inspector during the review, the (T, E, χ_o, S) model can highlight

[1] The review report acts as a 'simulated inspector'.

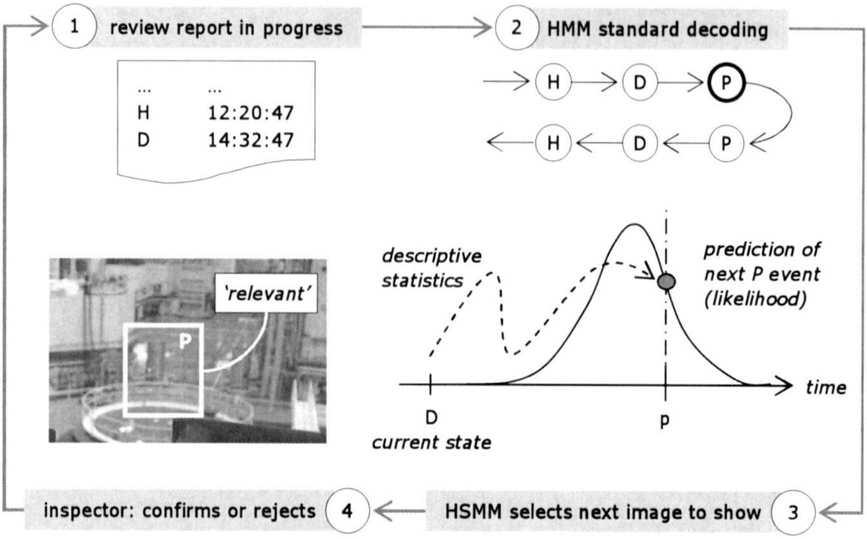

Fig. 9 Using the HSMM as a filter for assisting a review

the next most likely relevant image among those selected for motion events by scene change detection (SCD). The inspector decides whether to annotate the proposed image with a label corresponding to an event, e.g., H, D or P, or to reject it. The interaction is repeated with the next image until the end of the review (Fig. 9).

To illustrate the use of the MM tool as filter during a review, let us consider a mid-point during the review, when the inspector's past annotations form a sequence ν of k events (Step 1 of Fig. 9):

$$\nu = \{(o_1, t_1), (o_2, t_2), \ldots, (o_k, t_k)\}. \tag{5}$$

The pairs (o_j, t_j) indicate that the inspector has confirmed that an event o_j has occurred at time t_j. The information o_j specifies the type of event, expressed as one of the symbols of the emission alphabet. For the example of Fig. 5:

$$o_j \in \{H, D, P\}. \tag{6}$$

We assume the inspector to be a fully reliable in the labeling of events, so that ν is true with probability 1:

$$P(\nu) = 1. \tag{7}$$

Given ν, the MM tool estimates the current state χ of the plant processing (Step 2 in Fig. 9). Under the assumption (7), the decoding of ν via the HSMM detailed in [11] becomes superfluous and we are allowed to use the traditional HMM decoding [27] on the sequence $\{o_1, o_2, \ldots, o_k\}$ with an immediate advantage in terms of computational complexity.

The MM review tool uses the current state χ to select the next image to show to the inspectors on the graphical user interface (Step 3 in Fig. 9). To select this image I_{t*}, we first compute the likelihood of every future image selected by SCD given χ and S, for every symbol y in the emission alphabet:

$$L_y(I_t) = \sum_{j=1}^{M} \chi_j \left(\frac{S_j^*(t)}{\sum_t S_j^*(t)} \right). \tag{8}$$

In (8), we apply the total probability rule on the M states of the HSMM on the sum $S_j^*(t)$ of the duration probability distributions that pertain to symbol y. $S_j^*(t)$ is computed based on the assumption of stochastic independence of the processes of different flasks, as follows:

$$S_j^*(t) = \sum_{i \in \Gamma(j,y)} S_i\left(t - t_i^o\right). \tag{9}$$

In (9), S_i is the duration model for case i, t_i^o is the time of the previous event of case i, and $\Gamma(j, y)$ is the set of single-flask stories compatible with the real state j that can generate a symbol y if a transition occurs. Our choice of projecting the likelihood of duration models for one step only in the future is justified by the fact that predictions for multiple steps would involve a convolution of duration models, which quickly flattens the joint probability distribution and hence deteriorates the filtering effect. The image I_{t*} is then selected as follows:

$$t^* = \min\left(\arg\max_t \left(L_y(I_t)\right)\right). \tag{10}$$

In words, for each symbol y of the emission alphabet (of size A) we identify the image exhibiting maximum likelihood. Then, within this set of A images we select the image that comes first in order of time. This is to preserve, as much as possible, a sense of precedence in event sequences for inspectors when more than one motion event occurs as a consequence of multi-flask processing.

Finally, I_{t*} is presented to the inspector on the graphical interface of the MM review tool (Step 4 in Fig. 9). If the inspector rejects this candidate, $L_y(I_{t*})$ is set to zero and (10) is recomputed, and so on, until the inspector accepts the proposed image as a positive occurrence of an event. When an image is accepted and annotated, its symbol o_{k+1} is added to ν and the procedure restarts.

6.4 Discussion about the MM Review Tool

It can happen that events are missed because of the Bayesian nature of the MM review tool, giving rise to undesired false negatives. The informed, model-based nature of the filtering intrinsically carries this risk if the model does not match the data distribution in the image stream being reviewed. To possibly recover from these cases, our review tool implements some *heuristics* and *case-based reasoning* to avoid false negatives.[2]

[2]Recovery techniques are not covered here, because they are out of the scope of this chapter.

One case of false negative occurs when a flask's processing step is longer than usual. In this case, the duration model attributes a null probability of event to the image falling after the end of the distribution support function. To avoid this problem, we use duration functions with infinite support (Gamma functions). This practice may lower the filtering efficiency of the MM review tool, in that more images are shown to the inspectors, also those exhibiting a very low likelihood of event, but it can be useful in applications where a false negative is costly.

A second case of false negative is generated in multi-flask processing plants, when flasks follow the same process 'in pair' (i.e., flask 1 is moved from H to D, flask 2 follows the same movement, then flask 1 is moved from D to P, flask 2 follows, and so on). In this situation, an event referring to one flask may fool the HSMM for what regards the processing of other flasks. If the sequence ν is *inconsistent* with regards to the HSMM, i.e., the estimated current state χ is null, we adopt the heuristics of suppressing in turn the confirmed symbols $(o_k, t_k), (o_{k-1}, t_{k-1}), (o_{k-2}, t_{k-2})$, etc., until a *consistent* sequence is found (i.e. a sequence giving a non-null estimation for χ). The prediction procedure (Step 3 of Fig. 9) is then applied from that point. Instead, a missed detection giving birth to a sequence ν that is 'consistent' will not be discovered unless, as the review progresses, the sequence becomes 'inconsistent'. This fact nullifies any guarantee of null false detection rate of the HSMM.

A fundamental remark on MM is that it is intrinsically not designed to discover images of *irregular behavior. MM is useful as a way to establish where the regular events are*, and this can limit the application of this approach in some contexts.

Outside a review context, a further possible use of MM is for *consistency checks of review reports or operators' activity declarations*. Because MM uses probabilistic models, it can assign a likelihood score to a report or declaration and generate an alert in case of a low likelihood. This procedure may be applied just after a review has been completed so as to trigger a double-checking when needed. Or it may be applied during audits of review reports to direct the auditing inspector towards those 'more far away from normality'.

The MM review tool can be run only after the list SCD events has been computed for a given image stream. When MM is started, the first image selected for review is shown on the image viewer and the corresponding event is highlighted in the SCD event list (Fig. 10). Based on the image qualification provided by the inspector (positive or negative), MM proposes a second image highlighting the corresponding SCD event, and so on until the end of the review. To evaluate MM's performance, we keep track on the total number of highlighted events at the end of a review (see Sect. 7 on a benchmark).

From a user's viewpoint, a MM-based review has a peculiar aspect: images are presented in a likelihood-driven order on the basis of duration models and not in chronological order. If the maximum-likelihood image is refused, the second most likely image may happen to refer to a previous instant, and so the inspector sees a backward leap in the image stream (Fig. 11). Given that duration likelihoods are projected in the future for one step only, these jumps are generally localized in time around the segment of the image stream currently under review.

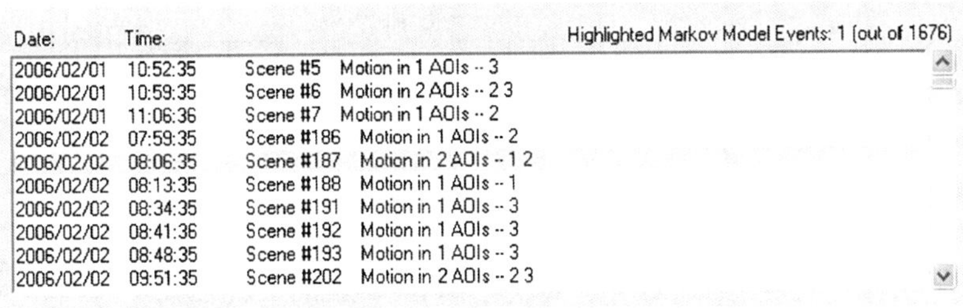

Date:	Time:	Highlighted Markov Model Events: 1 (out of 1676)
2006/02/01	10:52:35	Scene #5 Motion in 1 AOIs -- 3
2006/02/01	10:59:35	Scene #6 Motion in 2 AOIs -- 2 3
2006/02/01	11:06:36	Scene #7 Motion in 1 AOIs -- 2
2006/02/02	07:59:35	Scene #186 Motion in 1 AOIs -- 2
2006/02/02	08:06:35	Scene #187 Motion in 2 AOIs -- 1 2
2006/02/02	08:13:35	Scene #188 Motion in 1 AOIs -- 1
2006/02/02	08:34:35	Scene #191 Motion in 1 AOIs -- 3
2006/02/02	08:41:36	Scene #192 Motion in 1 AOIs -- 3
2006/02/02	08:48:35	Scene #193 Motion in 1 AOIs -- 3
2006/02/02	09:51:35	Scene #202 Motion in 2 AOIs -- 2 3

Fig. 10 SCD events proposed for review are highlighted on the event list

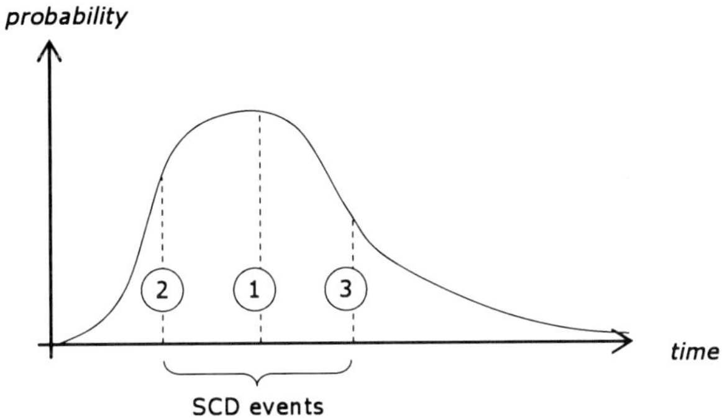

Fig. 11 The order of presentation of SCD events is not strictly chronological because it is likelihood-driven

Fig. 12 MM's log file

```
#1464: -;
#1465: p;    state: [3 5]
#1615: -;
#1614: -;
#1613: -;
#1603: d;    revision ; mid positive
#1602: -;
#1601: p;    state: [4 5]
#1795: -;
#1796: -;
#1797: h;    state: [5 6]
#1798: -;
#1838: -;
#1839: -;
#1840: h;    branching starts
#1841: d;    branching ends; state: [2 5] ; mid positive
#2002: -;
#2003: h;    state: [2 6] ; mid positive
#2004: -;
#2005: -;
```

To avoid confusion, a visualization of the back and forth jumps would be desirable. This could be part of a future interface that shows the evolution of the flask process and makes MM reasoning explicit to the user. The basis for this could be the existing MM's log file (Fig. 12). This file includes MM's dialogue with the inspector and its reasoning on the flask process. In each line, the log keeps track of the image number proposed for review (#⟨*integer*⟩), the annotation provided by the inspector ('-' for irrelevant image; h, d, p for events), and a number of possible messages shown in Table 2.

Table 2 Log messages explain MM reasoning about flask processes

Message type	Meaning
State	Estimated current flask configuration resulting after the decoding procedure.
Mid positive	A positively annotated image does not correspond to the predicted class for the event.
Revision (start)	The decoding procedure cannot reconstruct the flask configuration advancement with certainty. A positively qualified image does not fit the sequence of events. Probably a relevant image has been skipped: backtracking starts.
Branching (start)	The decoding procedure cannot reconstruct the flask configuration advancement with certainty. More that one hypothesis explain the sequence of events: multi-hypothesis tracking starts.

As we will see from the benchmark of Sect. 7, employing the MM filter gives a strong advantage in terms of *workload reduction*. MM is completely embedded in the state-of-the-art review workflow: its use does not require extra input from the user other than what is already needed to browse and annotate the SCD events. Hence, the gain of using MM can be measured by the number of SCD false positives spared at review time, when all true positives have been detected.

The filter runs in the background and can be deactivated at anytime. Further, its computational load is completely manageable by a standard laptop processor, thanks in particular to the possibility of substituting HMM-like decoding in place of the more complex HSMM decoding.

7 Benchmarking Image Review Filters

In the following, we focus on experimental results obtained by running scene change detection (SCD) and the Markov models (MM) review tool on grayscale images acquired by a safeguards surveillance system [5] in two different plants, A and B.

A is a single-flask plant ($F = 1$), while B has the capacity to process two flasks ($F = 2$) with the constraint of a single crane ($K = 1$) (Sect. 6.2).

The tests aimed at measuring the performance of SCD and MM in the detection of flask events: events on the hatch area (H), on decontamination (D) and over the pond (P).

7.1 Image Sets

Table 3 provides information about the image sets used for tests. Sets $A1$, $A2$ and $A3$ stem from plant A, while $B1$ and $B2$ stem from plant B. Each image set spans over several months of plant activity. For each set, the number of target events to identify

Table 3 Image sets used in the benchmark. For each set, the table lists the number of images in the set, and the number of events to be detected over the hatch (H), decontamination (D) and pond (P) areas

Image set	Images	H	D	P
$A1$	20160	17	17	17
$A2$	15661	1	1	1
$A3$	16022	–	–	–
$B1$	16020	30	30	30
$B2$	15446	12	12	12

is shown in the table. This ground truth information has been derived from inspectors' review reports. As anticipated in Sect. 3, the number of safeguards-relevant events is exiguous compared to the number of images in each set. Also, it is not unusual to have image sets where no safeguards-relevant activity needs to be reported by the inspectors as in $A3$.

7.2 Performance Metrics

To evaluate the performance of the image review tools, we adopt two measures.

The *first* is the classical performance evaluation for classifiers used in information retrieval, namely *recall* and *precision* [29]. For a given classification method M and benchmark containing R^* true events, the retrieval indexes are defined as:

$$recall_M = \frac{CR^*_M}{R^*}, \tag{11}$$

$$precision_M = \frac{CR^*_M}{CR_M}, \tag{12}$$

where:

- CR_M is the number of images classified by M as relevant.
- CR^*_M is the number of relevant images correctly classified by M.

The classification is optimal in terms of *recall* and *precision* when both indexes have value 1. *Recall* equals 1 when there are no false negatives, i.e., no relevant image is classified as irrelevant. *Precision* equals 1 when there are no false positives, these being defined as:

$$FP_M = CR_M - CR^*_M. \tag{13}$$

In our benchmark, we apply *precision* in a strict way, i.e., counting as correctly classified only the images annotated in review reports. In a sense, this is too strict for surveillance reviews where the goal is to highlight images in the *proximity* of the one that will be annotated: users always explore surrounding images to enhance their understanding on what happened in the scene. For this reason, *precision* has to be read as an indicative measure of performance, good enough to rank retrieval algorithms.

The *second* performance index is application-oriented [16] and by far more meaningful from the user point of view. Given that SCD is the default filter used in safeguards reviews, we measure the user advantage to filter SCD events by a second classifier M (provided that $recall_{\mathrm{SCD}}$ and $recall_M$ equal 1) by:

$$workload\_reduction_M = \frac{FP_{\mathrm{SCD}} - FP_M}{FP_{\mathrm{SCD}}}. \tag{14}$$

If FP_M equals FP_{SCD}, M does not bring any advantage and the *workload reduction* is 0. If FP_M is 0, M brings a large advantage and the *workload reduction* is 1.

7.3 Experimental Results

SCD has been parameterized in this benchmark to guarantee that *all* events are detected by optimal thresholds for each area of interest (AOI), i.e., the thresholds that provide $recall_{\mathrm{SCD}}$ equal to 1 and minimize FP_{SCD}.

MM is trained by the procedures described in Sect. 6.2 on roughly the first half of the events contained in $A1$ and $B1$. The remaining events test MM's generalization performance.

Since *in these experiments MM detected all ground truth events* (i.e., $recall = 1$), the evaluation is focused on *precision* and *workload reduction*.

Concerning *precision* (12), Fig. 13 (striped bars) shows that SCD produces many false positives, even when it is parameterized to work at its best. In particular, to detect all decontamination events, a low detection threshold had to be set. Figure 13 (black bars) shows that the *use of an informed technique after SCD pays off*: MM scores a significant increase in *precision* on all image sets while not missing events.

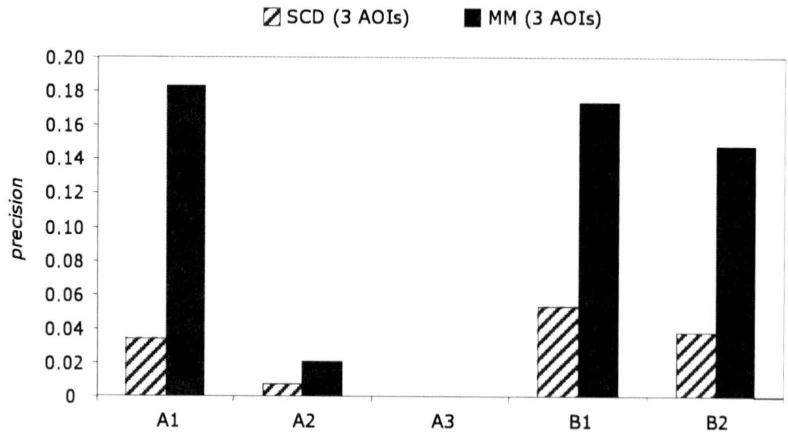

Fig. 13 *Precision* provided by two filters measured on five image sets: SCD (scene change detection) run on 3 AOIs, followed by MM (Markov models)

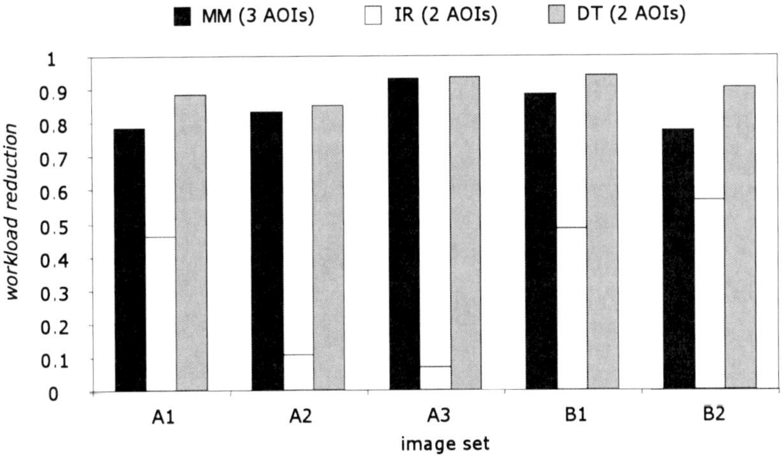

Fig. 14 *Workload reduction* provided by three filters measured on five image sets: MM (Markov models) run after SCD on 3 AOIs, IR (image retrieval by nearest neighbors) run after SCD on 2 AOIs, and DT (decision trees) run after SCD on 2 AOIs

The real advantage of using MM after SCD is measured *by the workload reduction* index (14) shown in Fig. 14 (black bars). The high peaks reached by MM indicate that it manages to reduce the number of false positives generated by SCD by a very significant amount, while retaining the true events. From the user point of view, the work reduction due to MM *implies at least* 75% *of images less to review*. At the same time, the use of MM does not require machine learning expert knowledge by the user, a design constraint to be respected for this technology to be accepted by users. Overall MM copes well with the task of tracking the whole flask-processing chain (i.e., over 3 AOIs) despite the noisy stream of SCD events.

As terms of comparison for MM, Fig. 14 provides the performance of two *alternative informed techniques*, namely image retrieval (IR, white bars) and decision tree based classification (DT, gray bars) [37]. Differently from MM, IR and DT exploit *regularities in the visual appearance of events* instead of regularities on the sequence and timing of events exploited by MM. The choice of focusing the comparison on DT and IR is that, like MM, these techniques require no expert machine learning knowledge. They can be used by any person, provided a pre-analysis work has been performed by machine learning experts on 'ML settings' that can work reasonably in a number of situations, in our case different plants and camera points of view.

IR is a nearest-neighbor classifier [4] that uses one example image per class of events to be detected, coupled with suitable similarity thresholds. Thresholds are learnt automatically based on annotations of events in past review reports. Images sufficiently close to the example (in a given feature space that describes the image content) are considered 'relevant' and displayed for review.

DT is implemented as a set of decision trees [26] (one per AOI) trained on positive and negative examples of events to be detected. Like IR, DT operates decision rules on image features describing the image content. An image is classified 'relevant' and proposed for review if at least one decision tree classifies it as relevant. DTs are trained interactively by inspectors based on relevance feedback cycles. Past review reports act as ground truth to help measure the classification performance. Further examples become available from the user online in an iterative way. This machine learning context is related to that of *active learning* [33, 34]. In active learning an algorithm selects data points which should receive a label by the user. This selection is driven, for instance, by the uncertainty associated to data classification as measured by the algorithm. Typically data points that fall close to decision boundaries of a classifier are the most uncertain, therefore presented with priority to the user to receive a label. Active learning is a lively area of research, very relevant to image retrieval. Still, it appears that the use of these techniques requires a level of expertise in machine learning to set-up a working system that surpasses the ordinary user's capabilities [32]. Hence, for DT, we followed a user-centered approach where the user decides which examples to label to improve the classification. These decisions are driven by application-oriented considerations (e.g., what is more relevant for safeguards purposes) and by an interface which makes the classifier's performance as transparent as possible to the end-user. Finally, we note that despite large image sets like ours, testing DT is fast because trees learn to focus classification rules on a reduced set of image features. Since the induced classifiers are very compact, they are suitable for interactive testing with online users [8].

IR and DT are examples of video annotation techniques introduced in Sect. 2. They learn to search images by content based on annotations (the training examples) made by the inspectors at previous reviews (Properties P1 and P2 in Table 1). Differently from MM, IR and DT classify each image 'on its own', i.e., outside the temporal context. They do not rely on a smooth transition in the visual appearance of objects of interest, and are applicable to low frame rate image streams like ours. Generally, these techniques are effective when classes of events exhibit distinctive, recurrent visual signatures.

In our context, not all classes of events satisfy this condition. As noted, events of type D are hard to distinguish by automatic means due to this area of interest be far away from the camera. Hence, in the experiments we limited the use of IR and DT to the detection of events H and P, because on D events *recall* for these techniques is definitely below 1. *Workload reduction* is then computed with respect to SCD run on 2 AOIs (H and P) for IR and DT, and on 3 AOIs for MM.

Figure 14 shows that MM outperforms simple informed search by content techniques such as IR, and is comparably effective as more sophisticated search by content techniques such as DT. On the other hand, MM is capable of detecting by temporal means all classes of relevant events, while DT is not.

Interestingly, analyzing the detection performance per type of event, we found that MM and DT have complementary detection skills. For instance, DT can detect well events of type H due to the camera favorable position with respect to the hatch area. By contrast, H entry events are hard to detect for MM because the start of a new flask processing cycle is *independent* of past cycles. For this reason, when no flask is in the plant for processing, MM fetches SCD events over the hatch until the inspector annotates that a new flask has entered the facility (see Sect. 6.1 on sojourn times). On the positive side, D events (the weak point for DT and IR) are well predicted by MM because they exhibit a strong regularity in the temporal dimension. Another source of workload reduction for MM is linked to the fact that, given a state of flasks processing, MM focuses its attention to SCD events generated over the AOIs where events are expected. For example, in a single flask plant, a flask entering the pond generates the expectation of the next event to be a 'pond exit' one. Hence, all SCD events happening on AOIs H and D are ignored until a P event is confirmed by the inspector. By contrast, IR and DT need to classify as relevant of irrelevant each single SCD event generated over the entire image stream because they do not 'follow' the flask process.

Since DT and MM have complimentary strong points and weaknesses, an integrated system of the two would be a natural proposal to improve the overall detection precision.

8 Discussion

Machine learning for the analysis of image streams in fields related to video surveillance is becoming a central topic of research within the computer vision community. To be effective, the range of techniques proposed need to address well the specificities of the applicative context envisaged. Part of this is reflected in the choice of the camera technology to be employed and how this is configured Today, camera systems are evolving in diverse directions, ranging from high-speed, to low-power devices and smart cameras, just to name a few.

The contribution presented in this chapter was focused on machine learning as a way to capture field knowledge from image annotations made by surveillance end-users. We have addressed an application field, that of nuclear safeguards, having the following characteristics: (i) very low frame rate due to the cameras working as standalone systems storing images on local memory only; (ii) objects of interest with distinct visual features for some classes of events of interest, contrasted by low resolution of the same objects in temporally following images. These characteristics contrast with the assumption of traceability of objects underlying many computer vision studies.

As specific technical contribution, we presented a combination of filters based on scene change detection (SCD) and Markov models (MM) to search streams of surveillance images. A design constraint for this study was to consider search techniques which do not require machine learning or computer vision expertise from the end-user.

SCD is an uninformed search technique. It does not use priors on the visual appearance of events to detect them. SCD is triggered by any change within areas of interest defined by the user. There are pros and cons to SCD. A clear advantage of SCD is its 'generality'. By setting conservative SCD thresholds, one has good guarantees that all events of interest will be included among SCD events. A second advantage of SCD lies in its 'programmability'. The only parameters to be specified by the user—who is not supposed to be an expert of computer vision—are the areas of interest and the associated detection thresholds. On the down side, SCD presents as events a large number of false positives.

MM is an informed search technique. It uses past event annotations made by end-users to facilitate the recognition of related events in new image streams to be reviewed. Specifically, MM's priors relate to temporal meta-data (i.e., the images' time-stamps). MM is totally 'blind'. All it does is reorganize the visit order of SCD events by their temporal likelihood. In doing so, a significant number of SCD false positives are spared from review. The choice of focusing MM on temporal aspects brings several advantages. *First*, when objects of interest are not traceable by image content (e.g., because of too weak visual signatures of events), a temporal filter allows to track sequences in a more robust way. *Second*, a temporal filter is easily embedded in the review cycle at no additional interaction cost for the user: with MM, the inspector performs all and only the usual operations necessary to annotate the SCD events when proposed by MM for review. Then, MM takes advantage of the 'user-in-the-loop': a freshly available annotation is immediately taken into account to recompute the next most likely relevant SCD event. Short term predictions are more accurate when based on solid information provided by the user. This information comes for free because the image review report is built progressively over time. *Third*, the MM model is *process-bound* instead of *camera-bound*. This means that if the camera is repositioned to provide a better view on the sequence of events, MM models are still valid, i.e., they require no retraining. For the same reason, MM models can be used to filter images from redundant cameras overlooking the scene. All this does not apply to search by content techniques which depend on visual regularities as perceived from a camera specific point-of-view.

Table 4 Best and worst cases of operative conditions for MM

	Best case	Worst case
SCD performance	SCD detects all events and scores few false positives (SCD thresholds well calibrated).	SCD does not detect all events, because of bad thresholds calibration or of bad camera position.
Availability of review reports	Past review reports documenting at least 10–15 complete processings are available for the plant under analysis.	Less than 3–4 complete processings or only incomplete processings are documented on the review reports available for training.
Regularity of event sequences	All events in the plant follow regular sequences according to what documented on past review reports.	One or more events happen in a logical sequence never documented on past review reports.
Regularity of event durations	All processing durations have a mono-modal Gamma distribution, i.e., occur with limited variability.	One or more events occur with a duration that is largely different from the mean duration documented on past review reports.

Temporal filters like MM can be applied when events of interest are recurrent over time and follow specific patterns as in surveillance for safeguards. Table 4 summarizes the 'best' and the 'worst' operative conditions for MM. The approach can be extended to other surveillance scenarios. The most compelling is industrial process monitoring, when video systems do not meet costly high frame rate surveillance standards.

References

1. Ahmed Elgammal, A., Duraiswami, R., Harwood, D., Davis, L.S.: Background and foreground modeling using nonparametric kernel density estimation for visual surveillance. Proc. IEEE **90**(7), 1151–1163 (2002)
2. Akyildiz, I.F., Su, W., Sankarasubramaniam, Y., Cayirci, E.: Wireless sensor networks: a survey. Comput. Netw. **38**(4), 393–422 (2002)
3. Brdiczka, O., Langet, M., Maisonnasse, J., Crowley, J.L.: Detecting human behavior models from multimodal observation in a smart home. IEEE Trans. Autom. Sci. Eng. **5**, 1 (2003)
4. Cover, T., Hart, P.: Nearest neighbor pattern classification. IEEE Trans. Inf. Theory **13**, 21–27 (1967)
5. DCM14. http://www.canberra.com/pdf/Products/Systems_pdf/DCM-14-SS.pdf
6. Dorado, A., Calic, J., Izquierdo, E.: A rule-based video annotation system. IEEE Trans. Circuits Syst. Video Technol. **14**(5), 622–633 (2004)
7. Ewerth, R., Freisleben, B.: Semi-supervised learning for semantic video retrieval. In: Proc. 6th ACM International Conference on Image and Video Retrieval, pp. 154–161 (2007)
8. Fails, J.A., Olsen, D.R.: A design tool for camera-based interaction. In: Human Factors in Computing Systems, CHI '03, pp. 449–456. ACM, New York (2003)
9. Ferrando, S., Gera, G., Massa, M., Regazzoni, C.: A new method for real time abandoned object detection and owner tracking. In: IEEE International Conference on Image Processing, pp. 3329–3332 (2006)
10. GARS. http://www.canberra.com/pdf/Products/Systems_pdf/GARS-SS.pdf
11. Guédon, Y.: Estimating hidden semi-Markov chains from discrete sequences. J. Comput. Graph. Stat. **12**(3), 604–639 (2003)
12. Hu, W., Tan, T., Wang, L., Maybank, S.: A survey of visual surveillance of object motion and behaviors. IEEE Trans. Syst. Man Cybern. C **34**(3), 334–352 (2004)
13. Iyengar, G., Nock, H.J.: Discriminative model fusion for semantic concept detection and annotation in video. In: Proc. 11th ACM International Conference on Multimedia, pp. 255–258 (2003)
14. Liu, Y., Yao, H., Gao, W., Chen, X., Zhao, D.: Nonparametric background generation. In: Proc. 18th International Conference on Pattern Recognition, vol. 4, pp. 916–919 (2006)
15. Lombardi, P., Versino, C.: Tracking nuclear material at low frame rate and numerous false detections. In: ICVS2007 Vision Systems in the Real World: Adaptation, Learning, Evaluation—Conference Proceedings, Bielefeld, Germany (2007)
16. Lombardi, P., Versino, C., Gonçalves, J.G.M., Heppleston, M., Tourin, L.: MM: the Markov model tool for image reviews. In: Proc. of the Institute of Nuclear Materials Management (2008)
17. Makris, D., Ellis, T.: Learning semantic scene models from observing activity in visual surveillance. IEEE Trans. Syst. Man Cybern. B **35**(3), 397–408 (2005)
18. Markou, M., Singh, S.: Novelty detection: a review—part 1: statistical approaches. Signal Process. **83**, 2003 (2003)
19. Markou, M., Singh, S.: Novelty detection: a review—part 2: neural network based approaches. Signal Process. **83**, 2499–2521 (2003)

20. Mcivor, A.M.: Background subtraction techniques. In: Proc. of Image and Vision Computing, pp. 147–153 (2000)
21. Messelodi, S., Modena, C.M., Zanin, M.: A computer vision system for the detection and classification of vehicles at urban road intersections. Pattern Anal. Appl. **8**(1), 17–31 (2005)
22. Mitra, S., Acharya, T.: Gesture recognition: a survey. IEEE Trans. Syst. Man Cybern., Part C, Appl. Rev. **37**(3), 311–324 (2007)
23. Morris, B.T., Trivedi, M.M.: A survey of vision-based trajectory learning and analysis for surveillance. IEEE Trans. Circuits Syst. Video Technol. **18**(8), 1114–1127 (2008)
24. Murphy, K.: Hidden semi-Markov models (HSMMs). Available online: http://people.cs.ubc.ca/~murphyk/Papers/segment.pdf (2002)
25. Qi, G.-J., Hua, X.-S., Rui, Y., Tang, J., Mei, T., Zhang, H.-J.: Correlative multi-label video annotation. In: Proc. 15th International Conference on Multimedia, pp. 17–26 (2007)
26. Quinlan, J.R.: C4.5: programs for machine learning. Morgan Kaufmann, San Mateo (1993)
27. Rabiner, L.R.: A tutorial on hidden Markov models and selected applications in speech recognition. Proc. IEEE **77**(2), 257–286 (1989)
28. Safeguards Techniques and Equipment IAEA—2003 Edition. http://www-pub.iaea.org/MTCD/publications/PDF/NVS1-2003_web.pdf
29. Smeulders, A.W.M., Worring, M., Santini, S., Gupta, A., Jain, R.: Content-based image retrieval at the end of the early years. IEEE Trans. Pattern Anal. Mach. Intell. **22**(12), 1349–1380 (2000)
30. Stauffer, C., Grimson, E.: Learning patterns of activity using real-time tracking. IEEE Trans. Pattern Recognit. Mach. Intell. **22**(8), 747–757 (2000)
31. Sullivan, G.J., Wiegand, T.: Video compression—from concepts to the H.264/AVC standard. Proc. IEEE **93**(1), 18–31 (2005)
32. Tao, D., Tang, X., Li, X., Wu, X.: Asymmetric bagging and random subspace for support vector machines-based relevance feedback in image retrieval. IEEE Trans. Pattern Recognit. Mach. Intell. **28**(7), 1088–1099 (2006)
33. Tong, S., Chang, E.: Support vector machine active learning for image retrieval. In: Proc. of ACM International Conference on Multimedia, pp. 107–118 (2001)
34. Tong, S., Koller, D.: Support vector machine active learning with applications to text classification. In: 17th International Conference on Machine Learning, pp. 401–412 (2000)
35. Toyama, K., Krumm, J., Brumitt, B., Meyers, B.: Wallflower: principles and practice of background maintenance. In: Proc. 7th IEEE International Conference on Computer Vision, pp. 255–261 (1999)
36. Trucco, E., Plakas, K.: Video tracking: a concise survey. IEEE J. Ocean. Eng. **31**(2), 520–529 (2006)
37. Versino, C., Stringa, E., Gonçalves, J.G.M., Heppleston, M., Tourin, L.: Towards the integrated review of safeguards surveillance data. In: 27th ESARDA Symposium on Safeguards and Nuclear Material Management (2005)
38. Wang, M., Hua, X.-S., Hong, R., Tang, J., Qi, G.-J., Song, Y.: Unified video annotation via multigraph learning. IEEE Trans. Circuits Syst. Video Technol. **19**(5), 733–746 (2009)
39. Yilmaz, A., Javed, O., Shah, M.: Object tracking: a survey. ACM J. Comput. Surv. **38**(4), 13 (2006)

Discriminative Multiple Target Tracking

Xiaoyu Wang, Gang Hua, and Tony X. Han

Abstract In this chapter, we introduce a metric learning framework to learn a single discriminative appearance model for robust visual tracking of multiple targets. The single appearance model effectively captures the discriminative visual information among the different visual targets as well as the background. The appearance modeling and the tracking of the multiple targets are all cast in a discriminative metric learning framework. We manifest that an implicit exclusive principle is naturally reinforced in the proposed framework, which renders the tracker to be robust to cross occlusions among the multiple targets. We demonstrate the efficacy of the proposed multiple target tracker on benchmark visual tracking sequences, and real-world video sequences as well.

1 Introduction

Visual tracking of multiple targets has been very active research in the past years [5, 18, 21, 22, 28, 29], largely due to its essentiality in video surveillance, and more emerging applications such as internet video annotation. To robustly track the multiple objects, firstly we need to *model* the visual targets, either based on contour shape or visual appearances. Then a matching algorithm match the image observation data with the models of the multiple targets. Appearance based modeling has induced a lot of attention due to its richness in representation.

X. Wang (✉) · T.X. Han
University of Missouri, Columbia, USA
e-mail: xw9x9@mail.missouri.edu

T.X. Han
e-mail: hantx@missouri.edu

G. Hua
Nokia Research Center, Santa Monica, USA
e-mail: ganghua@gmail.com

L. Wang et al. (eds.), *Machine Learning for Vision-Based Motion Analysis*, Advances in Pattern Recognition,
DOI 10.1007/978-0-85729-057-1_6, © Springer Verlag London Limited 2011

For visual appearance modeling of the multiple visual targets, one may model the different visual target separately, for example, either a generative model is built for each visual target to capture the visual variation [8, 14, 20–22, 27], or a discriminative model is built for each target to discriminate it from the background [1–3, 6]. Typical generative model for modeling visual target include appearance based subspace model [8, 14, 20] obtained using embedding methods such as principal component analysis [8, 20], or Gram–Schmidt decomposition [14], as well as Gaussian mixture model [17] learned from the Expectation-Maximization (EM) algorithm [9].

On the other hand, discriminative appearance models leveraged supervised learning algorithms to training a classification function to differentiate the appearances of the visual targets from the background. For example, support vector machine (SVM) and Boosting cascade classifier is adopted in [1] and [5], respectively, for training discriminative visual models, an ensemble classifier based on Boosting is leveraged in [2], a linear discriminative classifier is employed by [6], and a multiple instance Boosting classifier is utilized in [3]. A set of positive examples representing the target object and a set of negative examples representing the background are needed to train the discriminative model.

Compared to generative models, discriminative models aim directly on differentiating the visual target from the background clutter, hence they may be more desirable for robust visual tracking. However, separating the discriminative appearance modeling efforts for the multiple targets is problematic because each model is only focusing on differentiating the associated target with the background where the target presented. The discriminative information among the different visual targets themselves are totally ignored. Effectively capturing the discriminative information among the different visual targets may be vital in dealing with cross occlusions incurred among the multiple targets.

In this chapter, we present a formulation to learn a joint discriminative appearance model for discriminating the multiple visual targets from the background, as well as discriminating the multiple targets themselves. This formulation is cast under a discriminative metric learning framework proposed by Globerson and Roweis [11]. A nice property of this formulation is that the learning of the joint model only needs to optimize a convex function using gradient descent, where the optimal solution is guaranteed. Moreover, in our formulation, the visual matching process to track the multiple targets is optimizing the same objective function as what we used to learn the visual model.

The visual matching process in our tracking algorithm can be efficiently performed by gradient based optimization using any modern nonlinear optimization packages, such as the one proposed in [31]. This put our multiple target tracking algorithm into the literature of gradient based visual tracking algorithms [7, 12, 13, 26, 30]. Gradient based tracking algorithms directly match the visual model with the image observations based on the gradient of the objective function w.r.t. the motion parameters. It does not make any additional assumptions of the motion and observation models, which may often required in visual tracking algorithms based on hypothesis generation and observation verification, such as Kalman filter (KF) [19], probabilistic data association filter (PDAF) [4], and particle filter [16].

Due to the mutual discrimination of the appearance models of the different visual targets reinforced in the learning process, and the joint optimization of multiple motions, our tracking algorithm reinforces an implicit *exclusive principle* [21]. Exclusive principle, which is firstly defined by MacCormick and Blake [21] states that no two visual targets shall account for the same image observation, which is vital to handle and being robust to occlusions when dealing with multiple objects tracking. Notice that our proposed formulation for discriminative visual modeling of multiple visual appearances may not be utilized for tracking multiple identical objects in a visual scene. Nevertheless, exact identical multiple objects are scarce in real-world videos. Therefore, this limitation may not hinder the general applicability of the proposed multiple target tracking algorithm.

When compared with previous methods for multiple target tracking algorithms, the proposed modeling and matching framework presents three advantages. Firstly, it presents a discriminative formulation to simultaneously model the appearances of multiple objects, which not only discriminates the visual target from the background, but also seeks for mutual discrimination among the different visual targets. Secondly, in our formulation, an exclusive principle is naturally reinforced, which renders it robust to handle cross occlusions among the different visual target. Thirdly, our proposed framework is easily adapted for online model updating, which is supported by a principled criterion derived from the objective function to select the optimal set of visual examples for online modeling and matching.

2 Appearance and Motion Model of Multiple Targets

2.1 *Metric Learning Framework*

We cast our discriminative appearance and motion model of multiple targets by leveraging a metric learning framework similar to Globerson and Roweis [11]. Suppose we have a set of labeled training examples $\mathcal{X} = \{\mathbf{x}_{i,j} \in \mathbb{R}^N, o_{ij}\}_{j=1}^{n_i}$, where $o_{ij} = 0$ indicates background, and $o_{ij} = 1, \dots, K$ indicates the visual samples of each of the K visual targets we are intending to track. N is the dimension of examples. Let $\mathcal{S}_0 = \{(\mathbf{x}_{0j}, o_{0j} = 0)\}_{j=0}^{n_0}$, and also let $\mathcal{S}_i = \{(\mathbf{x}_{ij}, o_{ij} = i)\}_{j=0}^{n_i}$ for any $i = 1, \dots, K$, such that $n = \sum_{i=0}^{K}(n_i + 1)$ and $\mathcal{X} = \bigcup_{i=0}^{K} \mathcal{S}_i$. $\mathbf{x}_{ij}$ means the jth example for tracking target i ($i = 0$ implies background). In our experiments, each $\mathbf{x}_{ij}$ is usually a $w \times h$ image patch and $N = w \times h$.

We further denote $\forall i > 0$, $\mathbf{x}_{i0} = \mathbf{I}(\mathbf{m}_i)$, which indicates each of the K visual targets we want to track in the current frame where $\mathbf{m}_i \in \mathbb{R}^L$ is the motion parameters we want to recover. $\mathbf{I}(\mathbf{m}_i)$ is a mapping which maps the motion parameters, affine transformation parameters for example, to image patch. Obviously, the label o_{i0} of $\mathbf{I}(\mathbf{m}_i)$ is i, since it represents the ith visual target. For convenience, we will either use $\mathbf{x}_{i0}$ or $\mathbf{I}(\mathbf{m}_i)$ in our presentation depending on if we are learning for the appearance model or performing the visual matching for tracking of the multiple target. Following Globerson and Roweis [11], we propose to learn a Mahalanobis form metric, i.e.,

$$d_{\mathbf{A}}(\mathbf{x}_{ij}, \mathbf{x}_{kl}) = (\mathbf{x}_{ij} - \mathbf{x}_{kl})^T \mathbf{A}(\mathbf{x}_{ij} - \mathbf{x}_{kl}), \tag{1}$$

to achieve our unified formulation, where $\mathbf{A}$ is a positive semi-definite matrix we need to learn from data. Define, for each $\mathbf{x}_{ij} \in \mathcal{X}$, a conditional probability

$$p_{\mathbf{A}}(\mathbf{x}_{kl}|\mathbf{x}_{ij}) = \frac{1}{Z_{ij}} e^{-d_{\mathbf{A}}(\mathbf{x}_{ij}, \mathbf{x}_{kl})} = \frac{e^{-d_{\mathbf{A}}(\mathbf{x}_{ij}, \mathbf{x}_{kl})}}{\sum_{p \neq i \vee q \neq j} e^{-d_{\mathbf{A}}(\mathbf{x}_{ij}, \mathbf{x}_{pq})}}. \tag{2}$$

The ideal distribution of the optimal $\mathbf{A}$ shall collapse samples from the same class to be a single point. Specifically, the ideal distribution shall take the following form,

$$p_0(\mathbf{x}_{kl}|\mathbf{x}_{ij}) = \begin{cases} \frac{1}{n_c} & o_{ij} = o_{kl} = c, \\ 0 & o_{ij} \neq o_{kl}, \end{cases} \tag{3}$$

where $c \in \{0, 1, \ldots, K\}$. Recall that $\mathbf{x}_{i0} = \mathbf{I}(\mathbf{m}_i)$. Denote $\mathcal{M} = \{\mathbf{m}_1, \mathbf{m}_2, \ldots, \mathbf{m}_K\}$, we define

$$\begin{aligned} f(\mathbf{A}, \mathcal{M}) &= \sum_{i=0}^{n} \mathrm{KL}\big(p_0(\mathbf{x}_{kl}|\mathbf{x}_{ij}) \| p_{\mathbf{A}}(\mathbf{x}_{kl}|\mathbf{x}_{ij})\big) \\ &= C + \sum_{i=0}^{K} \sum_{j \neq k=1}^{n_i} \frac{1}{n_i} \big(d_{\mathbf{A}}(\mathbf{x}_{ij}, \mathbf{x}_{ik}) + \log Z_{ij}\big), \end{aligned} \tag{4}$$

where $C = \sum_{y_{ij}=y_{kl}=c} \frac{1}{n_c} \log \frac{1}{n_c}$ is a constant. To have $p_{\mathbf{A}}(\mathbf{x}_{kl}|\mathbf{x}_{ij})$ to be as close to $p_0(\mathbf{x}_{kl}|\mathbf{x}_{ij})$ as possible, we only need to proceed to minimize $f(\mathbf{A}, \mathcal{M})$. More formally, we formulate the following optimization problem,

$$\min \quad f(\mathbf{A}, \mathcal{M}) \tag{5}$$

$$\text{s.t.} \quad \forall \mathbf{a} \in \mathbb{R}^N, \quad \mathbf{a}^T \mathbf{A} \mathbf{a} \geq 0, \tag{6}$$

where the constraint in (6) confines $\mathbf{A}$ to be a positive semi-definite matrix (PSD). Solving the above optimization problem would allow us to jointly obtain the optimal discriminative appearance models for all of the multiple visual targets defined by $\mathbf{A}$, and track the motions of all of them as well, which is defined by $\mathbf{m}$. We solve both by efficient gradient based search, as we shall detail in the following subsections.

2.2 Joint Appearance Model Estimation

In formulation, discriminative appearance modeling refers to identifying the optimal $\mathbf{A}$ to define the metric between visual samples. Assume that the motion parameter $\mathbf{m}$ is fixed, following [11], it is easy to figure out that $f(\mathbf{A}, \mathbf{m})$ is a convex function of $\mathbf{A}$. Taking the derivative of $f(\mathbf{A}, \mathcal{M})$ with respect to $\mathbf{A}$, we have

$$\frac{\partial f(\mathbf{A}, \mathcal{M})}{\partial \mathbf{A}} = \sum_{i=0}^{K} \sum_{j=0}^{n_i} \sum_{k=0}^{K} \sum_{l=0}^{n_k} \omega_{ij}(kl)(\mathbf{x}_{kl} - \mathbf{x}_{ij})(\mathbf{x}_{kl} - \mathbf{x}_{ij})^T, \tag{7}$$

where

$$\omega_{ij}(kl) = p_O(\mathbf{x}_{kl}|\mathbf{x}_{ij}) - p_{\mathbf{A}}(\mathbf{x}_{kl}|\mathbf{x}_{ij}). \tag{8}$$

Similar to [11], we take a gradient projection algorithm [23] to obtain the optimal $\mathbf{A}$. Specifically the following two steps are performed:

1. Gradient Descent: $\mathbf{A} = \mathbf{A} - \epsilon \frac{\partial f(\mathbf{A},\mathbf{m})}{\partial \mathbf{A}}$, where ϵ determines the step length for gradient descent.
2. PSD Projection: Compute the eigen-value decomposition of $\mathbf{A}$, i.e., $\{\lambda_k, \mathbf{u}_k\}_{k=1}^N$ such that $\mathbf{A} = \sum_{k=1}^N \lambda_k \mathbf{u}_k \mathbf{u}_k^T$, set $\mathbf{A} = \sum_{k=1}^N \max(\lambda_k, 0)\mathbf{u}_k \mathbf{u}_k^T$.

The first step above performs gradient descent, and the second step reinforces the constraint to make $\mathbf{A}$ to be a positive semi-definite matrix. These two steps are iterated until convergence. Since $f(\mathbf{A}, \mathcal{M})$ is a convex function of $\mathbf{A}$ fixing $\mathcal{M}$, the iteration of these two steps is guaranteed to find the optimal solution of $\mathbf{A}$.

2.3 Motion Parameter Optimization

In this subsection, we fix the discriminative appearance model $\mathbf{A}$, and develop the gradient descent search for the motion parameters $\mathcal{M}$. Not losing any generality, we assume that each $\mathbf{m}_i$, $\forall i \in \{1, 2, \ldots, K\}$ is a linear motion model, that is,

$$\begin{bmatrix} x_i \\ y_i \end{bmatrix} = \begin{bmatrix} a_i & b_i \\ c_i & d_i \end{bmatrix} \begin{bmatrix} x'_i \\ y'_i \end{bmatrix} + \begin{bmatrix} e_i \\ f_i \end{bmatrix}, \tag{9}$$

where $[x'_i, y'_i]^T$ is the canonical coordinates for the labeled examples, and $[x_i, y_i]^T$ is the coordinates in the target video frame. This linear motion model covers a wide variety of visual motions such as translation, scaling, similarity, as well as full affine motion. We proceed to derive the gradient based search for the full affine motion model.

Recall that $\mathbf{x}_{i0} = \mathbf{I}(\mathbf{m}_i)$ is the only term that involves the motion parameter $\mathbf{m}_i$, $\forall i \in \{1, 2, \ldots, K\}$, according to chain rule, we have

$$\frac{\partial f(\mathbf{A}, \mathcal{M})}{\partial \mathbf{m}_i} = \frac{\partial f(\mathbf{A}, \mathcal{M})}{\partial \mathbf{x}_{i0}} \frac{\partial \mathbf{x}_{i0}}{\partial \mathbf{m}_i}. \tag{10}$$

With some mathematical manipulations, it can be shown that

$$\frac{\partial f(\mathbf{A}, \mathcal{M})}{\partial \mathbf{x}_{i0}} = \frac{4}{n_i} \sum_{j=1}^{n_i} \mathbf{A}(\mathbf{x}_{i0} - \mathbf{x}_{ij}) - 2 \sum_{k=1}^{K} \sum_{l=0}^{n_k} \beta_{i0}(kl)\mathbf{A}(\mathbf{x}_{i0} - \mathbf{x}_{kl}), \tag{11}$$

where

$$\beta_{i0}(kl) = p_{\mathbf{A}}(\mathbf{x}_{kl}|\mathbf{x}_{i0}) + p_{\mathbf{A}}(\mathbf{x}_{i0}|\mathbf{x}_{kl}). \tag{12}$$

For any parameter $\xi_i \in \mathbf{m}_i$, again, applying chain rule, we have

$$\frac{\partial \mathbf{x}_{i0}}{\xi_i} = \frac{\partial \mathbf{I}(\mathbf{m}_i)}{\partial \xi_i} = \frac{\partial \mathbf{I}(\mathbf{m}_i)}{x_i} \frac{\partial x_i}{\partial \xi_i} + \frac{\partial \mathbf{I}(\mathbf{m}_i)}{y_i} \frac{\partial y_i}{\partial \xi_i}, \tag{13}$$

where $\frac{\partial \mathbf{I}(\mathbf{m}_i)}{x_i}$ and $\frac{\partial \mathbf{I}(\mathbf{m}_i)}{y_i}$ represents the image gradient in the target frame in horizontal and vertical directions, respectively. For ease of notation, we denote them as $\mathbf{I}_{x_i}$ and $\mathbf{I}_{y_i}$, respectively. Following (13), we have, $\forall i \in \{1, 2, \ldots, K\}$

$$\frac{\partial \mathbf{x}_{i0}}{\partial a_i} = \mathbf{I}_{x_i} x_i', \frac{\partial \mathbf{x}_{i0}}{\partial b_i} = \mathbf{I}_{x_i} y_i', \tag{14}$$

$$\frac{\partial \mathbf{x}_{i0}}{\partial c_i} = \mathbf{I}_{y_i} x_i', \frac{\partial \mathbf{x}_{i0}}{\partial d_i} = \mathbf{I}_{y_i} y_i', \tag{15}$$

$$\frac{\partial \mathbf{x}_{i0}}{\partial e_i} = \mathbf{I}_{x_i} \frac{\partial \mathbf{x}_{i0}}{\partial f_i} = \mathbf{I}_{y_i}. \tag{16}$$

Therefore, we may easily calculate the gradient of $f(\mathbf{A}, \mathcal{M})$ with respect to $\mathbf{m}_i$ by applying (16) to (10). Then we can take a gradient descent step to recover the optimal motion parameter $\mathbf{m}_i$, $\forall i \in \{1, 2, \ldots, K\}$, that is,

$$\mathbf{m}_i = \mathbf{m}_i - \eta \frac{\partial f(\mathbf{A}, \mathbf{m}_i)}{\partial \mathbf{m}_i}, \tag{17}$$

where the step length η could be estimated, for example, by a quasi-Newton method such as L-BFGS [31].

3 Online Matching and Updating Multiple Models

Another challenge in appearance model based multiple target tracking is to robustly adapt the model to the visual environment. This adaptation may be indispensable for robust tracking since the target objects may go through drastic visual changes caused by environmental conditions such as extreme lighting, occlusions, casting shadows, and pose and view changes. The metric learning formulation we proposed in (5) enables us to naturally fulfill this task. We proceed to present it in a more formal way.

Extended from the notation of Sect. 2, let $\mathcal{X}^{(t)} = \bigcup \mathcal{S}_i^{(t)}$ be the set of n labeled examples we maintain at time instance t. We also let $\mathbf{A}_t$ be the current discriminative appearance model, and $\mathcal{M}_t = \{\mathbf{m}_i^{(t)}\}_{i=1}^K$ be the motion parameters we need to recover. Hence, we have $\mathbf{x}_{i0}^{(t+1)} = \mathbf{I}^{(t+1)}(\mathbf{m}_i^{(t)})$, which means the new positive example depends on current image and motion parameters corresponding to the previous frame. At each time instant t, given $\mathcal{X}^{(t)}$ and $\mathbf{A}_t$, we run the gradient descent optimization algorithm outlined in Sect. 2.3 to obtain the optimal motion parameters $\hat{\mathbf{m}}_i^{(t)}$, $\forall i \in \{1, 2, \ldots, K\}$. This fulfills our visual matching and tracking task. Then we perturb each $\hat{\mathbf{m}}_i^{(t)}$ in turn to generate a set of α background samples $\mathcal{S}_{0\alpha}^{(t+1)}$ to replace the oldest α samples subset $\mathcal{S}_{0\alpha}^{(t)}$ in $\mathcal{X}^{(t)}$. In practice, we sample examples around the current tracked target with a relative bigger distance to replace old background examples. This results in the new labeled examples $\mathcal{X}^{(t+1)}$, that is,

$$\mathcal{X}^{(t+1)} = \left(\mathcal{X}^{(t)} \setminus \mathcal{S}_{0\alpha}^{(t)}\right) \cup \mathcal{S}_{0\alpha}^{(t+1)}. \tag{18}$$

With $\mathcal{X}^{(t+1)}$, we can then run the gradient projection optimization algorithms outlined in Sect. 2.2 to obtain the optimal $\mathbf{A}_{t+1}$. To proceed with the next matching step, we need to retire one example for each visual target in the current $\mathcal{X}^{(t+1)}$ to update the example set, we propose a least consistent criterion based on the contribution of each of the target examples to the unified cost function $f(\mathbf{A}_{t+1}, \mathcal{M}_t)$. Indeed, fixing $\mathbf{A}_{t+1}$ and $\mathcal{M}_t$, $f(\mathbf{A}_{t+1}, \mathcal{M}_t)$ is a function of $\mathcal{X}^{(t+1)}$, that is, $f(\mathbf{A}_{t+1}, \mathcal{M}_t) = g(\mathcal{X}^{(t+1)})$. We can similarly define a $g(\cdot)$ function for any subset of $\mathcal{X}^{(t+1)}$ based on (4). Therefore, for each $\mathbf{x}_{ij} \in \mathcal{X}^{(t+1)}$, a consistent criterion can be defined as

$$c(\mathbf{x}_{ij}) = g\big(\mathcal{X}^{(t+1)}\big) - g\big(\mathcal{X}^{(t+1)} \setminus \{\mathbf{x}_{ij}\}\big). \tag{19}$$

It is easy to understand that the larger $c(\mathbf{x}_{ij})$ is, the more contribution $\mathbf{x}_{ij}$ has made to $f(\mathbf{A}_{t+1}, \mathcal{M}_t)$. If the label $o(x_{ij}) = i$, a larger $c(\mathbf{x}_{ij})$ indicates that $\mathbf{x}_{ij}$ is not very compatible to the rest of the visual samples of target i, and hence should be retired from the sample set. More formally, we select

$$\mathbf{x}_i^\star = \mathop{\arg\max}_{\mathbf{x}\in\mathcal{X}^{(t+1)}, o(\mathbf{x})=i} c(\mathbf{x}) \tag{20}$$

to retire from $\mathcal{X}^{(t+1)}$, for each $i \in \{1, 2, \ldots, K\}$. In real operation, we only need to change the numbering of $\mathbf{x}_{i0}^{(t+2)} = \mathbf{I}^{t+2}(\mathbf{m}_{t+1})$ to the numbering of $\mathbf{x}_i^\star$, then we reset $\mathbf{x}_{i0}^{(t+2)} = \mathbf{I}^{t+2}(\mathbf{m}_i^{(t)})$, $\forall i \in \{1, 2, \ldots, K\}$, which are initialized to kick off the matching process to recover the optimal motion parameter $\mathcal{M}_{t+1}$.

The above steps will be iterated from time instant t to time instant $t+1$. Therefore we track the multiple visual targets and estimate the joint discriminative visual appearance model in an online fashion, which are all based on efficient gradient based optimization. Most previous approaches resort to heuristics or the oldness of visual samples to select the optimal set of online training examples. While our proposed selection criterion for positive examples in (20) is derived directly from the objective function of the proposed formulation in a principled fashion, it manifests another benefit of our proposed metric learning framework for discriminative appearance modeling and matching of multiple visual objects.

To initialize the tracking algorithm, we can either run an object detector if it applies, such as a face detector [24] or a human detector [25], if we are tracking a number of faces or persons, or request the users to manually specify the tracking rectangles for the multiple visual target. Then the initialized tracking rectangles are perturbed to form the initial set of labeled examples $\mathcal{X}^{(1)}$. More specifically, perturbed rectangles with sufficient overlap with the initial rectangles are regarded as the visual samples of the corresponding targets, while those perturbed rectangles which are deviated too much from the initial rectangles are deemed as visual samples of the background. This bootstraps learning for the optimal discriminative appearance model $\mathbf{A}_2$, which is then adopted to obtain the optimal motion parameter $\mathcal{M}_2$. This processes will be repeated as described above.

Last but not least, when maintaining the labeled example set $\mathcal{X}^{(t)}$, we fix a small set of β background and β visual examples extracted from the initialization frame for each of the visual target in the working set, that is, we never replace them with new examples. This treatment is very important to keep our discriminative appearance model stable and avoid it to be drifted too drastically in the visual tracking process.

4 Discriminant Exclusive Principle

We argue that the proposed joint formulation for multiple object tracking naturally incorporates an exclusive principle [22] in the matching process. Therefore, it is robust to handle occlusions among the different visual objects. The exclusive principle states that no two visual tracker shall occupy the same image observation. Our proposed algorithm naturally achieved it because of the joint discriminative appearance model **A**, which reinforces the mutual discrimination of the appearances between two visual targets $\mathbf{I}(\mathbf{m}_i)$ and $\mathbf{I}(\mathbf{m}_j)$. To see this more clearly, given an optimal **A**, if $\mathbf{I}(\mathbf{m}_i)$ and $\mathbf{I}(\mathbf{m}_j)$ occupy similar image regions (a.k.a., $\mathbf{m}_j \doteq \mathbf{m}_i$), and thus have very similar visual appearances, the mutual discriminative information encoded in **A** would incur a large value for $f(\mathbf{A}, \mathcal{M})$. Therefore, $\mathbf{m}_j = \mathbf{m}_i$ is not an optimal solution to $\mathcal{M}$. In other words, the optimal motion parameter $\mathcal{M}$ is more likely to occur when $\forall 1 \leq i < j \leq K$, $\mathbf{m}_j \neq \mathbf{m}_i$. Therefore, the exclusive principle among the different visual targets are naturally reinforced, which makes our proposed framework for multiple target tracking to be more robust to cross occlusions among the different visual targets.

5 Experiments

5.1 Visualization of Learned Appearance Model

The appearance model **A** defines an discriminative embedding to differentiate the multiple visual objects from the background. Each eigenvector of **A** is corresponding to one basis vector of the embedding. To have a better understanding on how the appearance model **A** functions, in Fig. 1, we visualize the top 12 eigenvectors of an

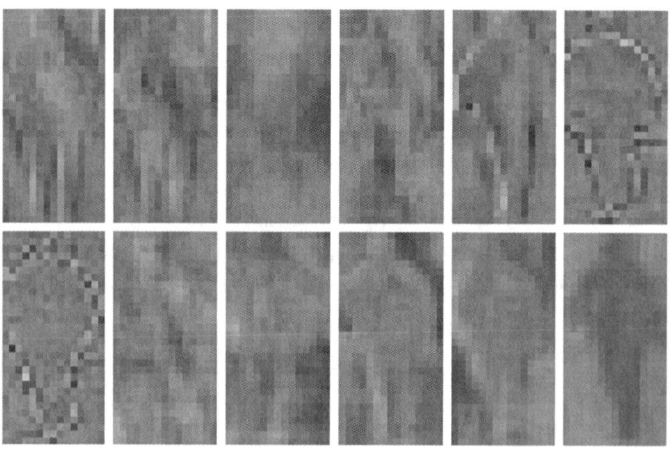

Fig. 1 The top 12 eigenvectors (with the descent order *from top left to bottom right*) for the discriminative matrix A

optimal **A** estimated at frame 512 when tracking three persons in the CAVIAR sequence. As clearly observed, they encode the contour and shape information of the target objects. It is quite sensible because A is used in our discriminative framework for discriminating the multiple objects from the background,and also reinforce the mutual discriminations among the different objects. The shape information is probably the most reliable one for achieving that.

5.2 *Multiple Target Tracking for Different Video Sequences*

We dub the name *TUDAMM* to the Tracker with Unified Discriminative Appearance Modeling and Matching (TUDAMM) for our tracker. We evaluate the tracker on two datasets: the CAVIAR [15] videos and the ETH Mobile Scence (ETHMS) [10]. For each single object, we randomly extract 20 positive examples to form a positive set tightly around the initial bounding box of the object. The number of negative examples around a single object is also set to be 20. We will have $20 * N$ negative examples for each object, supposing that we have N objects to be tracked. A confliction solving procedure is employed to avoid extracting a positive example from one object as a negative of another. After obtaining the matched patch in the current frame, negative examples would be generated by randomly selecting patches with a minimum and maximum distance toward the positive. The motion parameters (affine parameters) are kept the same in this step. Half of the positive and negative examples would be kept without updating to help the tracker recover from big changes and occlusions. The normalized pixel intensity is used as the feature. We downsample the image patches to 20×20, regardless of their original dimensions (the feature dimension for each object must be the same to fit into the metric framework). This procedure is implemented by solving a warping equation instead of directly sampling the image patch, which will provide a smoother objective function for the gradient descent optimization in the second step of the iteration.

Figure 2 shows the tracking results for a video from CAVIAR in which three persons walk on the corridor with big scale changes and occlusions. The objects encounter big occlusion by a crossing person from key frame 816. We present the sample results obtained by our tracker, the ILT [20] tracker and the Meanshift [7]. Our tracker shows quite robust responses. The ILT tracker loses the target when it's occluded by a person crossing the corridor. The Meanshift tracker shifts because it cannot deal with big scale change.

In order to give quantitative performance comparison with these two works, we employ a criterion called Average Tracking Precision (ATP) to do the evaluation, enlightened by the PASCAL grand challenge. More formally, for each tracking task, a ground truth mask for the object of interest is labeled in each frame j. The mask is represented as a point set $\mathcal{G}_j$ which is a collection of all points in the ground truth bounding box. The tracking result is represented as a point set $\mathcal{T}_j$ at frame j. $(x_i, y_i) \in \mathcal{G}_j$ or $\mathcal{T}_j$ indicates that the pixel at (x_i, y_i) is associated with the target. For an ideal tracker, $\forall j, \mathcal{G}_j = \mathcal{T}_j$.

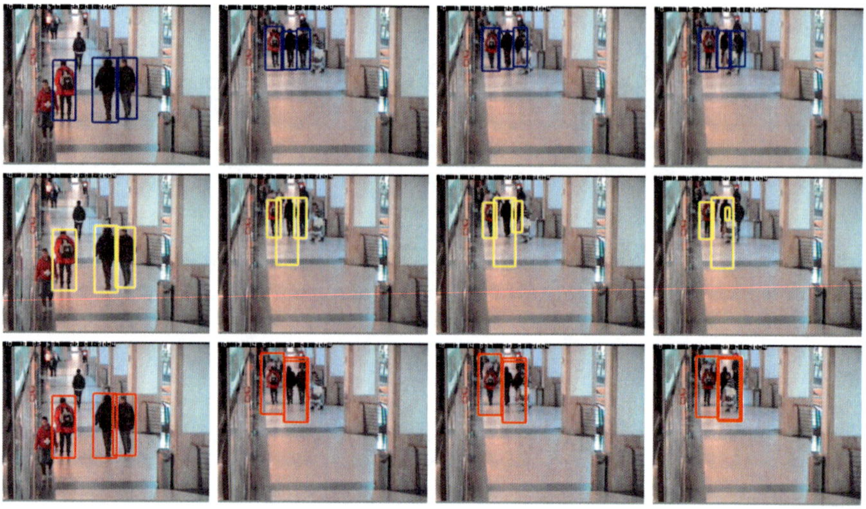

Fig. 2 Sampled multiple object tracking results on the Caviar dataset. Key frame NO.:513,809,817,828. The *first row*: our tracking results; the *second row*: tracking results of ILT; the *third row*: tracking results of Meanshift

For each frame j, the tracking precision r_j is defined as: $r_j = |\mathcal{G}_j \cap \mathcal{T}_j|/|\mathcal{G}_j \cup \mathcal{T}_j|$. Noticing that $r_j \in [0, 1]$, the ATP for a tracker of an object in a video clip is defined as:

$$\text{ATP} = \frac{1}{N}\sum_{j=1}^{N} r_j = \frac{1}{N}\sum_{j=1}^{N} \frac{|\mathcal{G}_j \cap \mathcal{T}_j|}{|\mathcal{G}_j \cup \mathcal{T}_j|}, \tag{21}$$

where N is the running length of the video clips in frame number. For an ideal tracker, ATP $\equiv 1$. We use it as the exclusive quantitative measure to compare the performance of the TUDAMM with other state-of-the-art trackers.

Because neither of the other two algorithms support multiple object tracking, we track the objects independently to obtain results from the two trackers. Figure 3 shows the ATP curve. The TUDAMM tracker gives the best performance, with an ATP above 0.7. Recall that in PASCAL grand challenge, a detection with an overlap bigger than 0.5 with ground truth would be treated as a true detection. The ATP value 0.7 implies perfect tracking performance.

Figure 4 presents sampled key frames from the result of tracking three persons on a street [10]. The person with red coat is occluded by a tree during the tracking. Our tracker shows robust tracking. Figure 5 presents sampled tracking results for a video from CAVIAR dataset [15].

Figure 6 shows the tracking result for a horse racing video in which cross occlusion happens frequently. Our tracker shows excellent performance. The ILT tracker cannot locate the object very well and the fifth (left to right) horse is completely lost during the tracking process. The Meanshift tracker is not good at solving cross occlusions and the bounding box shifts drastically. As we can clearly observe, our

Fig. 3 Tracking performance comparison using ATP. *Red curve*: TUDAMM; *blue curve*: ILT; *black curve*: Meanshift

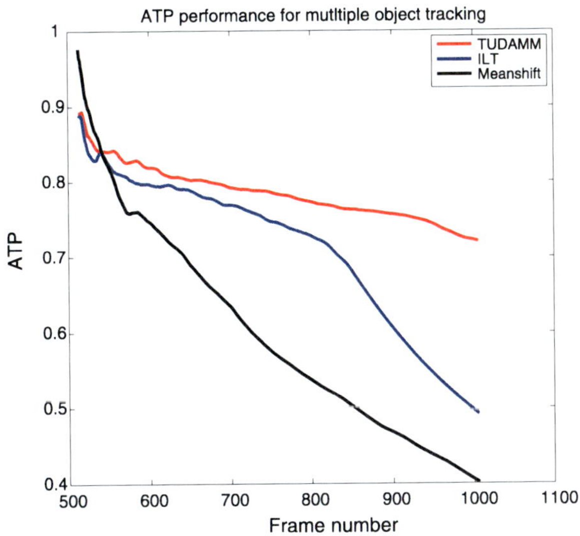

Fig. 4 Multiple Tracking result on the ETH dataset

Fig. 5 Multiple Tracking result on the Caviar dataset

Fig. 6 Multiple Tracking results for a horse racing video. Key frame No: 3026,3143,3202,3341. The *first row*: TUDAMM tracker; the *second row*: the Meanshift tracker; the *third row*: the ILT tracker

discriminative multiple targets tracker presents very robust tracking results under drastic visual variations induced by illumination changes, scale changes, pose articulations, as well as mutual occlusions.

6 Discussions, Conclusion and Future Work

We proposes a discriminative metric learning framework for robust tracking of multiple targets. It not only seeks for appearance models to discriminate the multiple foreground targets from the background, but also try to recover subtle discriminations between two different visual targets. Our experiments on a set of challenging real-world video sequences demonstrated the robustness of the proposed tracking algorithms in dealing with large visual variations and cross-occlusions.

Future work may include further exploration of different type of filters for further improving the robustness of the tracker under the same formulation. Meanwhile, as discussed above, this framework may encounter problem if objects are nearly identical. We will further investigate this issue and explore means of mitigating this issue. It may be addressed by posing strong dynamic models learned online, we will defer all this to our future work.

References

1. Avidan, S.: Support vector tracking. In: CVPR (2001)
2. Avidan, S.: Ensemble tracking. In: CVPR (2005)
3. Babenko, B., Yang, M.-H., Belongie, S.: Visual tracking with online multiple instance learning. In: CVPR (2009)
4. Bar-Shalom, Y.: Tracking and Data Association. Academic Press, San Diego (1987)
5. Cai, Y., de Freitas, N., Little, J.J.: Robust visual tracking for multiple targets. In: The 9th European Conference on Computer Vision, Graz, Austria, vol. 4, pp. 107–118 (2006)
6. Collins, R.T., Liu, Y.: On-line selection of discriminative tracking features. In: ICCV, vol. 1, pp. 346–352 (2003)
7. Comaniciu, D., Ramesh, V., Meer, P.: Real-time tracking of non-rigid objects using mean shift. In: CVPR, vol. 2, pp. 142–149 (2000)
8. Cootes, T., Edwards, G., Taylor, C.: Active appearance models. In: ECCV, pp. 484–498 (1998)
9. Dempster, A., Laird, N., Rubin, D.: Maximum likelihood from incomplete data via the em algorithm. J. R. Stat. Soc., Ser. B **39**(1), 1–38 (1977)
10. Ess, B.A., Gool, L.: Depth and appearance for mobile scene analysis. In: ICCV (2007)
11. Globerson, A., Roweis, S.T.: Metric learning by collapsing classes. In: NIPS (2005)
12. Hager, G.D., Belhumeur, P.N.: Efficient region tracking with parametric models of geometry and illumination. IEEE Trans. Pattern Anal. Mach. Intell. **20**(10), 1025–1039 (1998)
13. Hager, G.D., Dewan, M., Stewart, C.V.: Multiple kernel tracking with ssd. In: CVPR, vol. 1, pp. 790–797 (2004)
14. Ho, J., Lee, K.-C., Yang, M.-H., Kriegman, D.: Visual tracking using learned subspaces. In: CVPR, vol. 1, pp. 782–789 (2004)
15. http://homepages.inf.ed.ac.uk/rbf/caviardata1
16. Isard, M., Blake, A.: Contour tracking by stochastic propagation of conditional density. In: ECCV, vol. 1, pp. 343–356 (1996)
17. Jepson, A.D., Fleet, D.J., El-Maraghi, T.F.: Robust online appearance models for visual tracking. In: CVPR, pp. 415–422
18. Khan, Z., Balch, T., Dellaert, F.: Mcmc data association and sparse factorization updating for real time multitarget tracking with merged and multiple measurements. IEEE Trans. Pattern Anal. Mach. Intell. **28**(12), 1960–1972 (2006)
19. Lee, J.W., Kim, M.S., Kweon, I.S.: A kalman filter based visual tracking algorithm for an object moving in 3d. In: Proc. of IEEE/RSJ International Conference on Intelligent Robots and Systems, vol. 1, pp. 342–347 (1995)
20. Lim, J., Ross, D., Lin, R.-S., Yang, M.-H.: Incremental learning for visual tracking. In: NIPS, pp. 801–808 (2005)
21. MacCormick, J., Blake, A.: A probabilistic exclusion principle for tracking multiple objects. In: ICCV, pp. 572–587 (1999)
22. MacCormick, J., Isard, M.: Partitioned sampling, articulated objects, and interface-quality hand tracking. In: ECCV, pp. 3–19 (2000)
23. Rosen, J.B.: The gradient projection method for nonlinear programming. Part I. Linear constraints. J. Soc. Ind. Appl. Math. **8**(1), 181–217 (1960)
24. Viola, P., Jones, M.J.: Robust real-time face detection. Int. J. Comput. Vis. **57**(2), 137–154 (2004)
25. Wang, X., Han, T.X., Yan, S.: An HOG-LBP human detector with partial occlusion handling. In: ICCV (2009)

26. Wu, Y., Fan, J.: Contextual flow. In: CVPR (2009)
27. Yang, M., Wu, Y.: Tracking non-stationary appearances and dynamic feature selection. In: CVPR (2005)
28. Yu, T., Wu, Y.: Collaborative tracking of multiple targets. In: Proc. of IEEE Conference on Computer Vision and Pattern Recognition, Washington, DC, vol. I, pp. 834–841 (2004)
29. Yu, Q., Medioni, G., Cohen, I.: Multiple target tracking using spatiotemporal Markov chain Monte Carlo data association. In: Proc. of IEEE Conference on Computer Vision and Pattern Recognition, Minneapolis, Minnesota (2007)
30. Zhao, Q., Brennan, S., Tao, H.: Differential EMD tracking. In: ICCV (2007)
31. Zhu, C., Byrd, R.H., Lu, P., Nocedal, J.: Algorithm 778: L-BFGS-B: Fortran subroutines for large-scale bound-constrained optimization. ACM Trans. Math. Softw. **23**(4), 550–560 (1997)

A Framework of Wire Tracking in Image Guided Interventions

Peng Wang, Andreas Meyer, Terrence Chen, Shaohua K. Zhou, and Dorin Comaniciu

Abstract This chapter presents a framework of using computer vision and machine learning methods to tracking guidewire, a medical device inserted into vessels during image guided interventions. During interventions, the guidewire exhibits non-rigid deformation due to patients' breathing and cardiac motions. Such 3D motions are complicated when being projected onto the 2D fluoroscopy. Furthermore, fluoroscopic images have severe image artifacts and other wire-like structures. Those factors make robust guidewire tracking a challenging problem. To address these challenges, this chapter presents a probabilistic framework for the purpose of robust tracking. We introduce a semantic guidewire model that contains three parts, including a catheter tip, a guidewire tip and a guidewire body. Measurements of different parts are integrated into a Bayesian framework as measurements of a whole guidewire for robust guidewire tracking. For each part, two types of measurements, one from learning-based detectors and the other from appearance models, are combined. A hierarchical and multi-resolution tracking scheme based on kernel-based measurement smoothing is then developed to track guidewires effectively and efficiently in a coarse-to-fine manner. The framework has been validated on a testing

P. Wang (✉) · T. Chen · S.K. Zhou · D. Comaniciu
Siemens Corporate Research, Siemens Corporation, 755 College Road E., Princeton, NJ, USA
e-mail: peng-wang@siemens.com

T. Chen
e-mail: terrence.chen@siemens.com

S.K. Zhou
e-mail: shaohua.zhou@siemens.com

D. Comaniciu
e-mail: Dorin.Comaniciu@siemens.com

A. Meyer
Siemens Healthcare, Siemensstr. 1, Forchheim, Germany
e-mail: andreas.am.meyer@siemens.com

L. Wang et al. (eds.), *Machine Learning for Vision-Based Motion Analysis*, Advances in Pattern Recognition,
DOI 10.1007/978-0-85729-057-1_7,

set containing 47 sequences acquired under clinical environments, and achieves a mean tracking error of less than 2 pixels.

1 Background

Computer vision and machine learning methods have been applied to medical image analysis and have gained success because of their effectiveness in identifying autonomy structures [6, 21, 22]. Among their applications in medical imaging areas, the image guided intervention [3, 12] is where the real-time object detection and tracking methods have found their particular importance. In the image guided intervention, three-dimensional images or two-dimensional videos are acquired not only for diagnosis and for treatment strategy planning, but also for conducting surgeries and minimal invasive interventions. Figure 1 shows an interventional room equipped with an X-ray modality dedicated for image guided interventions. In a typical image guided intervention, the preoperative data of patient, usually acquired from three-dimensional tomography, is collected before the intervention for treatment planning. During interventions, clinicians insert medical devices, for example, catheter, balloon, and stent. Such devices are used for various purposes, such as imaging the internal body structure, delivering drugs, and treating vascular diseases. The positions of such medical devices need to be continuously localized to provide guidance for clinicians during interventions. For example, based on the devices localization, real-time acquired fluoroscopy images can be overlayed onto preoperative CT data, and provide real-time guidance for clinicians.

In this chapter, we present a framework of tracking a guidewire in 2D X-ray fluoroscopy for image guided interventions. The guidewire is a medical device that is inserted into vessels through a guiding catheter for various tasks, such as stent delivery, and balloon inflation. Robust guidewire tracking is essential for many applications in image guided interventions, for example, real-time assessment of guidewire

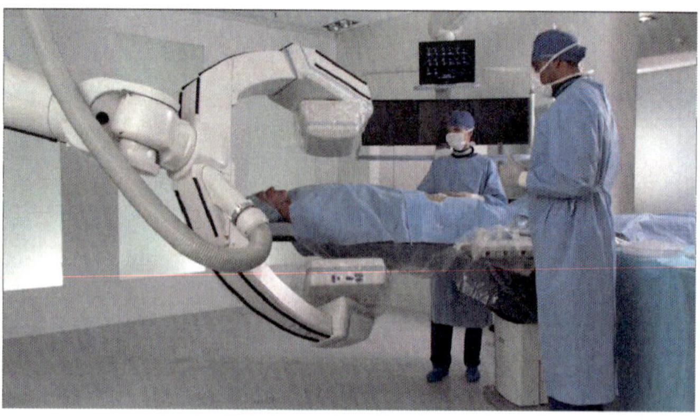

Fig. 1 An interventional procedure performed with Siemens AXIOM Artis Zeego system

position and shape, the visibility enhancement of guidewires, and guidance of coregistration between 2D real-time X-ray fluoroscopic images and 3D preoperative images.

A guidewire usually starts from a tip of a guiding catheter (thicker tubes in the images), and ends at a guidewire tip. Some exemplar guidewires in fluoroscopy are shown in Fig. 2. The figure demonstrates the challenges of guidewire tracking. First, guidewires are thin and have low visibility in fluoroscopic images because the fluoroscopy images usually have low dose of radiations thus low signal-to-noise ratio. Sometimes, parts of guidewires are barely visible in noisy images. Such weak and thin wire structures in noisy images make the robust tracking challenging. Second, guidewires exhibit large variations in their appearances, shapes, and motions. The shape deformation of a guidewire is mainly due to a patient's breathing and cardiac motions in 3D. The 3D motions become more complicated when being projected onto a 2D image space. Third, there exist many other wire-like structures, such as guiding catheters and ribs, as shown in Fig. 2. Some of the structures are close to the guidewire, could distract guidewire tracking and finally lead to tracking failures. All the aforementioned factors, along with robustness and speed requirements for interventions, render guidewire tracking a challenging task.

Due to the unique characteristics of the guidewire in fluoroscopy, conventional tracking methods would encounter difficulties and cannot deliver desired speed, accuracy, and robustness for interventions. Since a guidewire is thin, the tracking methods that use regional features such as holistic intensity, textures, and color histogram [20], cannot track it well. Active contour [9, 13, 23] and level set based methods, heavily rely on intensity gradients, and are easily attracted to image noise and other wire-like structures in fluoroscopy. There is some work on guidewire detection [4] and tracking [2, 11]. Barbu et al. [4] present a learning-based method to automatically detect guidewires in fluoroscopic sequences. The method aims at detecting a guidewire in individual frames, not continuously tracking the guidewire in a sequence. Beyar et al. [11] use a filter based method to identify a guidewire in an X-ray image, and then use the Hough transform to fit a polynomial curve to track a guidewire. There are no quantitative reports on tracking performance in their papers. The method of Baert et al. [2] tracks a guidewire enhanced by image subtraction and coherence diffusion. However, their experiments show that only a part of guidewire, not the whole guidewire, has been tracked.

We present in this chapter a probabilistic framework for robust guidewire tracking. An earlier version of this work has been presented in [17, 18]. The tracking method is based on 2D fluoroscopic images, because a 3D guidewire model and associated 3D projection matrices are not always available in a clinical practice. This framework makes three contributions to address the aforementioned challenges in the guidewire tracking:

1. This method introduces a semantic guidewire model, based on which a probabilistic method is presented to integrate measurements of three guidewire parts, that is, a catheter tip, a guidewire body and a guidewire tip, in a Bayesian framework to track a whole guidewire. This tracking framework is robust to measurement noises at individual guidewire parts.

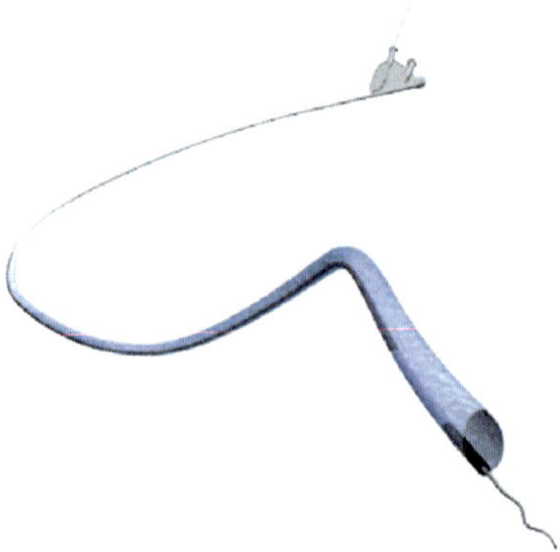

(a)

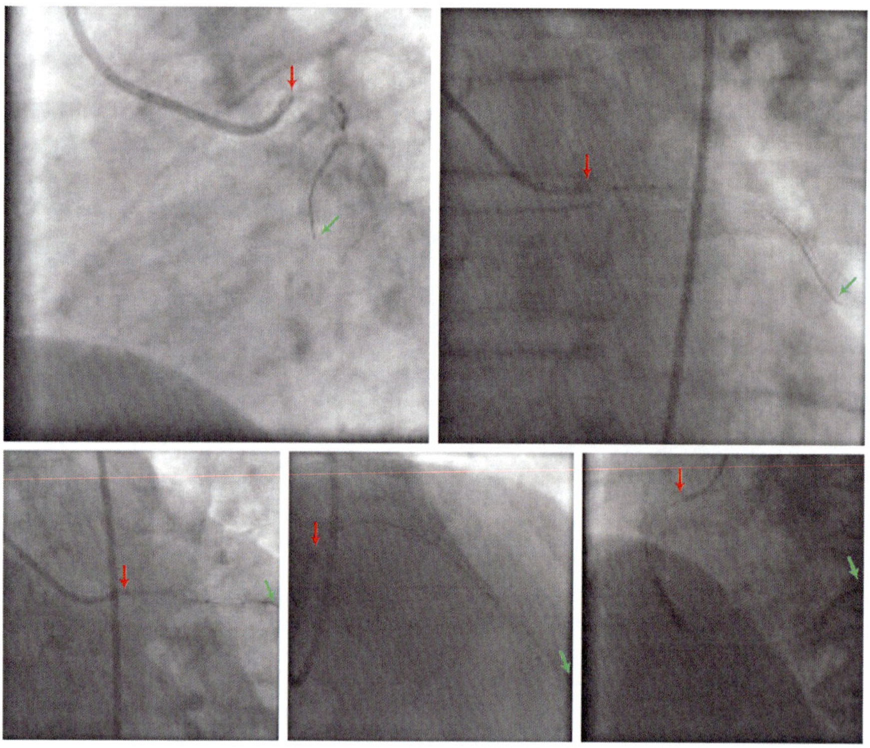

(b)

Fig. 2 Guidewires. (**a**) A guidewire is a catheter system (Acumen Spirit Navigable Lead Delivery Catheter System). (**b**) Some examples of guidewires in fluoroscopic sequences. The guidewires exhibit low visibility, with a variety of shapes and appearances. For better visualization, two ends of guidewires are marked with *red* and *green arrows* respectively

2. Learning based measurement models are used in our method to track the guidewire. The learning-based measurement models are trained from a database of guidewires, in order to detect and track guidewire parts in low-quality images. Our method further incorporates online measurement models, which are based on guidewire appearances, as a complementary to learning based measurements to improve the tracking robustness.
3. We develop a hierarchical and multi-resolution scheme to track a deforming guidewire. By decomposing the guidewire motion into two major components, the hierarchical tracking starts from a rigid alignment, followed by a refined non-rigid tracking. At each stage, we apply a multi-resolution searching strategy by using variable bandwidths in a kernel-based measurement smoothing method, to effectively and efficiently track the deforming guidewire.

Compared with the previous work [18], the framework combines both measurements from offline learned detectors and online appearance models, and provides a principled probabilistic tracking framework. We validate the guidewire tracking framework on a test set containing 47 sequences that are captured in real-life interventional scenario. Quantitative evaluation results show that the mean tracking error is less than 2 pixels, that is, 0.4 mm. This demonstrates the great potential of our method for clinical applications.

The rest of the chapter is structured as follows. We first introduce a guidewire model, and present the probabilistic formalization of guidewire tracking in Sect. 2.1. Details on measurement models used for guidewire tracking are provided in Sect. 2.2, and the hierarchical tracking scheme in Sect. 2.3. The quantitative evaluations of our method are presented in Sect. 3. Section 4 concludes the chapter.

2 Guidewire Tracking Method

Before presenting algorithm details, we explain the notations used. The regular font represents a scalar variable or a function, and the bold font represents a vector, for example, the $\mathbf{x}$ as a 2D location and the $\mathbf{u}$ as guidewire motion parameter. The $\mathbf{Z}$ is used to denote image observation, and Γ represents a guidewire curve.

2.1 Method Overview

2.1.1 A Guidewire Model

As shown in Fig. 2(a), a guidewire contains several parts: the catheter tip, the guidewire body, and the guidewire tip. Accordingly, we can establish a semantic model for the guidewire, as in Fig. 3. All the three parts are connected together, but each part has slightly different appearances in fluoroscopic images. For example, as shown in Fig. 2(b), a guiding catheter is a tube containing the guidewire, and has

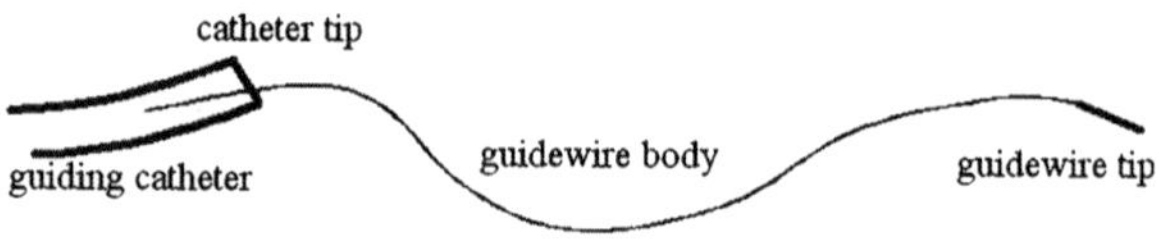

Fig. 3 A semantic model of guidewire. It contains a catheter tip, guidewire body, and a guidewire tip

better visibility than the guidewire body in images. But sometimes the catheter tips can be occluded by contrast material injected during interventions. Some guidewire tips could be thicker than the guidewire body, and also show more flexible deformations. The guidewire body has the least visibility in images, but its deformations are more constrained than tips. It is therefore highly desirable to have a method to handle the appearance differences among different parts, and also to allow for flexible movements of each part of the guidewire.

To represent all the parts along the guidewire, a spline model is used. Assuming that there are M control points $\mathbf{x}_i^c, i = 1, \dots, M$ that can decide the shape of a guidewire, other points on the guidewire can be interpolated from the control points, as (1):

$$\Gamma(\mathbf{x}) = \{\mathbf{x} = (\gamma^x(\lambda), \gamma^y(\lambda)) | 1 \le \lambda \le M\}, \quad (1)$$

where $\gamma^x(\lambda)$ and $\gamma^y(\lambda)$ are cubic spline functions, and $\lambda \in [i-1, i]$ means that $\mathbf{x}$ is interpolated between control points $\mathbf{x}_{i-1}^c$ and $\mathbf{x}_i^c$. In the spline model, two controls point, $\mathbf{x}_1^c$ and $\mathbf{x}_M^c$, represent the catheter tip and the guidewire tip, respectively. A general form of a cubic spline interpolation can be found in [5]. Using the spline control points to represent the whole wire can significantly reduce the number of the shape parameters, therefore reducing the complexity of guidewire tracking. A guidewire spline is resampled at each frame, and usually has 10–20 control points depending on the spline length. The control points on cubic spline splines pass through the spline, which is desirable in the guidewire shape model. Furthermore, the occlusion and overlapping of an individual control point will not affect the tracking of a whole guidewire.

2.1.2 A Probabilistic Guidewire Tracking Framework

The tracking problem can be modeled in a Bayesian framework in which unknown states are inferred from sequential data [8, 20]. We also formalize the guidewire tracking in this probabilistic inference framework, to maximize the posterior probability of a tracked guidewire given fluoroscopic images. In the framework, a guidewire hypothesis at the tth frame is a guidewire deformed from a previous frame, denoted as $\Gamma_t(\mathbf{x};\mathbf{u})$:

$$\Gamma_t(\mathbf{x};\mathbf{u}) = T(\Gamma_{t-1}(\mathbf{x}), \mathbf{u}_\mathbf{x}), \quad (2)$$

where T is a guidewire shape transformation function, and $\mathbf{u}_\mathbf{x}$ is the motion parameter. $\Gamma_{t-1}(\mathbf{x})$ is a tracked guidewire at a previous frame and is used as a template for the tracking at the tth frame. For the simplicity of notations, a guidewire candidate is denoted as $\Gamma_t(\mathbf{x})$. Therefore, the posterior probability $P(\Gamma_t(\mathbf{x})|\mathbf{Z}_t)$ is given in (3)

$$P(\Gamma_t(\mathbf{x})|\mathbf{Z}_t) \propto P(\Gamma_t(\mathbf{x}))P(\mathbf{Z}_t|\Gamma_t(\mathbf{x})). \quad (3)$$

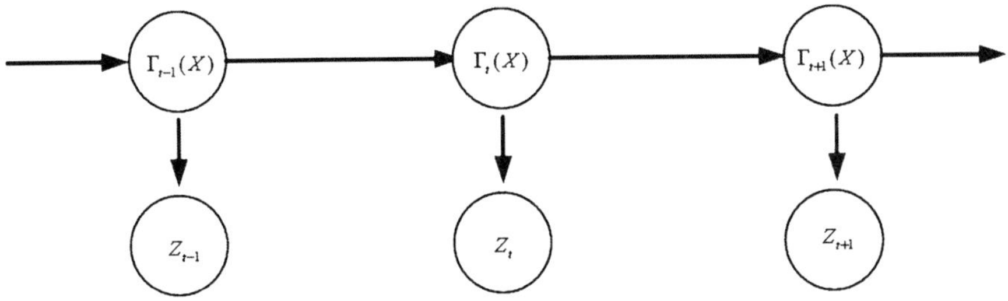

Fig. 4 The graphical model for the guidewire tracking

The graphical model corresponding to the tracking problem is shown in Fig. 4. The tracked guidewire $\hat{\Gamma}_t(\mathbf{x})$ is estimated as the guidewire candidate that maximizes the posterior probability, that is,

$$\hat{\Gamma}_t(\mathbf{x}) = \underset{\Gamma_t(\mathbf{x})}{\arg\max}\, P\big(\Gamma_t(\mathbf{x})|\mathbf{Z}_t\big). \tag{4}$$

In (3), $P(\Gamma_t(\mathbf{x}))$ is a prior probability, which can be propagated from previous tracking results. We model the guidewire prior probability as:

$$P\big(\Gamma_t(\mathbf{x})\big) = \frac{1}{\sqrt{2\pi}\sigma_\Gamma} \exp\left(\frac{-|D(\Gamma_t(\mathbf{x}), \Gamma_{t-1}(\mathbf{x}))|^2}{2\sigma_\Gamma^2}\right), \tag{5}$$

where $D(\Gamma_t(\mathbf{x}), \Gamma_{t-1}(\mathbf{x}))$ is the average of the shortest distances from points on a guidewire candidate $\Gamma_t(\mathbf{x})$ to the guidewire shape template $\Gamma_{t-1}(\mathbf{x})$. A large kernel size σ_Γ (a typical value is 60 in this method) is chosen to allow a large guidewire deformation. Another component, the likelihood measurement model $P(\mathbf{Z}_t|\Gamma_t(\mathbf{x}))$, plays a crucial role in achieving robust tracking results. Given a guidewire represented by N points $\Gamma_t(\mathbf{x}) = \{\mathbf{x}_1, \mathbf{x}_2, \ldots, \mathbf{x}_N\}$ that are interpolated from control points, the guidewire $\Gamma_t(\mathbf{x})$ is in an N-dimensional space, which make the measurement model $P(\mathbf{Z}_t|\Gamma_t(\mathbf{x}))$ difficult to represent. To simplify the model, we assume the measurement independency among different parts along a guidewire, that is, $P(\mathbf{Z}_t|\mathbf{x}_i, \Gamma_t(\mathbf{x})) = P(\mathbf{Z}_t|\mathbf{x}_i)$. By this assumption, we can decompose the measurement model $P(\mathbf{Z}_t|\Gamma_t(\mathbf{x}))$ into measurements at individual guidewire points, as (6):

$$P\big(\mathbf{Z}_t|\Gamma_t(\mathbf{x})\big) = \sum_{\mathbf{x}_i} P(\mathbf{Z}_t|\mathbf{x}_i) P\big(\mathbf{x}_i|\Gamma_t(\mathbf{x})\big), \tag{6}$$

where $P(\mathbf{Z}_t|\mathbf{x}_i)$ is the measurements at individual points on a guidewire, and $P(\mathbf{x}_i|\Gamma_t(\mathbf{x}))$ is the weights of individual points on a guidewire. Since the two ending points at the guidewire model represent the catheter tip and the guidewire tip, respectively, the measurement model $P(\mathbf{Z}_t|\Gamma_t(\mathbf{x}))$ is rewritten as a combination of three parts of measurements in (7):

$$P\big(\mathbf{Z}_t|\Gamma_t(\mathbf{x})\big) = \omega_1 P(\mathbf{Z}_t|\mathbf{x}_1) + \omega_N P(\mathbf{Z}_t|\mathbf{x}_N) + \frac{1-\omega_1-\omega_N}{N-2} \sum_{i=2}^{N-1} P(\mathbf{Z}_t|\mathbf{x}_i), \tag{7}$$

where ω_1 and ω_N are the weights of the catheter tip and the guidewire tip, respectively. Usually the tips have more distinguishing characteristics than the guidewire body, so they are assigned higher weights (the tip weights are empirically set between 0.05 and 0.2 in our algorithm). All other points on a guidewire are assigned equal weights $\frac{1-\omega_1-\omega_N}{N-2}$.

The decomposition of the measurement model $P(\mathbf{Z}_t|\Gamma_t(\mathbf{x}))$ as the form of (7) allows for independent measurements of different guidewire parts, while their integration on the guidewire $\Gamma_t(\mathbf{x})$ provides a unified measurement model for the whole guidewire. The probabilistic framework provides the flexibility to track deformable guidewires, and also ensures robustness to the measurement noise at individual parts. Another advantage of this framework is its capability of introducing different types of measurement $P(\mathbf{Z}_t|\mathbf{x}_i)$, at individual parts, making this framework general enough to fuse multiple measurement modalities.

2.2 Guidewire Measurement Models

Robust measurement models $P(\mathbf{Z}_t|\mathbf{x}_i)$ are crucial to addressing the difficulties encountered in guidewire tracking. In our method, learning-based methods are applied for robust measurements of guidewire parts. Guidewire part detectors are learned, from off-line collected training data, to model a large variety of guidewires, especially for guidewire body and guidewire tips. Another measurement modality, online measurement model based on guidewire appearance, is combined with the learning-based measurements. The integration of two types of measurements can correct failures caused by one measurement modality, such as false or missing detections of learning-based measurements, and drifting of appearance-based models, therefore is able to robustly track guidewires under various environments.

2.2.1 Learning-Based Guidewire Measurements

Learning-based detectors recently have been widely used in object detection and tracking [1, 6]. The reason behind their increasing popularity is their robustness to noises and their capability of handling objects with large variations. Different from traditional measurements based on low-level features such as edges and ridges, learning-based measurement models can be trained from a set of off-line collected data, thus being able to model objects with large variations. Since the training data also includes non-objects, the trained measurement models can distinguish objects from background noise. For guidewire tracking, we use the probabilistic boosting tree (PBT) [14] to construct the guidewire part detectors. PBT is a tree based general form of AdaBoost classifiers, and has a nice property of modeling a complex distribution of a class of objects. For more details on and PBT, please refer to [7, 14].

During tracking, the trained detectors can identify if an image patch at given location $\mathbf{x}_i$ belongs to a class of objects, that is, one of three guidewire parts. The output of an AdaBoost classifier, denoted as $f(\mathbf{z}, \mathbf{x}_i)$, is a combination of outputs from a collection of learned weak classifiers $H_k(\mathbf{z}, \mathbf{x}_i)$ with associated weights α_k. The numeric outputs can be further interpreted into probabilistic measurements, seeing (8):

$$f(\mathbf{z}, \mathbf{x}_i) = \sum_k \alpha_k H_k(\mathbf{z}, \mathbf{x}_i),$$
$$P^d(\mathbf{Z}_t|\mathbf{x}_i) \propto \frac{e^{f(\mathbf{Z}_t, \mathbf{x}_i)}}{e^{-f(\mathbf{Z}_t, \mathbf{x}_i)} + e^{f(\mathbf{Z}_t, \mathbf{x}_i)}}. \tag{8}$$

In this method, we train one detector for each part of the guidewire. The guidewire body detector mainly identifies line segments of the guidewire. To train the guidewire body detector, we collect line segments from annotated guidewires as positive samples, and randomly sample the image outside guidewire as negative samples. Similarly, the guidewire tip detector and the catheter tip detector are trained. All the learning-based models are built on Haar features [15], as their computational efficiency is favorable in interventional applications. Therefore, three learning-based measurement models, $P^d_{\text{cath}}(\mathbf{Z}_t|\mathbf{x}_1)$, $P^d_{\text{gw}}(\mathbf{Z}_t|\mathbf{x}_i)$, and $P^d_{\text{tip}}(\mathbf{Z}_t|\mathbf{x}_N)$, are obtained for the catheter tip, guidewire body, and guidewire tip, respectively. Figures 5(a) and 5(b) show the detected tip candidates and the line segment candidates of the guidewire body in a frame.

2.2.2 Appearance-Based Measurements

As shown in Fig. 5, learning-based measurements may suffer from missing or false detections due to the noises in the guidewire appearances. This motivates us to integrate another type of measurements, appearance-based measurement, to improve the robustness of tracking. Different from the learning-based measurement model, an appearance-based measurement model aims at modeling the online appearance of a specific guidewire being tracked. In our method, the appearance-based model takes the form in (9):

$$P^a(\mathbf{Z}_t|\mathbf{x}_i) \propto \exp\left\{-\frac{\sum_{\mathbf{x}' \in S(\mathbf{x}_i)} |\rho(\mathbf{Z}_t(\mathbf{x}') - I^0(\mathbf{x}'); \sigma_a)|^2}{2\sigma_a^2}\right\}, \tag{9}$$

where $\mathbf{Z}_t(\mathbf{x}')$ is the image intensity at the tth frame, and $I^0(\mathbf{x}')$ is the corresponding image intensity in a guidewire template, which is updated from the tracked guidewire at a previous frame. $S(\mathbf{x}_i)$ represents the geometric shape of the template, centered at the point $\mathbf{x}_i$. For example, $S(\mathbf{x}_i)$ is an ellipse for the catheter tip, and a segment of guidewire for the guidewire body and tip. As defined in (10), ρ is a robust function that is used to measure the intensity differences between current observations and the guidewire template, with removal of outliers.

$$\rho(y; \sigma_a) = \begin{cases} y, & \text{if } |y| \le 3\sigma_a, \\ 3\sigma_a, & \text{if } |y| > 3\sigma_a. \end{cases} \tag{10}$$

Similar to learning-based measurement models, we build an appearance-based model for each part of the guidewire. So, three appearance-based measurements

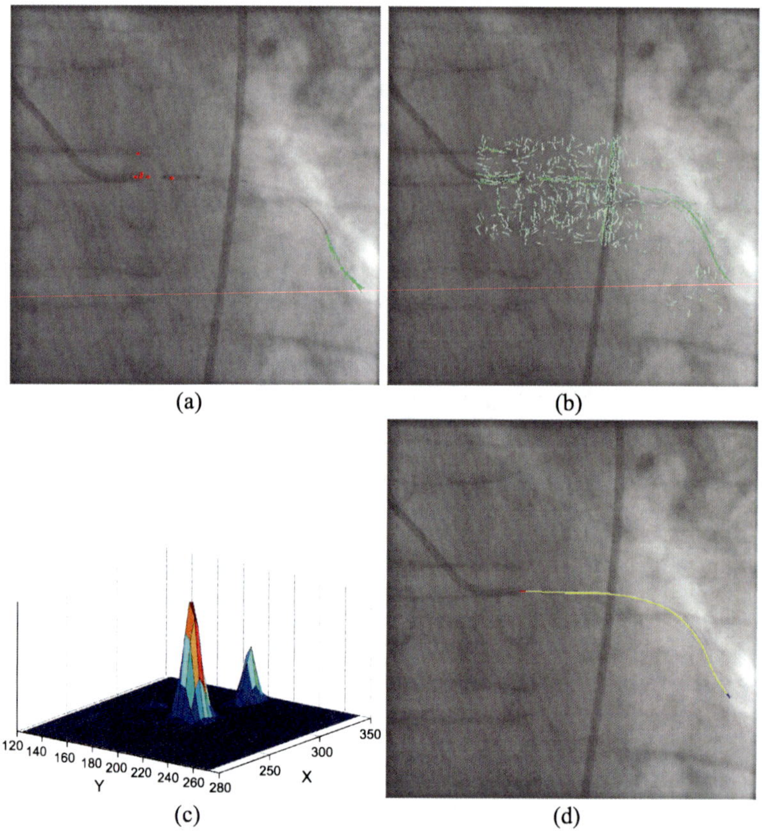

Fig. 5 Guidewire measurements for robust tracking. (**a**) Detected catheter tip candidates (*red blocks*) and guidewire tip candidates (*green lines*) in a frame. (**b**) The detected line segment candidates (in a region around the guidewire body) are shown in *green*. (**c**) A distribution of measurements of the catheter tip after combining learning-based and appearance-based measurements. (**d**) The tracked guidewire. The figures are best viewed *in color*

models, $P^a_{\text{cath}}(\mathbf{Z}_t|\mathbf{x}_1)$, $P^a_{\text{gw}}(\mathbf{Z}_t|\mathbf{x}_i)$, and $P^a_{\text{tip}}(\mathbf{Z}_t|\mathbf{x}_N)$, are obtained for the catheter tip, guidewire body, and guidewire tip, respectively.

2.2.3 Fusion of Multiple Measurements

The fusion of multiple measurements has been demonstrated to provide more robust tracking results than using a single measurement [16, 19]. In our method, two types of measurement models are integrated into one measurement model:

$$P(\mathbf{Z}_t|\mathbf{x}_i) = P^d(\mathbf{Z}_t|\mathbf{x}_i)P_d + P^a(\mathbf{Z}_t|\mathbf{x}_i)P_a, \tag{11}$$

where P_d and P_a are corresponding priors for two types of measurement models. An example of integrated measurements at the catheter tip is shown in Fig. 5(c), where the measurements have a peak at the catheter tip with false detections being suppressed.

2.3 *Hierarchical and Multi-resolution Guidewire Tracking*

The guidewire exhibits large variations in shape and motion, especially due to projections from 3D to 2D. Since a 3D guidewire model and a 3D projection matrix are not always available in a clinical practice, our method does not impose any assumptions that depends on 3D information. Instead, this method tries to handle guidewire motions that could be captured from arbitrary directions. For this purpose, our method decomposes the guidewire motion into two major steps: rigid and nonrigid motions, as the guidewire motion caused by the breathing motion can be approximated as a rigid motion in 2D, and the cardiac motion is nonrigid. The decomposed motions can be effectively and efficiently recovered in a hierarchical and coarse-to-fine manner, based on a kernel-based measurement smoothing method.

2.3.1 Kernel-Based Measurement Smoothing

We here present a kernel-based measurement smoothing method for multi-resolution guidewire tracking. To obtain measurements at each point $\mathbf{x}$ is computationally expensive, and is prone to measurement noises at individual points. For example, the measurements at points that are classified by detectors as non-guidewire parts are not reliable and therefore ignorable in $P^d(\mathbf{Z}_t|\mathbf{x}_i)$. Guidewire measurements can be more robust and more efficient to compute by using kernel-based estimation (or smoothing).

In the kernel-based estimation, measurements are made at a set of sampled locations $\mathbf{x}_j^s$, instead of a whole image. For learning-based measurements, $\mathbf{x}_j^s$ are those points classified as guidewire parts, and for appearance-based measurements, $\mathbf{x}_j^s$ are uniformly sampled points. We can conveniently assume the Markov conditional independence that the observations at sampling points $\mathbf{x}_j^s$ are independent with the unsampled points $\mathbf{x}_i$, i.e., $P(\mathbf{Z}_t|\mathbf{x}_i,\mathbf{x}_j^s) = P(\mathbf{Z}_t|\mathbf{x}_j^s)$. Therefore, the kernel-based measurement estimation is represented as (12):

$$P(\mathbf{Z}_t|\mathbf{x}_i) = \sum_j P\left(\mathbf{Z}_t|\mathbf{x}_j^s\right) G_\sigma\left(\mathbf{x}_j^s,\mathbf{x}_i\right), \tag{12}$$

where $P(\mathbf{x}_j^s|\mathbf{x}_i) = G_\sigma(\mathbf{x}_j^s,\mathbf{x}_i)$ is a Gaussian kernel with a bandwidth σ. The kernel-based measurement estimation can obtain smooth measurements in a neighborhood, reduce computations of measurements, and also allow for multi-resolution searching during rigid and nonrigid tracking by varying bandwidths in kernels.

2.3.2 Rigid Tracking

Rigid tracking aims at recovering the rigid motion of a guidewire between two successive frames. In rigid tracking, the motion parameter $\mathbf{u_x}$ in (2) contains only

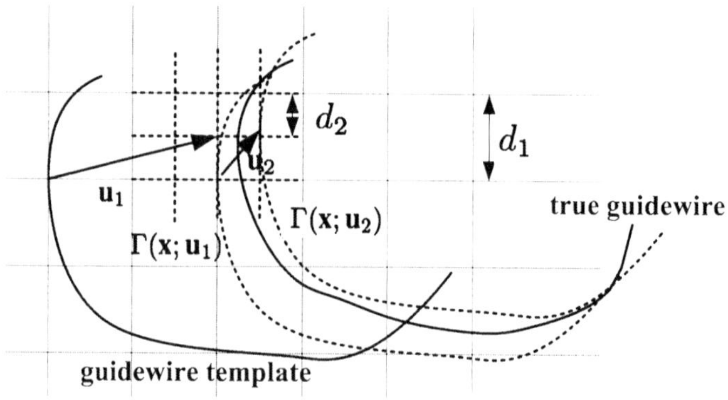

Fig. 6 Multi-resolution rigid tracking with incrementally decreased search intervals and kernel bandwidths. The rigid tracking started from a template, and ends at position close to the true guidewire, up to an error caused by deformable motions. The *dotted curves* represent intermediate tracking results

global translation and rotation, that is, $\mathbf{u_x} = \mathbf{u} = (c, r, \theta)$, where c, r, and θ are the translation and rotation parameters. Therefore, the rigid tracking is formulated as maximizing the posterior probability under a rigid motion of the guidewire, that is, maximizing $E(\mathbf{u})$ as below:

$$E(\mathbf{u}) = P\big(\Gamma_t(\mathbf{x})\big) \sum_{\mathbf{x}_i} P\big(\mathbf{x}_i | \Gamma_t(\mathbf{x};\mathbf{u})\big) P(\mathbf{Z}_t|\mathbf{x}_i). \tag{13}$$

Tracking the rigid motion can be efficiently implemented using variable bandwidths in kernel-based measurement smoothing. As illustrated in Fig. 6, the rigid tracking is performed at multiple resolutions, with decreased search intervals $\{d_1 > d_2 > \cdots > d_T\}$. During the multi-resolution tracking, the corresponding bandwidth in (12) varies accordingly, denoted as σ_i. At coarse resolutions, we use larger kernel bandwidths to avoid missing tracking caused by larger sampling intervals; and at fine resolutions, we use smaller kernel bandwidths to obtain finer tracking results.

Furthermore, the rigid tracking is performed at both global and local scales. At a global scale, the whole guidewire is tracked, while at a local scale, a whole guidewire is divided into several local segments for rigid tracking, which follows the same formalization as (13). By the two-stage tracking, a guidewire is roughly aligned at the current frame. A rigid tracking result is shown as the red curve in Fig. 7(b).

2.3.3 Nonrigid Tracking

After the rigid tracking, a guidewire is further refined by the nonrigid tracking, where the guidewire motion parameter $\mathbf{u_x}$ is point dependent. Different from rigid

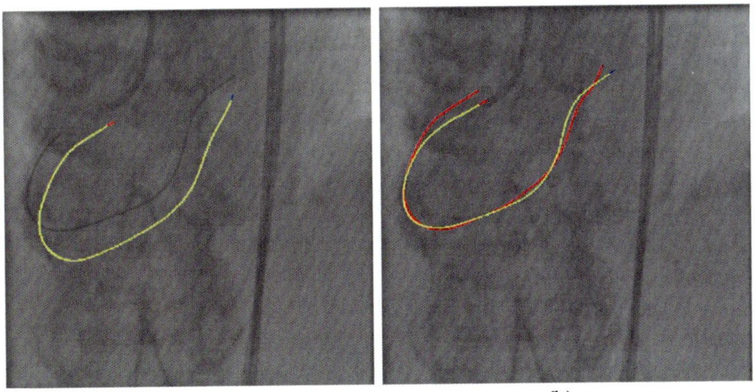

(a) (b)

Fig. 7 Hierarchical tracking. (**a**) Tracking starts from a tracked guidewire from the previous frame (*yellow line*). (**b**) The tracked guidewire after rigid tracking (*red line*) and after nonrigid tracking (*yellow line*)

tracking, the nonrigid tracking imposes a prior from guidewire smoothness constraints, as (14):

$$E(\mathbf{u_x}) = P\big(\Gamma_t(\mathbf{x};\mathbf{u_x})|\mathbf{Z}_t\big) + \alpha \int \left|\frac{d\Gamma_t(\mathbf{x};\mathbf{u_x})}{ds}\right|^2 ds + \beta \int \left|\frac{d^2\Gamma_t(\mathbf{x};\mathbf{u})}{ds^2}\right|^2 ds. \quad (14)$$

The two additional terms, $\int |\frac{d\Gamma_t}{ds}|^2 ds$ and $\int |\frac{d^2\Gamma_t}{ds^2}|^2 ds$, are integrals of the first-order and second-order derivatives of guidewire curves, and act as guidewire smoothness priors to prevent over-deformations of guidewires. The weights, α and β, are then used to balance the smoothness constraints and probabilities scores. Such α and β are empirically set (a typical value is between 0.05 and 0.2), but the tracking performance is not sensitive to the parameter settings as observed from our experiments. Although (14) looks similar to the formalization in active contour based methods such as Snakes [9], they have fundamental differences. In our method, the tracking is based on robust probabilistic measurements, while Snakes is mainly based on intensity gradients, and is prone to image noise in fluoroscopy. Our method maximizes a posterior probability, based on the novel fusion of learning-based measurements and online appearance measurements. Also, our method tracks an open curve of a guidewire, while Snakes and level set based methods [9, 13, 23] mainly handle closed object boundary. At last, our method integrates multiple measurements from different guidewire parts, making it more suitable to track a wire structure.

The search space of the nonrigid guidewire motion in (14) is high dimensional. To reduce the dimensionality of the searching space, we deform control points on the guidewire body along normal directions, and two tips along both the tangent and normal directions, as illustrated in Fig. 8. But still, to exhaustively explore such a deformation space is formidable considering computational complexity. For example, if there are 20 control points, and each control point has 10 deformation candidates,

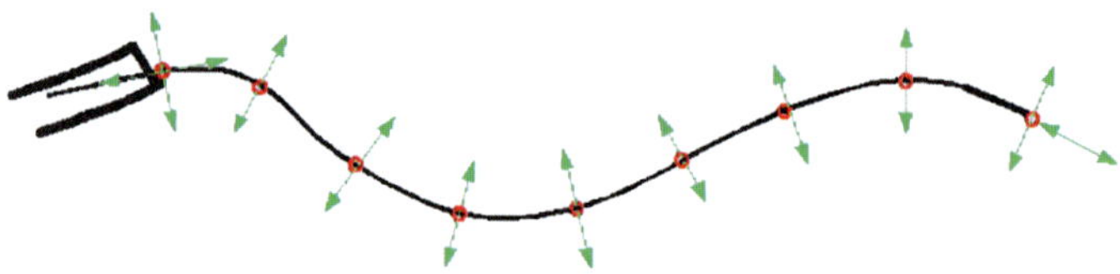

Fig. 8 Nonrigid guidewire tracking. The control points (the *red dots*) on the guidewire body deform along normal directions, and the catheter tip and guidewire tip deform along both the normal and tangent directions

the searching space contains 10^{20} candidates. Instead of parallel searching or sampling the search space, our method searches guidewire deformation sequentially. At each step, only one control point deforms to achieve a maximum $E(\mathbf{u_x})$. The sequential deformation will iterate until the maximum number of iterations is reached or it converges. The same as rigid tracking, the multi-resolution searching strategy is applied during the nonrigid tracking. The sequential searching strategy in most cases leads to an optimal solution, because the rigid tracking has roughly aligned the guidewire near the true shape. An example of nonrigid tracking is shown as the yellow curve in Fig. 7(b).

3 Experiments

3.1 Data and Evaluation Protocol

The tracking algorithm is evaluated on a set of 47 fluoroscopic sequences. The frame size of each sequence is 512×512 or 600×600, with the pixel size between 0.184 mm and 0.278 mm. There are totally more than 1000 frames in the test set. The test sequences cover a variety of interventional conditions, including low image contrast, thin guidewire, and contrast injection. Some exemplar frames in the test set are displayed in Fig. 9. The guidewire part detectors are trained on a set of 500 guidewire images that are previously collected [4].

To establish ground truth for evaluation, we manually annotate the guidewires in the test set as the ground truth. An annotated guidewire starts from a guiding catheter tip, and ends at a guidewire tip. For the purpose of evaluating tracking performance, the annotation at the first frame of each sequence is used to initialize the guidewire tracker, and the rest of annotation is used for validations. In clinical applications, the guidewire can be automatically initialized at the first frame [4], or semi-automatically detected using the interactive detection method [10].

To comprehensively and quantitatively evaluate the performance of guidewire tracking, we define a set of performance metrics, including *overall guidewire tracking precision*, *guidewire body tracking precision*, *tip tracking precision*, and *missing and false tracking rate*, as follows.

1. The *overall guidewire tracking precision* is defined as the average of shortest distances from points on a tracked guidewire to the corresponding annotated

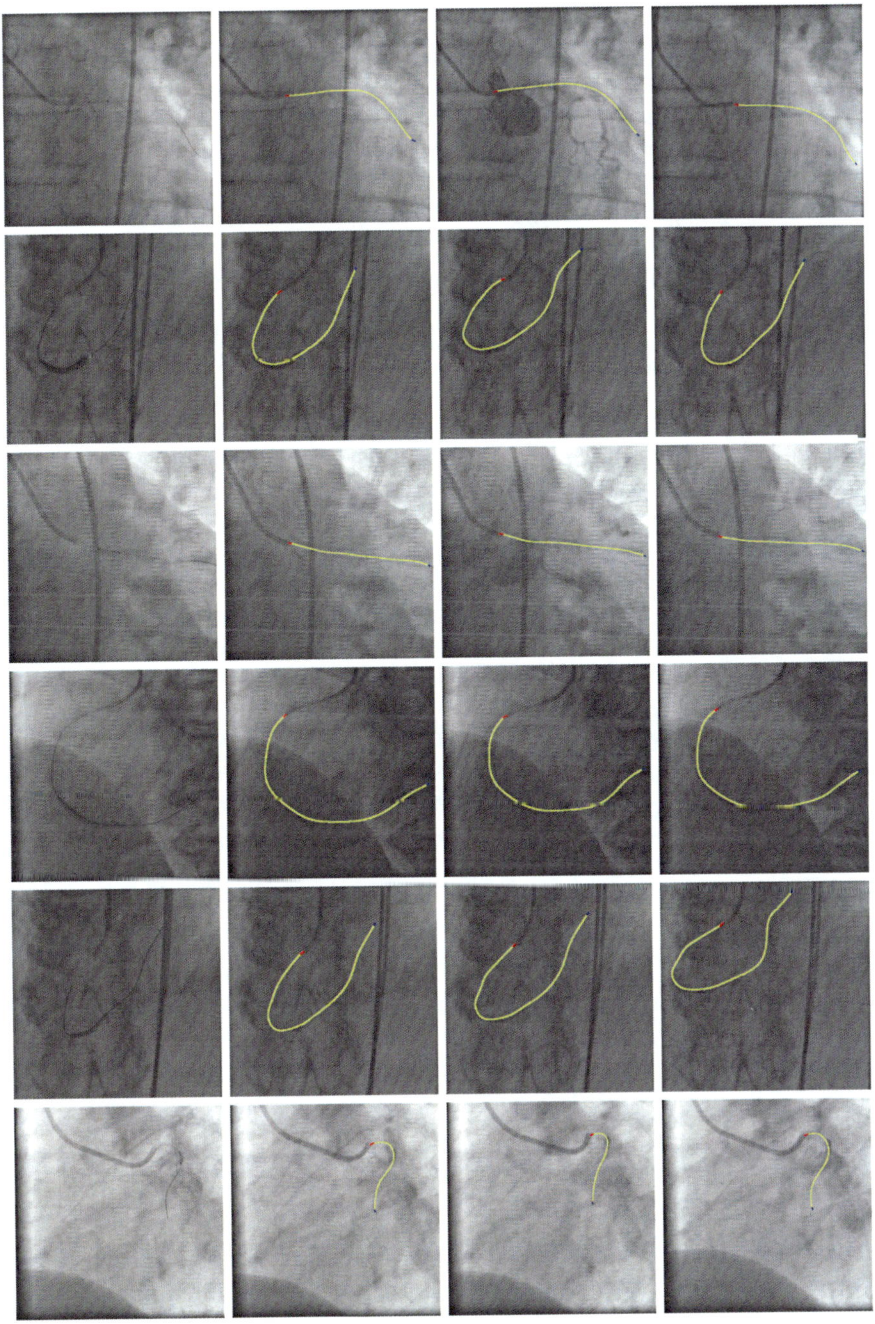

Fig. 9 Some guidewire tracking results (shown in *yellow curves*)

guidewire. Such a precision describes how close a tracked guidewire is to the ground truth.
2. The *guidewire body tracking precision* is the tracking precision specifically at the guidewire body. The tracking errors at tips are excluded in evaluating the guidewire body tracking precision.
3. The *tip tracking precision* is the tracking precision specifically at the catheter and guidewire tips.
4. The *missing and false tracking rates* describe the percentages of guidewire points that have not been successfully tracked. A miss-tracked guidewire point is the point on an annotated guidewire whose shortest distance to the tracked guidewire is greater than a pre-set threshold (e.g., a threshold of 3 pixels is used in this evaluation.) A false-tracked guidewire point is the point on a tracked guidewire whose tracking error is greater than the threshold. Missing and false tracking rates are the percentages of such failed guidewire points.

3.2 Quantitative Evaluations

Some guidewire tracking results in fluoroscopic sequences are shown in Fig. 9. Our method can successfully track the guidewire, even for those sequences with low visibility, background distraction, and contrast injection. For quantitative evaluations, the performance metrics defined in Sect. 3.1, are computed on the test set. Table 1 summarizes quantitative evaluations of our tracking methods, and also compares the performance of using combined measurements with tracking only using learning-based measurements. For each performance metric, including overall guidewire tracking precision, body tracking precision, tip tracking precision, and missing and false tracking rate, we compute its mean, standard deviation, and median. As shown in the table, the overall guidewire tracking precision is around 1.8 pixels, that is, less

Table 1 Quantitative evaluation of guidewire (GW) tracking

Measurements models	With combined models			With only offline learned models		
	mean	std	median	mean	std	median
Overall GW tracking prec. (in pixels)	1.80	3.41	0.95	7.53	27.37	1.05
GW body tracking prec. (in pixels)	1.70	2.96	0.95	1.79	4.02	0.97
GW tip tracking prec. (in pixels)	5.45	8.21	3.69	9.06	15.43	3.79
Cath. tip tracking prec. (in pixels)	11.62	12.76	7.18	68.33	76.85	35.85
GW missing tracking rate	9.88%	8.43%	8.27%	11.06%	13.97%	8.55%
GW false tracking rate	9.62%	6.92%	7.80%	11.87%	9.74%	9.33%

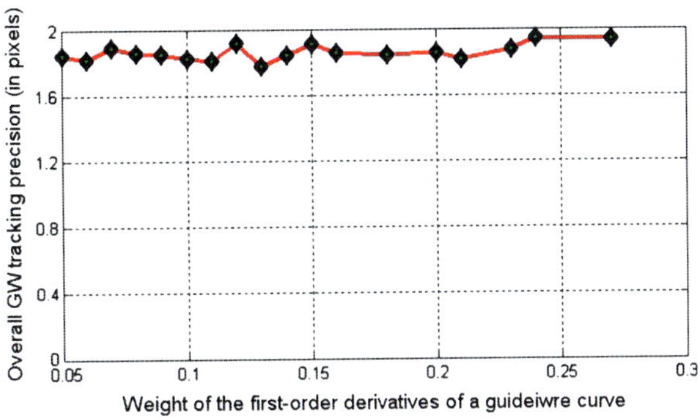

Fig. 10 The tracking accuracy and an algorithm parameter (the weight of the first-order derivative of guidewire *curves*)

than 0.4 mm. The tracking errors on the guidewire and catheter tips are greater than on the guidewire body, because the motions are larger on the tips, and the background distractions affect more on tips than on the guidewire body. Our method successfully tracks more than 90% points on the guidewire, with a false tracking rate lower than 10%.

Table 1 also shows the tracking accuracy when only using learning-based measurements. It demonstrates that using both measurements improves the tracking robustness, as the mean and standard deviation of tracking errors have been reduced, especially at the catheter tip where the appearance-based measurement plays a dominant role. The large tracking errors of learning-based measurements at catheter tips are mainly caused by image noises and occlusions by contrast injection during interventions. Another important observation is that the learning-based measurements provide fairly robust measurements on the guidewire body and guidewire tips, as the improvement of tracking precision on guidewire body and tip is smaller than on the catheter tip. This confirms the advantage of our framework that unifies multiple measurements from different guidewire parts in a principled way.

The presented method is not sensitive to parameter changes. Due to limited space, we present only one experiment in Fig. 10, which shows the tracking accuracy does not change much with a wide range of a parameter α, that is, the weight of the smoothness constraint based on the first-order derivative of guidewire curves. Validations on other parameters show similar results. We conclude from the quantitative evaluations that our probabilistic tracking method provides robust and accurate guidewire tracking results. This method currently runs at 2 frames per second at a Core 2 Duo 2.0 GHz computer, and can achieve a near real-time speed with an implementation optimization, such as multi-threading and GPU accelerations.

4 Conclusion

This chapter presents a probabilistic framework of robust guidewire tracking in fluoroscopy for image guided interventions. Our framework can track nonrigid guidewire motions under arbitrary projections. The validation on a test set of 47 real interventional sequences demonstrates that this method provides robust and accurate tracking results. The future work will be integrating guidewire tracking into clinical applications, such as breathing motion compensation.

References

1. Avidan, S.: Ensemble tracking. In: CVPR, pp. 130–136 (2005)
2. Baert, S.A.M., Viergever, M.A., Niessen, W.J.: Guide wire tracking during endovascular interventions. IEEE Trans. Med. Imaging **22**(8), 965–972 (2003)
3. Baim, D.S.: Cardiac Catheterization, Angiography, and Intervention, 7th edn. Lippincott William's and Wilkins, Philadelphia (2006)
4. Barbu, A., Athitsos, V., Georgescu, B., Boehm, S., Durlak, P., Comaniciu, D.: Hierarchical learning of curves application to guidewire localization in fluoroscopy. In: CVPR (2007)
5. Bartels, R.H., Beatty, J.C., Barsky, B.A.: An Introduction to Splines for Use in Computer Graphics and Geometric Modelling. Morgan Kaufmann, San Mateo (1998)
6. Carneiro, G., Georgescu, B., Good, S., Comaniciu, D.: Detection and measurement of fetal anatomies from ultrasound images using a constrained probabilistic boosting tree. IEEE Trans. Med. Imaging **27**(9), 1342–1355 (2008)
7. Friedman, J., Hastie, T., Tibshirani, R.: Additive logistic regression: a statistical view of boosting. Ann. Stat. **28**(2), 337–374 (2000)
8. Isard, M., Blake, A.: Condensation: conditional density propagation for visual tracking. Int. J. Comput. Vis. **29**(1), 5–28 (1998)
9. Kass, M., Witkin, A., Terzopoulos, D.: Snakes: active contour models. Int. J. Comput. Vis. **1**(4), 321–331 (1987)
10. Mazouer, P., Chen, T., Zhu, Y., Wang, P., Durlak, P., Thiran, J.-P., Comaniciu, D.: User-constrained guidewire localization in fluoroscopy. In: Medical Imaging: Image Processing. Proc. SPIE. SPIE, Bellingham (2009)
11. Palti-Wasserman, D., Brukstein, A.M., Beyar, R.: Identifying and tracking a guide wire in the coronary arteriesduring angioplasty from x-ray images. IEEE Trans. Biomed. Eng. **44**(2), 152–164 (1997)
12. Peters, T., Cleary, K.: Image-Guided Interventions: Technology and Applications. Springer, Berlin (2008)
13. Staib, L., Duncan, J.: Boundary finding with parametrically deformable models. IEEE Trans. Pattern Anal. Mach. Intell. **14**, 1061–1075 (1992)
14. Tu, Z.: Probabilistic boosting-tree: learning discriminative models for classification, recognition, and clustering. In: ICCV, pp. 1589–1596 (2005)
15. Viola, P., Jones, M.: Robust real-time object detection. Int. J. Comput. Vis. **57**(2), 137–154 (2004)
16. Wang, P., Ji, Q.: Robust face tracking via collaboration of generic and specific models. IEEE Trans. Image Process. **17**(7), 1189–1199 (2008)
17. Wang, P., Chen, T., Zhu, Y., Zhang, W., Zhou, S., Comaniciu, D.: Robust guidewire tracking in fluoroscopy. In: IEEE Computer Society Conference on Computer Vision and Pattern Recognition, pp. 691–698 (2009)
18. Wang, P., Zhu, Y., Zhang, W., Chen, T., Durlak, P., Bill, U., Comaniciu, D.: Hierarchical guidewire tracking in fluoroscopic sequences. In: SPIE: Medical Imaging, vol. 7258, pp. 72591L–72591L–8. SPIE, Bellingham (2009)
19. Wu, Y., Huang, T.S.: Robust visual tracking by integrating multiple cues based on co-inference learning. Int. J. Comput. Vis. **58**(1), 55–71 (2004)
20. Yilmaz, A., Javed, O., Shah, M.: Object tracking: a survey. ACM Comput. Surv. **38**(4), 13 (2006)

21. Zheng, Y., Barbu, A., Georgescu, B., Scheuering, M., Comaniciu, D.: Four-chamber heart modeling and automatic segmentation for 3d cardiac ct volumes using marginal space learning and steerable features. IEEE Trans. Med. Imaging **27**(11), 1668–1681 (2008)
22. Zhou, X., Comaniciu, D., Gupta, A.: An information fusion framework for robust shape tracking. IEEE Trans. Pattern Anal. Mach. Intell. **27**(1), 115–129 (2005)
23. Zhu, S.C., Yuille, A.L.: Forms: a flexible object recognition and modeling system. Int. J. Comput. Vis. **20**, 187–212 (1996)

Part III
Motion Analysis and Behavior Modeling

An Integrated Approach to Visual Attention Modeling for Saliency Detection in Videos

Sunaad Nataraju, Vineeth Balasubramanian, and Sethuraman Panchanathan

Abstract In this chapter, we present a framework to learn and predict regions of interest in videos, based on human eye movements. In our approach, the eye gaze information of several users are recorded as they watch videos that are similar, and belong to a particular application domain. This information is used to train a classifier to learn low-level video features from regions that attracted the visual attention of users. Such a classifier is combined with vision-based approaches to provide an integrated framework to detect salient regions in videos. Till date, saliency prediction has been viewed from two different perspectives, namely visual attention modeling and spatiotemporal interest point detection. These approaches have largely been vision-based. They detect regions having a predefined set of characteristics such as complex motion or high contrast, for all kinds of videos. However, what is 'interesting' varies from one application to another. By learning features of regions that capture the attention of viewers while watching a video, we aim to distinguish those that are actually salient in the given context, from the rest. The integrated approach ensures that both regions with anticipated content (top–down attention) and unanticipated content (bottom–up attention) are predicted by the proposed framework as salient. In our experiments with news videos of popular channels, the results show a significant improvement in the identification of relevant salient regions in such videos, when compared with existing approaches.

S. Nataraju (✉) · V. Balasubramanian · S. Panchanathan
Center for Cognitive Ubiquitous Computing, CUbiC, Arizona State University, Tempe, AZ 85287, USA
e-mail: snatara4@asu.edu

V. Balasubramanian
e-mail: vineeth.nb@asu.edu

S. Panchanathan
e-mail: panch@asu.edu

L. Wang et al. (eds.), *Machine Learning for Vision-Based Motion Analysis*, Advances in Pattern Recognition,
DOI 10.1007/978-0-85729-057-1_8,

1 Introduction

As growing numbers of videos are generated each day, there has been an equally increasing need to reliably identify appropriate regions of interest in videos. Saliency detection has been widely used in many computer vision problems, and has applications from several perspectives. In video analysis based tasks such as recognition and detection, identifying salient regions is often used as a preprocessing step to narrow down on potential areas of interest in the video frames. This reduces the computational cost of these algorithms by eliminating the need to analyze regions that are not of any relevance. It also improves the accuracy when images have excessive background clutter, by ensuring that only the regions of interest are processed by the algorithm. In addition, saliency detection has been used in video and image compression, where a higher bit-rate is used to encode regions that are of interest, thereby not compromising on the quality of regions that are relevant.

Existing approaches to identify salient regions in images and videos can be broadly categorized into two different schools of thought: *interest point detection* and *visual attention modeling*, each of which is described below.

1.1 Interest Point Detection

Interest point detection refers to identifying a set of locations in an image/video that have a sufficiently high response to a predefined saliency function (or a set of predefined saliency functions). Historically, interest point detection methods originated from the popular corner detection [25] and edge detection methods [45]. While most methods bear a direct relevance to images, approaches for interest point detection in videos include spatiotemporal methods such as Laptev's 3-D Harris corner detector [21], and the periodic detector presented by Dollar et al. in [1] (more details of such methods are presented in Sect. 2). These methods have also largely been vision-based, i.e., they rely on predefined filters that characterize regions of interest with a tendency to capture extrema in certain image characteristics. An inherent problem with such approaches is that they use the same set of filters to determine saliency in all types of videos.

While both these aforementioned kinds of approaches generally perform well to detect artifacts such as corners, edges and motion-based interest regions, '*interestingness*' is strongly dependent on an application under consideration, and it may not be appropriate to generalize this concept. For example, in surveillance applications, monitoring personnel would consider specific events such as a person carrying a gun as salient, although the background of this scene may be cluttered with other 'interesting' image artifacts based on edges or color. Similarly, in biomedical applications such as colonoscopy videos, a physician would specifically consider patterns corresponding to tumors as salient, rather than all complex patterns corresponding to normal characteristics in the body. This chapter presents a novel approach to detect saliency that is not only driven by image/video artifacts, but is also influenced by what is of interest for a given class of videos.

1.2 Visual Attention Modeling

Visual attention in humans, while viewing a scene, is understood to be driven by *bottom–up* (characterized by image features alone) and *top–down* (characterized by user's intent and context) approaches. In simpler terms, a user, while viewing a scene belonging to a specific application, looks out for regions that are of salience based on the context. For example, a security personnel looks for specific entities in a video, while monitoring video feeds from surveillance cameras. Such regions are said to correspond to top–down salience. However, in the same example, it is possible that there are other regions in the video having distinct features (such as a bird flying by) that unintentionally catch the user's attention and are task-independent. These regions are said to be of bottom–up salience.

Approaches to bottom–up saliency modeling [5, 13] detect regions having certain distinct features that stand out and capture the visual attention of the viewer. Itti et al. [11–13] have been the pioneers of modeling saliency from this perspective. These approaches are generally independent of the context of the scene and only aim to detect regions having features that may arouse the interest of a viewer. As a result, computational models for bottom–up saliency have been universal by nature, and the same model is applied to all kinds of images and videos. These methods are based on image processing and computer vision, and involve the use of filters that respond to certain predefined characteristics based on contrast, motion, texture and orientations. At a fundamental level, research in this domain has been focused on designing image filters that respond to features that biologically drive visual attention in humans. Saliency models based on these approaches typically are represented using *saliency maps*, which are two-dimensional gray scale plots indicating the visual saliency of each of the pixel locations in an image. A sample saliency map is illustrated in Fig. 1.

Bottom–up saliency modeling accounts for only one aspect of visual attention. While this type of saliency drives human eye-movements towards regions that have distinct features, they do not drive visual attention to regions that are relevant in a given scenario. This is where top–down attention comes into play. As stated earlier, attention is not only driven by features that are distinct in terms of its contrast or

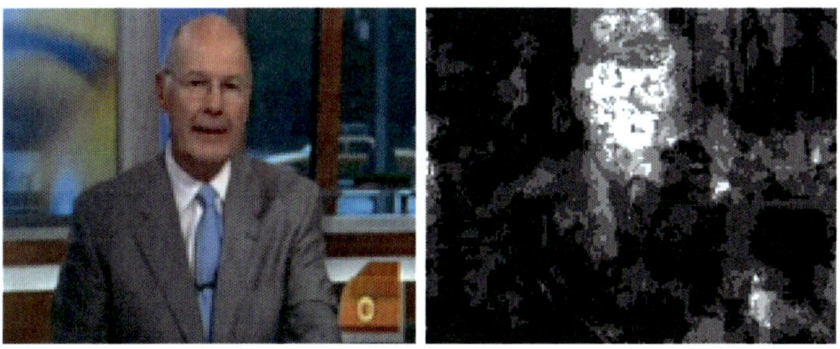

(a) (b)

Fig. 1 Example of a saliency map for an image frame

motion, but by features corresponding to specific regions that the users are trying to locate. In other words, top–down attention is application specific. Frameworks that model this kind of attention use the domain knowledge of the scene. A comprehensive framework to model saliency in a scene is one that combines both these approaches.

Most existing research in modeling attention has been directed towards detecting bottom–up saliency. However, there has been some limited work in modeling top–down attention. These approaches [4, 28] define top–down saliency as being task-driven (more details presented in Sect. 2). However, the 'task' or the 'goal', such as object recognition or detection is specified a priori. In this work, an alternate paradigm towards integrated top–down attention modeling is proposed. Instead of predetermining the task explicitly, human eye movements are used to indicate the relevance of regions in videos, and thus the 'task' is learnt rather than prespecified. This is done by recording the eye gaze of several users while they naturally watch videos belonging to a specific application or context. In the next stage, a classifier is trained to learn the features corresponding to the regions that were viewed. By doing so, the learnt features correspond to what was commonly viewed by most users since all the videos belong to a single domain, or context. Thus, in this work top–down saliency is defined as the saliency based on the context of the scenario in the videos.

Further, in order to provide a complete framework for visual attention, a novel probabilistic approach is presented to model the final combined saliency. Saliency maps are modeled as 2-D probability distribution functions. The final saliency is calculated as a joint distribution function derived from a graphical model. Such an approach is shown to be capable of handling top–down as well as bottom–up saliencies. This approach is useful in high risk scenarios where along with the regions that are relevant to that specific context, it might also be important to locate regions that 'pop out'. This approach ensures that all regions that are possibly of interest are highlighted, and the user is alerted to watch these regions.

1.3 Proposed Approach

Human eye movements comprise of saccades and fixations [3]. A fixation occurs when the eye-gaze is directed to a particular location. Their durations can vary from a few tens of a millisecond to several hundreds of milliseconds. Fixations point to regions in a scene that the user paid attention to, and thus help indicate the regions of interest in videos/images. Saccades, on the other hand, are rapid eye movements that occur between fixations. Human eye movements have been known to be good indicators of visual saliency. The approach of using human eye movements has two distinct advantages:

- Unlike video datasets that are typically captured under controlled conditions with a stationary background, real-world videos are often comprised of several regions, each of which has its own spatiotemporal characteristics. Each of these regions

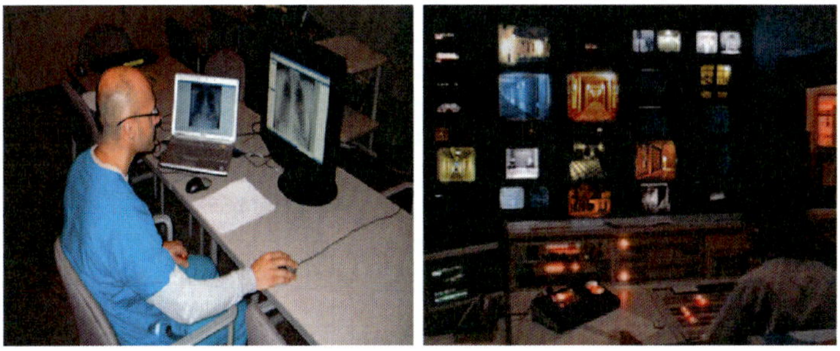

(a) (b)

Fig. 2 Eye trackers could be used to record the regions viewed by radiologists and surveillance personnel as they analyze videos

may or may not be relevant, depending on the application and the nature of videos being studied. Human eye movements are capable of indicating the relevant regions of interest in an application-specific manner.

- Supervised learning approaches that can possibly learn the relevance of regions need labeling of training samples. This is a very tedious and time-consuming process, where users would be required to manually select the regions that are of interest. This can be avoided by using human eye movements to label the regions. Users can be asked to watch videos naturally, and their eye movements can be recorded to subconsciously label regions of interest. This process proves to be an efficient and quick way of obtaining ground truth for a learning framework that can predict regions of interest in new videos.

Eye-trackers are used to record the eye gaze of users as they are viewing a computer screen (see Fig. 2). Eye-tracking technology has evolved over the years, and state of the-art eye-trackers can easily be integrated into a regular desktop monitor. They allow unobtrusive viewing, and are also robust to minor head movements. As mentioned above, the advantages of using eye-trackers in the proposed framework is that they provide a means of implicitly labeling training data without interfering with the user activity. They remove the necessity for manually labeling each and every frame of a video and thus automate the process. For example, as radiologists study medical videos as part of their daily work, they subconsciously label the data simultaneously without having to explicitly mark the regions of interest. Thus, the proposed framework can be used with ease in such applications without any explicit procedure for data capture, or labeling methodology that may be needed every time the model needs to be learnt or updated.

The main objectives and contributions of this chapter are as follows:

- *Develop a methodology to detect saliency by integrating context-specific (dominated by top–down) and bottom–up approaches in videos*: Pure vision based approaches to detect saliency are designed to identify regions that 'pop out' in the image. However, in many real world applications such as medical videos, it is required to find regions that are relevant to the context, such as tumors and polyps.

In order to do so, pure vision-based approaches alone are insufficient. It is imperative to use a supervised learning based approach to detect salient regions and be able to distinguish these regions from regions that only have distinct features and may not be relevant to the application. This chapter presents an approach to achieve this objective.

- *Use eye movements to indicate user interest rather than have a manual labeling process to capture what is salient for a given class of videos*: In applications that require skilled professionals to analyze videos, labeling training data becomes an expensive and a lengthy procedure. Eye trackers provide an automated way to overcome the manual labeling process, by recording the eye movements as the users analyze the videos.

The key features of the proposed approach are mentioned below:

1. This framework provides a supervised learning framework based on human eye movements for saliency detection in videos, and integrates this framework with a vision-based approach. Existing approaches have largely been focused on a vision-based approach.
2. This framework can handle pan and zoom camera movements, unlike existing methods that rely on motion in videos as an indicator of saliency.
3. The conceptual framework can be generalized to different classes of videos.
4. The approach has been validated with a real-world dataset with complex backgrounds. Most existing work in this regard use videos captured under highly controlled conditions.
5. A probabilistic formulation of saliency provides for a meaningful interpretation in real-world application contexts.

The remainder of the chapter is organized as follows. Prior background work and motivation for the proposed approach is discussed in Sect. 2. The conceptual framework to learn and predict saliency based on eye movements is presented in Sect. 3. The experiments and results for the learnt model along with suitable illustrations are described in Sect. 4. Section 5 summarizes the methodology, discusses possible applications and directions of future work.

2 Prior Work

Saliency is defined in this work as a measure of possible user interest on a single unit of video (for example, pixel on a video frame). As mentioned earlier, saliency detection has been studied from two different perspectives, visual attention modeling and interest point detection. Figure 3 provides a high-level illustrative summary of the different kinds of approaches that have been used in this context. Visual attention in humans is driven by bottom–up, as well as top–down approaches. Bottom–up attention is driven by regions having distinct features, and is independent of the task, or the context of the scenario. On the other hand, top–down saliency is specific to a context where the user is trying to locate something in particular. Approaches to detect interest points make use of predetermined filters that measure saliency based on

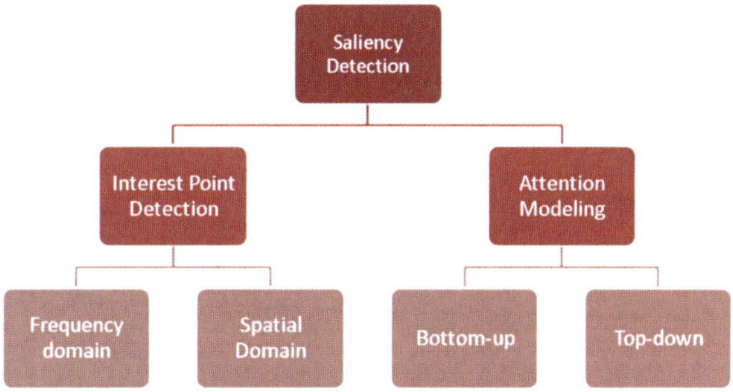

Fig. 3 Prior work in Saliency detection

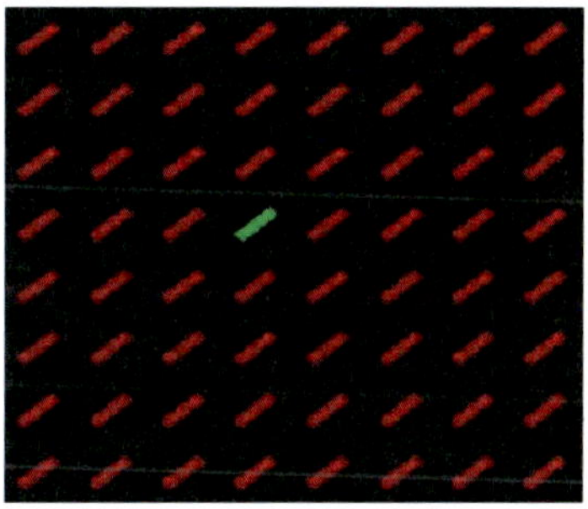

Fig. 4 Regions having distinct features are of bottom–up salience. *Illustration taken from* [5]

specific artifacts such as motion, and corners. This section reviews the related work from each of these perspectives, and also discusses prior work in the use of human eye movements. The drawbacks of these existing approaches, and the motivation for the proposed integrated approach are also illustrated.

2.1 Visual Attention Modeling Methods

2.1.1 Bottom–Up Saliency

Research in visual attention modeling has primarily focused on bottom–up saliency. Bottom–up attention is driven by regions having salient stimuli (see Fig. 4). Such approaches involve algorithms that detect regions in an image/video that have distinctive features, and is independent of the context of the video. Most computational frameworks that model bottom–up attention are based on the feature integration theory, as explained in [42]. This theory explains the visual search strategy in humans. It proposes that several features are used to obtain individual feature maps that are then integrated to provide the final measure of saliency. The most popular framework to model bottom–up attention was proposed by Itti et al. in [13], as illustrated in Fig. 5. This model was built on the architecture that was proposed by Koch and Ullman in [20] which is based on the concept of a saliency map that indicates the visual saliency of every pixel in an image.

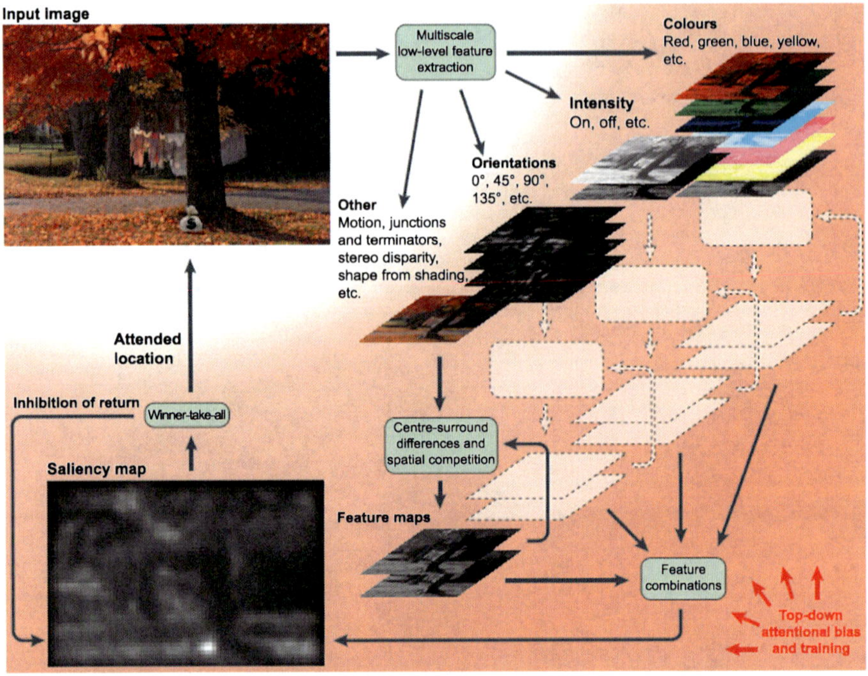

Fig. 5 Framework used by Itti and Koch in [13] to model bottom–up attention. *Illustration taken from* [12]

In Itti's seminal work [13], a Gaussian pyramid is constructed at multiple dyadic scales for a given image. Individual features are then extracted for each location based on color, orientation, intensity and motion. The operators used to extract these features are driven by the visual attention system in primates, which is sensitive to a small region surrounded by a region having characteristics that inhibit the neural response. Differences are then taken across scales to implement this center-surround to obtain the individual feature maps. The feature maps are then normalized so as to enhance the conspicuity of strong peaks and diminish the conspicuity of feature maps having numerous comparable peaks. The normalized feature maps are then combined along the various scales to obtain a single feature map for each individual feature. Finally, these are then linearly combined to obtain the final saliency map, that indicates the saliency at each pixel location.

Another approach was proposed by Gao and Vasconcelos in [5]. In their formulation, they equate saliency to discriminability. Although their approach also uses the concept of obtaining different feature maps and combining them into a single saliency map, the filters they use are more suited to locate regions that have discriminative features. Stentiford [39, 40] proposed a measure of saliency that depended on the dissimilarity between neighborhoods in an image. This was also linked to the notion of *self-similarity* or a fractals approach. A comprehensive survey of approaches to model bottom–up attention in humans that aim to extract such conspicuous re-

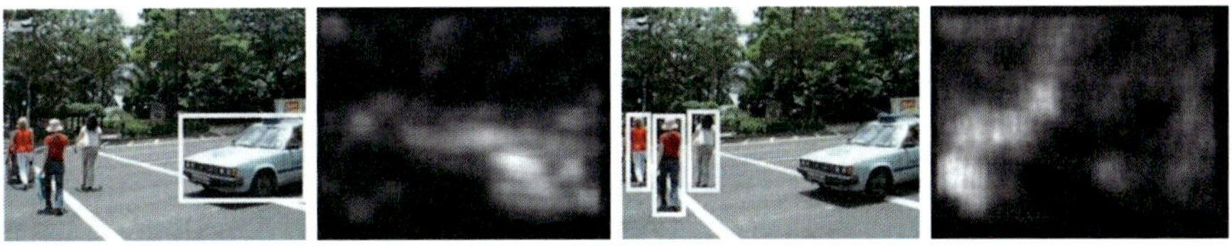

Fig. 6 Top–down saliency maps derived using recognition based approaches. *Illustration taken from* [6]

gions is presented in [11]. Bottom–up approaches have been well-studied over the last few decades, and hence, in the presented framework, the existing framework by Itti et al. is used to integrate bottom–up saliency with the context-based (which we use to represent the top–down component) saliency algorithm.

2.1.2 Top–Down Saliency

Unlike bottom–up saliency, top–down saliency is driven by user intent. There has been limited work done in this regard [4, 23, 28]. Existing approaches that model top–down saliency use a specific goal or task, such as object recognition. Figure 6 illustrates such an approach, where two different saliency maps are derived for a single image based on car and person recognition respectively.

Similar to their approach to bottom–up saliency, Gao and Vasconcelos in [4] presented a discriminative saliency based approach for top–down processing. Here, they defined saliency based on features that are best discriminative for a given class of objects. Navalpakkam and Itti, in [28] investigated the use of top–down strategies to select optimal cues in a predetermined search task for objects in a scene, and came up with a model to maximize the saliency of the target with respect to its distractors. In the proposed approach, top–down saliency is defined from a different perspective. As mentioned earlier, the task, or the goal, is subtly indicated by the regions fixated by the users while viewing the videos, rather than artificially defining a potential task in a scene.

2.1.3 Integrating Top–Down and Bottom–Up Saliency

There have been very few efforts in the past on developing models that integrate the top–down and bottom–up components for saliency. An approach to integrate top–down and bottom–up saliencies for target detection was proposed by Navalpakkam and Itti in [27]. In their framework, Itti's bottom–up saliency model [13] is used to obtain the individual feature maps. Subsequently, they used the prior statistical knowledge of the features based on the target as weights of these individual maps to linearly combine them into an overall saliency map. However, this approach is purely vision-based. Another approach proposed by Oliva et al. [31] and Torralba in [41] to model attention in images used a machine learning approach to integrate contextual information along with bottom–up image saliency. They defined bottom–up saliency based on the likelihood of finding certain features in an image. The contextual priming is learnt using a database of training images having known locations of the target objects. However, in both these approaches, a prespecified task such as object recognition is used to derive the top–down approach for saliency. In other words, the user intent is prespecified in these approaches, which restricts the generalizability of such approaches.

2.2 Interest Point Detection Methods

Interest point detection refers to identifying a set of pixel locations in images based on a certain saliency, or 'interestingness' measure. The filters or functions used to detect salient locations are chosen so as to respond to certain artifacts such as corners, textures, or motion. Different approaches use different sets of filters to measure the saliency of a pixel. In [24], Lowe proposed the SIFT algorithm to find 'key points' and their corresponding descriptors that are invariant to scale and orientation. Another popular approach to detect interest points is the Harris corner detector [9] that uses a measure based on a second moment matrix to compute the 'cornerness' of a pixel. The interest points detected using this approach are known to be very sparse, since the pixels need to be spatial as well as temporal corners. Kadir and Brady [15] proposed an approach that measures interestingness based on the information content. In their approach, Shanon entropy is used to measure the complexity of a pixel, using the probability distribution of a descriptor extracted from the region around it. This is evaluated across various scales. The final saliency measure of a pixel is calculated based on the scales that exhibit a peak in entropy, as well as a high gradient in the probability distribution.

Another set of pure vision-based approaches are directed towards detecting saliency in images from their Fourier spectrum. These approaches make use of the $1/f$ law [34, 38] which describes the statistics of natural images. It states that the amplitude spectrum of the Fourier Transform of an image is inversely proportional to the frequency, that is, on a log–log scale, the amplitude spectrum of images is approximately a straight line. An example for this is illustrated in Fig. 7. Hou and Zhang [10] use this property to define saliency based on the spectral residual of an image. This is calculated as a difference between the log spectrum of an image with its averaged spectrum. The saliency map is constructed using inverse Fourier Transform of this spectral residual. Wang and Li also make use of this property to detect saliency in color images in [44]. However, in their approach they propose a two step approach, where a coarse saliency map is obtained in the first step based on the spectral residual of the image. In the second step, this map is refined and thresholded to obtain a binary saliency map. Another approach in this regard [8] proposes the use of phase spectrum, over the amplitude spectrum of the Fourier Transform, and argue that it provides better results with lesser computations.

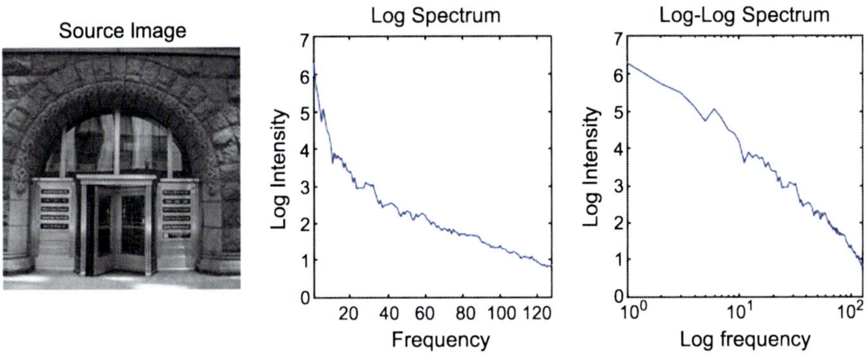

Fig. 7 Example of spectrum and log spectrum of an image. *Illustration taken from* [10]

In the spatiotemporal domain, there have been two categories of approaches to find interest points. One of them seeks to extend the algorithms that exist in 2-D spatial domain to the temporal domain. Examples of these approaches include the 3-D Harris corner detector [21], 3D SIFT descriptor [37], and the work of Oikonomopoulos et al. [30], which is an extension of the work done by Kadir and Brady in [15] into the third dimension. On the other hand, there have been approaches that have been specifically designed to identify interestingness in the spatiotemporal domain. A popular algorithm in this regard is the periodic detector, that was proposed by Dollar et al. in [1]. In their approach, a train of pixels in a temporal neighborhood is considered, and their response to a quadrature pair of 1-D Gabor filters in the temporal domain is used to measure saliency. Such an approach responds to any kind of complex motion in the video, to detect the salient points. Another approach suggested by Ke et al. in [16] makes use of volumetric features and video optical flow to detect motion. This is based on the rectangular features used by Viola and Jones [43].

Incidentally, most of these existing approaches to detect spatiotemporal interest points test the performance of their algorithms on human action classification [1, 22, 29, 36] using the KTH dataset [36]. The KTH dataset is artificially created, and has a simple stationary background, as shown in Fig. 8. All the motion in the videos correspond to those of human actions, thus enabling these approaches to perform well. However, in real life scenarios this is seldom the case, and there is a need to have a detector that is capable of classifying regions that are actually of interest in complex scenes based on the given context or application of the video, and not only detect points that respond to motion. Unlike these approaches, the proposed framework is validated on real world data which also include moving backgrounds and artifacts.

Though the interest point detectors use measures to determine the saliency of each location in an image, they differ from the attention modeling approaches in that the filters used may not necessarily be biologically driven by the human visual system. These detectors are approached purely from a computer vision perspective. The attention modeling frameworks, on the other hand, are influenced by neuro-

Fig. 8 Sample frames taken from the KTH dataset

logical and cognitive perspectives. Both these approaches, however, do not take the context of the scene into account. The 'interestingness' of a pixel, as in the case of bottom–up saliency, is solely based on the strength of its response on predefined filters. As a result, the saliency is only a measure of the distinctiveness of the neighborhood of a pixel in terms of texture, motion and other such features. Instead, the proposed integrated approach computes saliency as a measure of visual relevance to users for a given class of videos.

2.3 Human Eye Movement as Indicators of User Interest

Eye-tracking is the procedure of tracking the position of the eye gaze of a user. One of the earliest uses of eye-tracking was in the field of psychology in understanding how text is read. Researchers analyzed the variations in fixation and saccade durations with line spacing and difficulty of textual content. Eye-tracking was also used to understand scene and art perception. In more recent times, eye-tracking is being increasingly used in other commercial and research applications ranging from Human Computer Interaction (HCI) and medical research to marketing. In recent work on advertising and web applications, eye-trackers are used to monitor the eye-movements of consumers as they view specific websites. This data is statistically analyzed and used to determine the effectiveness of an advertisement, strategize the location of a product, learn the features that catch the attention of an eye, and so on. Duchowski [2] presented an exhaustive survey of various applications that have used eye tracking, with a specific focus on its use in interactive applications rather than its diagnostic use.

In HCI, the eye tracker has been used as a selection tool, where eye gaze can be used as a substitute for a computer mouse. It has also been used in gaze-contingent displays, where displays respond based on the user's fixations. Another application has been in evaluating and training novices or students as they are performing certain tasks. Since they can provide relevant feedback as to what a user is viewing, it has been used to distinguish novices from experts in the driving and aviation industry. This has been extended to study the visual behavior of drivers (or students), and warn them if they are not paying attention or if they are drowsy.

Eye-tracking has also been used in the field of computer vision and machine learning. Granka et al. in [7] used eye tracking to understand how users react to results provided by an internet search engine, and gain insight into user behavior. Salojarvi et al. [35] investigated the use of eye movements to study relevance in information retrieval. Oyekoya and Stentiford [32, 33] also conducted experiments in image retrieval that indicated the fact that eye-gaze is attracted by regions of interest in images. They found that eye tracking data can be used to retrieve images faster than random selection.

2.3.1 Use of Eye-Tracking in Related Work

There has been limited work in detecting interest points or regions based on human eye movements, almost all of which have been focused on images than videos. Kienzle et al. [17, 18] used human eye movements to learn a model to detect bottom–up saliency in images. They recorded eye-gaze data of users as they were viewing 200 natural images, and built a classifier to learn the image patterns that resulted in high visual saliency. Pixel intensities in rectangular image patches were used as the feature vectors in the for the classifier. The results indicated that the performance of their model was comparable to other bottom–up approaches. More recently, Judd et al. [14] used eye movements to learn saliency in images. They used a larger dataset of 1003 randomly selected landscape and portrait images to collect eye tracking data of users. As stated in their work, their methodology is very closely related to the work of Kienzle et al. In their approach, in addition to having low-level features descriptors such as color and contrast, the classifier is also trained on mid-level and high-level features. These include horizon line detectors since most objects are on the ground and humans look for objects, face and person detectors, as well as the distance from the center of the image.

For videos, Kienzle et al. [19] extended their work to detect spatiotemporal interest points in videos. The dataset used for the training comprised of arbitrary videos sampled from a movie. The eye movements of users were recorded as they watched these videos, and a classifier was trained to learn the features corresponding to regions viewed by the users. The feature descriptor for the learning model was based on the periodic detector [1], where a sequence of pixels in the temporal neighborhood of a pixel (pixels having the same spatial coordinates in neighboring frames) is used. However, in their approach, the filter coefficients of the temporal filters were learnt based on the eye movements of users, instead of using a 1-D Gabor filter. The classifier learnt using their approach was tested on the KTH dataset, which as described earlier, has a plain background and all the motion in the videos belong to the foreground.

Though the performance of such an approach is shown to be comparable to the other state-of-the-art approaches on such a dataset, this approach has limitations in other complex scenarios. Firstly, the above approach uses only temporal descriptors, and a temporal descriptor alone is insufficient to characterize a region. Eye tracking, however, provides spatial interest regions, and this is the most useful benefit of using human eye movements. However, this aspect is ignored in prior work, and the spatial characteristics are not taken into consideration while constructing the feature vector. As a result, all regions having motion are classified as salient (more illustrations are provided in the next subsection). Thus, such a classifier alone is insufficient to distinguish regions that are actually of salience from those simply having motion. Moreover, in the presence of any camera induced motion such as pan or tilt, all regions having edges or any sort of texture are classified as salient since the approach relies purely on motion. Although the approach performs well, it was only tested on videos having motion in the foreground alone without any background clutter.

2.4 Limitations of Existing Work

Most of the existing approaches that model saliency based on visual attention are directed towards bottom–up attention. Such approaches are modeled to be independent of the context of the videos, and detect regions having a standard set of features for all kinds of videos. Interest point detectors have drawbacks that are similar to the bottom–up visual attention models. Other existing approaches that use human eye movements are motion based, and are not capable of handling any kind of camera induced motion. The performance of some of these approaches on a frame taken from a real world news video is illustrated in Fig. 9. The frame has many regions with motion, and it can be seen that these approaches detect all regions having motion and texture, and are unable to distinguish between regions that are actually relevant for a news video from the background clutter.

Bottom–up approaches are designed to detect regions that are significantly distinct from the rest of the image. In images where such regions do not exist, they are not capable of highlighting the salient regions. In general, pure vision-based approaches do not take the application of the video into consideration. However, what is 'interesting' is dependent on the class of the videos. Regions that are salient in one class of videos may not be relevant for another class.

The use of eye-trackers allows regions of interest to be implicitly indicated as users view videos in a natural manner. Existing approaches to model top–down attention are done so for prespecified tasks, such as object or person recognition, that is known to represent user intent. Eye trackers eliminate the need for this since the 'task' is indicated by what users watch when they view videos. Moreover, most existing models require a manual process for labeling to train classifiers which is tedious and expensive. Eye trackers provide a mechanism to automate and speed up the process. In the rest of this work, the performance of the presented approach is compared against a 3D Harris corner detector, periodic detector, Itti's bottom–up attention model and Kienzle's approach in [19]. These are popular approaches

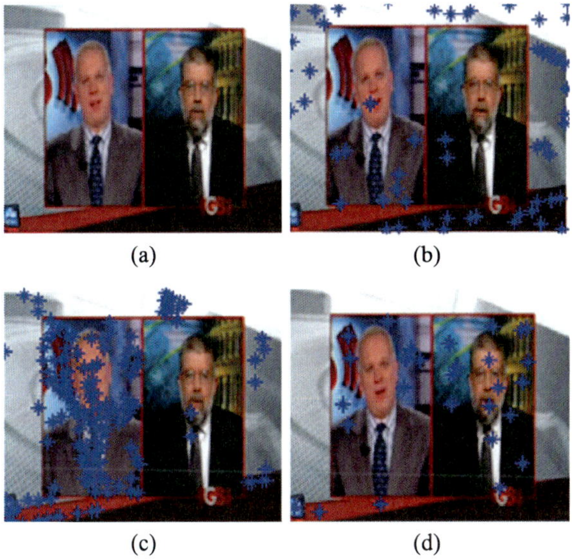

Fig. 9 Interest points detected on real world scenario (**a**), using (**b**) 3D Harris corner detector, (**c**) periodic detector, and (**d**) Itti's approach

that have been used so far in each of the categories of methods for saliency detection, and are representative of the underlying concepts behind each of the different perspectives discussed earlier in this section.

3 Learning Attention-Based Saliency: Conceptual Framework

The conceptual framework of the proposed approach is discussed in this section. The approach to learn top–down saliency is first presented, and subsequently, the probabilistic graphical model used to integrate the saliencies is described. It is to be noted that in the presented approach to integrate top–down and bottom–up saliencies, the bottom–up saliency map is obtained directly using one of the existing computational models, such as [13]. The reason for this, as mentioned earlier, is that bottom–up saliency has been well studied. Hence, this section is directed towards predicting the context-based saliency map, and devising a methodology to integrate this with a bottom–up saliency map that are derived from existing models.

The computational model for the proposed prediction framework to detect context-based saliency is also motivated by the biologically inspired, widely accepted Itti's saliency model [13] embedded into a machine learning framework based on human eye movements. The focal concept of obtaining saliency maps [20] from low-level feature maps is used. However, instead of thresholding generic filter responses of such low-level features, a learning framework is introduced that can capture the image characteristics of salient regions based on the labeled data that is obtained by recording the eye gaze of users as they are naturally watching videos for a specific application. The implementation details of the framework are presented as part of the experiments and results in Sect. 4.

3.1 Learning Context-Specific Saliency

A framework to learn the features corresponding to visual attention based on eye-gaze information for any given application is shown in Fig. 10. Firstly, eye movements of a number of users are recorded as they are naturally watching videos belonging to a single given application (the experimental protocols are described in the next section). As mentioned earlier, fixations correspond to regions that are of visual salience. These regions represent the positive samples for training the classifiers. Negative samples are then generated by randomly selecting regions in the videos from a uniform distribution. The samples are rejected if they happen to be in the neighborhood of a positive sample. The rationale behind this method of generating negative samples is that such regions do not drive the visual attention of any of the users. A sample frame with pixels corresponding to positive and negative labels is illustrated in Fig. 11. The positive samples are shown using green markers, and the negative samples are indicated by red markers.

In the next stage, descriptors for various features corresponding to these samples are extracted. Experiments demonstrated that motion, color, and orientations are good descriptors of regions. Individual classifiers are then trained using these feature descriptors. Subsequently, another classifier is learnt to determine the weights of the

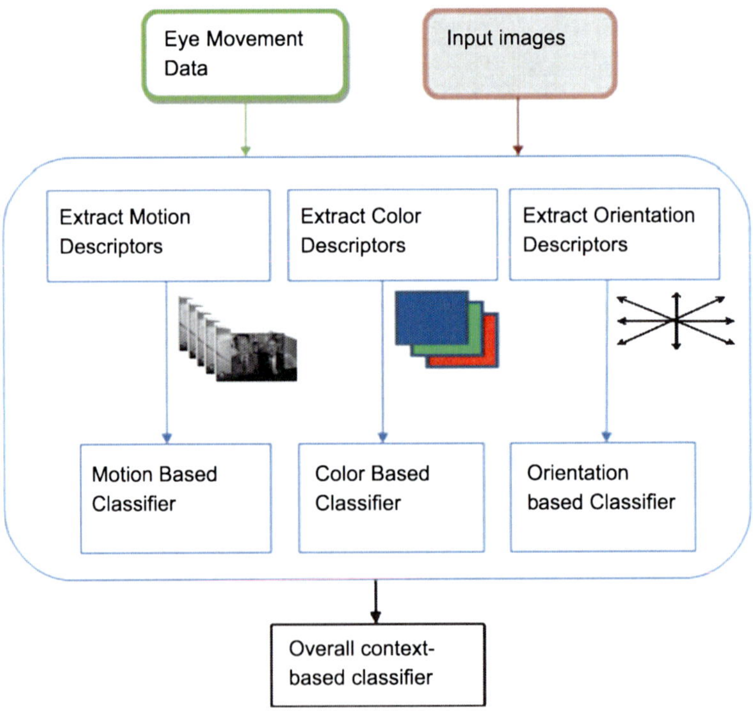

Fig. 10 Framework to learn classifiers for individual features based on human-eye movements

Fig. 11 A frame with positive and negative training samples. The positive samples are shown using *green markers*, and the negative samples are indicated by *red markers* (best viewed *in color*)

individual features in order to combine them into a single saliency map, which is interpreted as a 2-D probability distribution.

3.2 Predicting Context-Specific Saliency

The framework to predict context-specific saliency is shown in Fig. 12. For an incoming image frame, the various feature descriptors are extracted for each pixel location. The individual classifiers are used to predict the feature maps that indicate saliency based on their corresponding features. The outputs from these individual feature maps are used as inputs for the final classifier that results in the final saliency map. The saliency map can be interpreted as a weighted sum of individual feature maps, with the weights as determined by this classifier.

4 Experiments and Results

The proposed framework has been validated on real world news videos. So far, most existing approaches that detect saliency have been tested on artificially created datasets having stationary backgrounds. On the other hand, the news videos that are used in this experiment were downloaded from popular news broadcasting channels, and have significant motion in them apart from those of the news readers. Some of the distractions include flashing logos of the news channels, camera movements, traffic, and other moving patterns of the news channel in the background. The reason that such videos are used to demonstrate the performance of the algorithm is that they have regions corresponding to the news readers, which are salient for the given context, as well as other regions having distinct contrasting features.

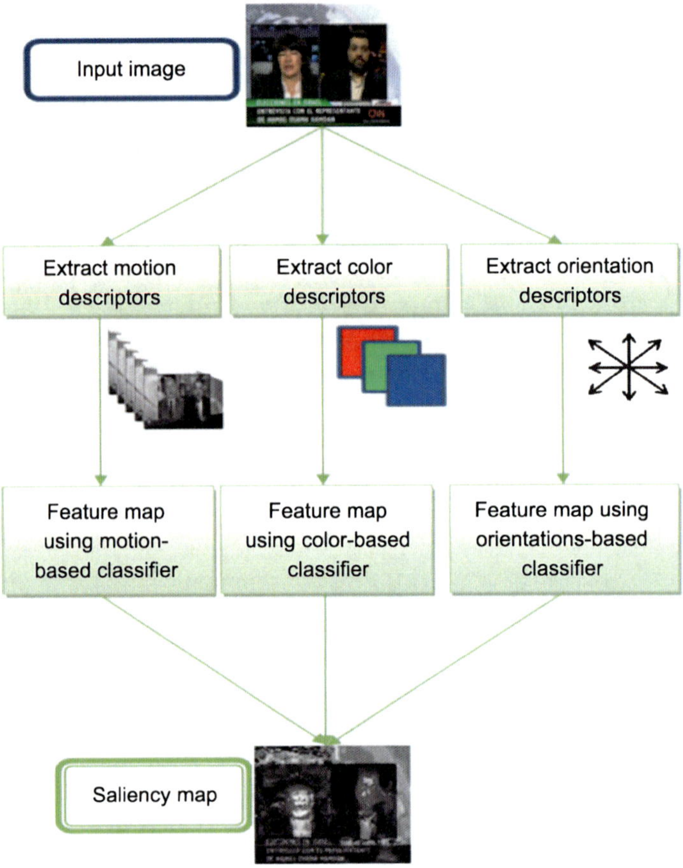

Fig. 12 Framework to obtain saliency map from the learnt classifiers

4.1 Experimental Setup

The Tobii 1750 eye tracker was used to record eye movements. This device, as shown in Fig. 13, is integrated into a 21 inch monitor. It tracks the eye gaze of the viewers while using the monitor with a sampling frequency of 50 Hz. The device has an accuracy of 0.5 degrees. The eye tracking procedure involves an initial calibration for each person. Once calibrated, the eye tracker is capable of compensating for the user's head movements, thereby allowing a natural uninterrupted viewing procedure.

Eleven volunteers participated in this experiment. Each of these volunteers were unaware of the experiment and were instructed to watch the news videos naturally. The users were asked to sit at a normal viewing distance from the monitor while watching the videos. The eye gaze information of these volunteers were recorded for thirty six different news video snippets. In between each of the videos, a blank screen was displayed for four seconds so that the initial eye movements during a

Fig. 13 Tobii 1750 eye tracker

video is not biased by the previous video. The positive and negative samples were obtained as discussed in the previous chapter, and used to train the learning framework.

4.2 Implementation

The videos were separated into a training and test dataset. The eye-tracking data from the first 28 videos was used to train the model. The bottom–up framework is not used in the training phase as all the parameters are independent of the context of the video. Feature vectors were extracted at the locations corresponding to the positive and negative training samples. For a sample p located at (x_p, y_p) in image I_p, the following features are constructed.

- *Color intensity histograms:* Three separate histograms of the intensities corresponding to red, green and blue are concatenated to obtain the color intensity histogram. Ten equally spaces bins were used for each of the histograms. Thus, the length of the concatenated color descriptor was 30. Thus, the feature vector, F_c, for color can be represented as:

$$F_c(p) = \left[H_r\left\{I_p(x_p, y_p)\right\}H_g\left\{I_p(x_p, y_p)\right\}H_b\left\{I_p(x_p, y_p)\right\}\right],$$

  where $H\{.\}$ is the intensity histogram for an individual color component at a given location in an image.
- *Gabor orientations:* To capture the orientation information of the sample, the 2-D Gabor responses at three scales and six orientations are taken. Thus, the length of the feature vector for orientation is 18. The Gabor feature vector, F_g is given by:

$$F_g(p) = G\left\{I_p(x_p, y_p), s, o\right\},$$

  where $G\{.\}$ represents the 2-D Gabor filter response. s and o represent varying scales and orientations, respectively.
- *Motion descriptor:* This is similar to that suggested in [19], and consists of the pixel intensities in the neighboring frames corresponding to the same spatial coordinates. Such a vector characterizes the temporal features well, as illustrated in their results. Pixels from the previous nine frames as well as the future nine frames were used (this choice was made based on preliminary experiments). Thus, the length of the temporal descriptor was 19. The temporal feature vector at frame n, F_t is given by:

$$F_t(p) = I_p(x_p, y_p, n-9:n+9).$$

Once the feature vectors, F, for all the samples are obtained, the saliency model used is captured using a feedforward neural network:

$$O_s = b_{s0} + \sum_{i=1}^{k_s} \alpha_{si} \tanh(F \cdot W_{si} + b_{si}). \tag{1}$$

Here, W_s and α are vectors representing weights of the first and second layer, respectively. b_{s_i} are their corresponding bias parameters. The final saliency, S_s is given by the logistic sigmoidal activation function applied to the output of the second layer of hidden nodes, O_s

$$S_s = 1/\big(1 + \exp(-O_s)\big). \tag{2}$$

The logistic function is known to produce good results for binary classification, and is hence used. The motivating factor in using such a function also arises from the fact that its inverse, O_s gives a probabilistic interpretation of the saliency. It is known as the *logit* function and represents the log of the ratios of the probabilities, $\ln(P(C = 1|F)/P(C = 0|F))$. Here $C = 1$ and $C = 0$ represent the positive and negative classes, respectively. On the other hand, most vision-based saliency functions have arbitrary ranges, and this problem makes their interpretation a difficult challenge.

Once the saliency with respect to each of these low-level features is obtained, we adopt a neural network ensemble approach to learn the final saliency, similar to the feature aggregation theory [13]. The final saliency function is obtained using a second neural network layer which computes the weighted sum of the outputs of the individual feature-specific neural networks. A single layer perceptron is trained to determine the weights of these individual features. This approach learns the relevance of each of the features for the given application of videos.

All of the neural networks were implemented in MATLAB using the Netlab toolbox [26]. Two layered feed-forward neural networks were used with the 'logistic' option to describe the output activation function in the neural network models. The weights were learnt using Scaled Conjugate Gradient (SCG) optimization. Fifteen hidden nodes were used for the neural network to train the color. Ten hidden nodes were used for the other two descriptors. The number of hidden nodes were empirically selected to maximize the performance of the classifiers. In each of the above implementations, the eye gaze of the users was used to label pixels as salient or otherwise. This corresponded to a label of '+1' or '−1', respectively. The feature vectors corresponding to the pixels were thus labeled in the training phase, to thereby provide for a supervised machine learning methodology to predict saliency in new videos. Hence, the input to the neural network is a pixel from a video frame, and the target function is the saliency value associated with the input pixel.

4.3 Results

The results for the experiments are presented from three perspectives. In the first set of results, the performance of the individual feature-specific neural networks is presented in terms of the eye-gaze prediction accuracy. In the second part, the performance of the learning framework is visually illustrated for frames taken from various videos and compared with other saliency detectors.

4.3.1 Eye Movement Prediction

As mentioned earlier, the eye gaze data for eleven participants was recorded for a set of 36 news videos. The samples from the first 28 videos were used to train the models. The positive and negative samples from the remaining eight videos were used to test the performance of the classifiers. Classification results for the individual neural networks that were discussed in the previous section are shown in Table 1. As mentioned earlier, a pixel that is classified as salient is said to have a label '+1' (or positive) and a pixel that is not salient bears the label '−1' (or negative). The performance of the motion descriptor also indicates the performance of the approach of Kienzle et al. [19]. The other methods mentioned in Sect. 2 were vision-based and did not use a supervised learning methodology. Hence, they could not be used for comparison in this study.

An important point to note is that the negative samples are classified much better in the final model. The classifier using Kienzle's approach is only able to detect true positives (TPR) with an accuracy comparative to the presented approach. The True Negative Rate (TNR), however, is low since it classifies all regions having motion as salient. The presented approach, on the other hand is shown to be able to distinguish between regions that are salient from those simply having motion or texture. This is clearly indicated by a high TPR as well as TNR. Another point worth mentioning is that the dominance of individual features is dependent on the class of videos. It can be inferred from the table that color is the dominant feature in news videos. This, however, does not indicate that the same feature may be most relevant in other classes of videos. The choice of these low-level features ensures that such a framework can generalize to other application contexts too.

Performances were also experimented with a concatenated vector, in which the individual feature descriptors were concatenated into a single vector, and a neural network was trained to predict the saliency outcome. However, the performance of such an approach was not comparable to that of the combined classifier and hence, is not reported in this work.

Table 1 Prediction accuracy of different neural networks on the samples from test videos

Neural network	True positive rate	True negative rate
Orientation descriptor	79.8%	63.8%
Motion descriptor	80.2%	60.6%
Color histogram	82.4%	77.3%
Final model	80.9%	83.1%

4.3.2 Context Specific Saliency Detection

Figures 14 and 15 illustrate the performance of the algorithm using two different examples. The images to the top left are the original frames. These frames have different regions with motion. The fixations of the users, as recorded by the eye tracker are shown in the images to the top right, that is, Figs. 14(b) and 15(b). Figures 14(c) and 15(c) show the points detected using the periodic detector. In order to better illustrate salient regions from the saliency maps, 'interest points' based on the saliency maps were plotted using a simple thresholding. Since saliency maps are probability distributions, thresholding merely allows us to select pixels that have a high probability of being salient. Figures 14(d) and 15(d) show the pixels predicted by Kienzle et al.'s algorithm [17]. It is evident that both these detectors, being motion-based, respond to all regions in the video frame that have movements.

Fig. 14 (**a**) Is the original frame from a video with moving patterns in the background, and (**b**) shows the fixations of users on the frame. Interest points detected using (**c**) periodic detector, (**d**) Kienzle et al.'s detector, (**e**) 3D Harris corner detector, (**f**) Itti's bottom up model, and (**g**) presented learned detector

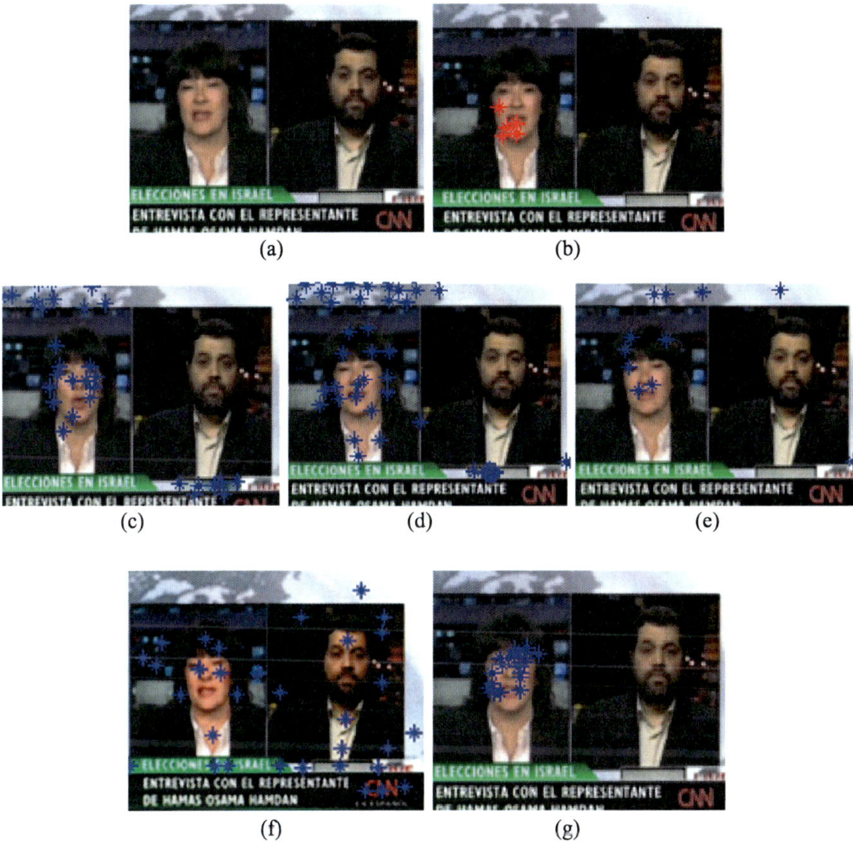

Fig. 15 (a) Is the original frame, and (b) shows the fixations of users on the frame. Interest points detected using (c) periodic detector, (d) Kienzle et al.'s detector, (e) 3D Harris corner detector, (f) Itti's bottom up model, and (g) presented learned detector

Figures 14(e) and 15(e) illustrate the results obtained using 3-D Harris corner detector. These points, as expected are sparsely distributed in regions having spatial and temporal corners. The results obtained using the presented approach in which the saliency map is obtained from individual feature maps are presented in Figs. 14(g) and 15(g). Clearly, this framework is able to distinguish the salient regions from the irrelevant ones having motion. As can be seen, the interest points detected by the presented approach bear a high correlation to the fixations of the users in both the examples. The presented detector is able to detect regions that are of visual interest to users, unlike the other existing approaches.

The frame shown in Fig. 16 is taken from a video that was captured using a moving camera. As expected, the pure motion-based approaches detect motion in all parts of the scene, and in turn predict interest points from all regions. Here, it can be seen that the learned detector is robust to the camera movements. It is evident from the examples that the proposed learning framework ensures interest points

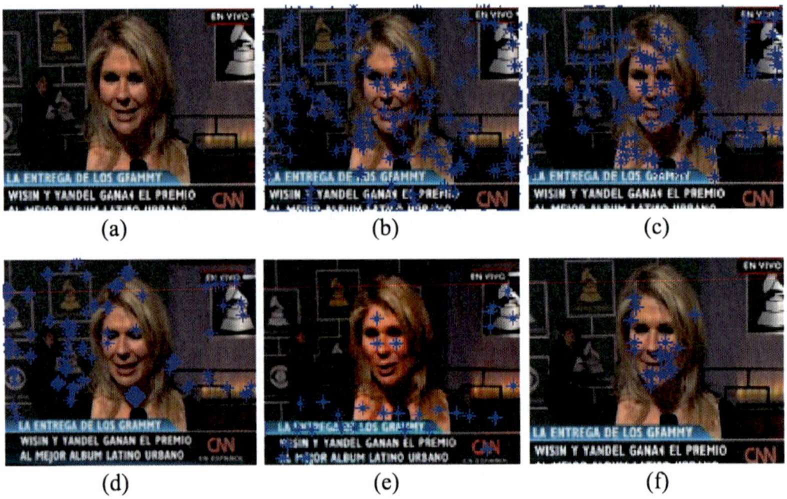

Fig. 16 (**a**) Is the original frame from a video taken using a moving hand-held video camera. Interest points detected using (**b**) periodic detector, (**c**) Kienzle et al.'s detector, (**d**) 3-D Harris corner detector and (**e**) Itti's bottom–up model, and (**f**) presented learned detector

have relevant spatial content along with temporal content. In addition, as mentioned briefly earlier, experiments were conducted with the proposed framework to design a single neural network that was built using a concatenated feature vector (using all the considered features). However, it was found that this approach did not perform as well as using individual neural networks for each feature. It is believed that this may have been due to the high dimensionality of the resulting concatenated feature vector. Another point to be noted is that the saliency model detects regions based on the regions that are commonly viewed by all the users of the experiment. For example, in this work, it was found that all the volunteers in this experiment viewed the faces of the newscasters during the video. However, there might be a case where some users view other regions (possibly the text on the screen). In such a scenario, the learning framework will learn features based on the most commonly viewed regions.

4.4 Discussion

In the case of news videos, the learnt model detects regions corresponding to talking faces as being salient since that is what most viewers look out for while watching news. While it can be argued that a face detector combined with a motion-based detector could achieve similar results, it should be noted that the overall framework presented in the work can be applied to any scenario. The training process remains the same in all scenarios. The eye movements are recorded as users analyze videos, and classifiers are trained so that the weights for each of the features are optimized for that class of videos.

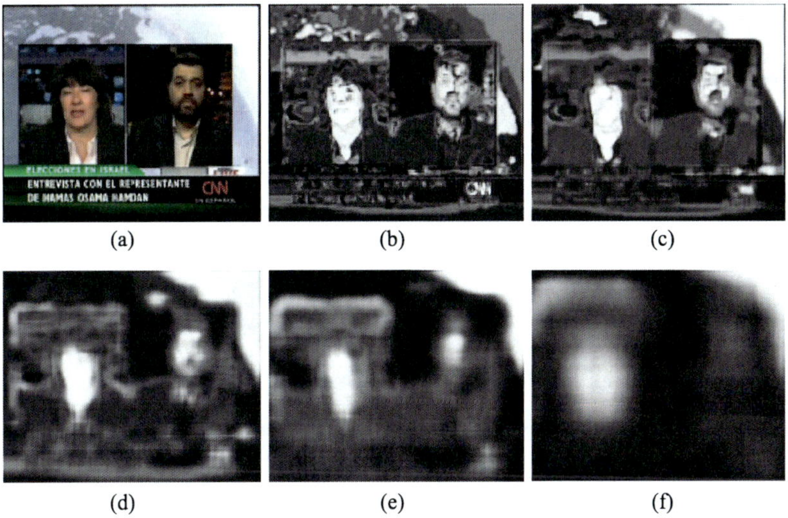

Fig. 17 Variation of color-specific saliency maps with increase in spatial window sizes *from left to right*. The window sizes include 3×3, 5×5, 11×11, 17×17 and 33×33

Some parameters that need to be addressed in this approach include the spatial window size and the length of the temporal descriptor. Figure 17 indicate the variations in color-based feature maps with increase in the size of spatial windows to extract the histograms. The window sizes used in the illustrations include 3×3, 5×5, 11×11, 17×17 and 33×33. It can be observed that the details are not captured while using larger windows. This is intuitive since using a larger number of pixel locations to calculate the histogram has a smoothening effect. Hence, aberrations in the histograms caused by individual pixel values are suppressed. Larger window sizes result in highlighting whole blobs of regions in image frames. On the other hand, histograms using smaller windows are extremely sensitive to individual pixel intensities. While this results in a more accurate feature map, it is also more sensitive to noise. In the presented work, a mid-sized window of size 11×11 pixels was used so as to compromise between the two extremes.

The nature of the temporal descriptor varies based on the application. In the current implementation of the system, the temporal descriptor used is noncausal. The pixels in the temporal neighborhood of the pixel under consideration in previous as well as future frames are used. However, in a real-time implementation, this is not feasible. Pixels from only the previous and current frames can be used to construct the descriptor. Figure 18 shows the variations in temporal-based feature maps. In the presented work, pixels from the previous nine frames as well as from the future nine frames are used in the temporal descriptor. This is illustrated in Fig. 18(e).

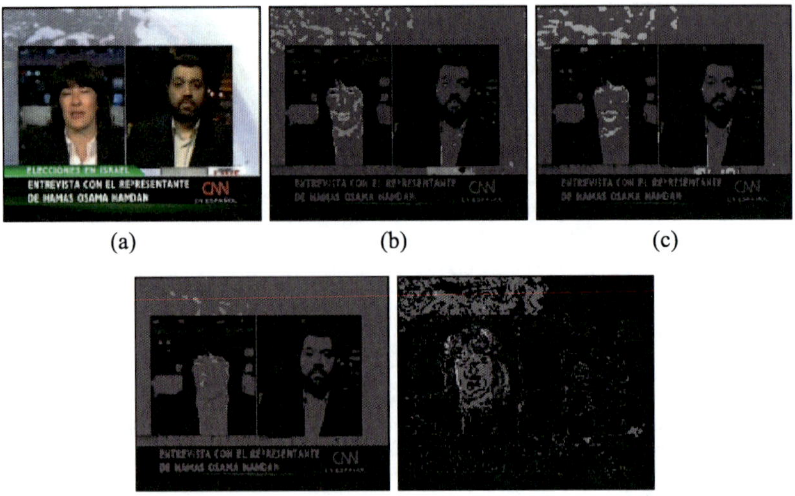

(a) (b) (c)

(d) (e)

Fig. 18 (**a**) Variation of temporal-specific saliency maps with varying lengths of temporal descriptor. (**b**) Uses pixels from previous five frames, (**c**) uses pixels from previous nine frames, (**d**) uses pixels from previous five frames and the future five frames, and (**e**) uses pixels from previous nine frames and the nine future frames

4.5 Integrating Bottom–Up Approaches: A Possible Extension

In high risk scenarios, it may not be sufficient to only highlight regions based on a classifier trained on what is viewed by the users. In addition to regions that are highlighted by the classifier, it may also be appropriate to indicate regions that are of high bottom–up salience. The need for such an approach arises since the classifier is trained only on regions viewed by a specific set of users. Unexpected events, indicated by the bottom–up approaches may not be detected since such regions are not common, and hence may not be learnt by the context-based saliency model alone. In order to do so, the presented framework can be extended to integrate a bottom–up approach along with the learnt model. One way to do so is by modeling the final saliency as a joint probability distribution.

4.5.1 Integration Using Probabilistic Framework

In this approach, as a preliminary approach to integrate top–down and bottom–up methods, saliency is interpreted as a 2-D random variable and the saliency map as a 2-D probability distribution function (PDF) that describes the saliency of each pixel location in an image frame. In other words, the saliency map is a mapping function that assigns a numerical value that corresponds to the probability of a particular pixel location being salient.

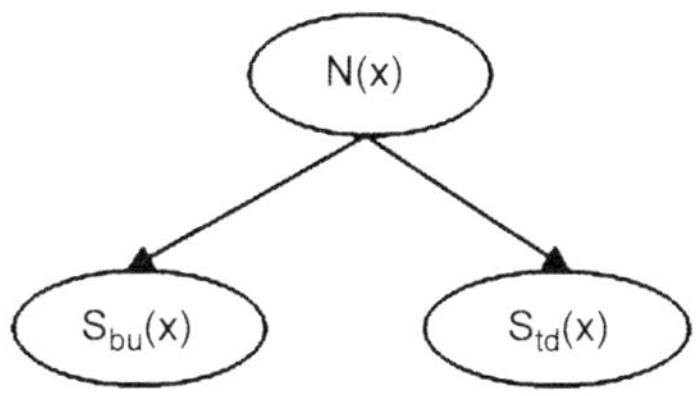

Fig. 19 Graphical model representing top–down and bottom–up saliencies as random variables

The top–down saliency of a pixel indicates its visual saliency obtained based on the context or the application of the video. The bottom–up saliency of a pixel indicates its visual saliency obtained using existing computational models such as [13]. The saliency maps obtained using the two approaches are converted to probability density functions to ensure that the sum is one, and that the measurements at each of the locations is nonnegative. This can be construed as a 2-D PDF, while still maintaining the shape of the distribution as output by the classifiers. (While we have defined this PDF in a straightforward manner in this work as proof-of-concept, we intend to study the generative relationship between the saliency of the pixel, surrounding pixels and image features in future work.)

The saliencies are represented as the nodes of a graphical model as shown in Fig. 19. In this figure, $N(x)$ represents the neighborhood of a pixel x. S_{bu} and S_{td} represent the bottom–up and top–down saliencies, respectively. Each pixel is represented by a feature vector that is obtained based on pixels in its spatial and temporal neighborhoods.

In order to have an attention model that captures both these saliencies, the proposed framework models the final saliency of a pixel, $S(x)$, as a joint distribution of the top–down and bottom–up saliencies.

$$S(x) = P\big(S_{bu}, S_{td}|N(x)\big) = P\big(S_{bu}/S_{td}, N(x)\big) * P\big(S_{td}/N(x)\big). \tag{3}$$

In the above graphical model, when $N(x)$ is observed, S_{bu} and S_{td} are conditionally independent, that is, $S_{bu} \perp S_{td} | N(x)$

$$S(x) = P\big(S_{bu}|N(x)\big) * P\big(S_{td}|N(x)\big). \tag{4}$$

Since $P(S_{bu}/N(x))$ and $P(S_{td}/N(x))$ are independent distributions, the final PDF for the saliency is a product of the individual PDFs corresponding to the bottom–up and top–down approaches. Intuitively as well, this makes sense since regions that are of high salience are those that are salient based on the context or the application of the videos, as well as have distinctive features that catch the attention of the user.

4.5.2 Results of the Integrated Framework

The results at each of the intermediate steps in the prediction process are illustrated in Fig. 20. The input image frame belongs to a news video from a popular news channel. It consists of a news reader speaking in front of a transparent window. There is moving traffic in the background, and a yellow car is seen speeding. The

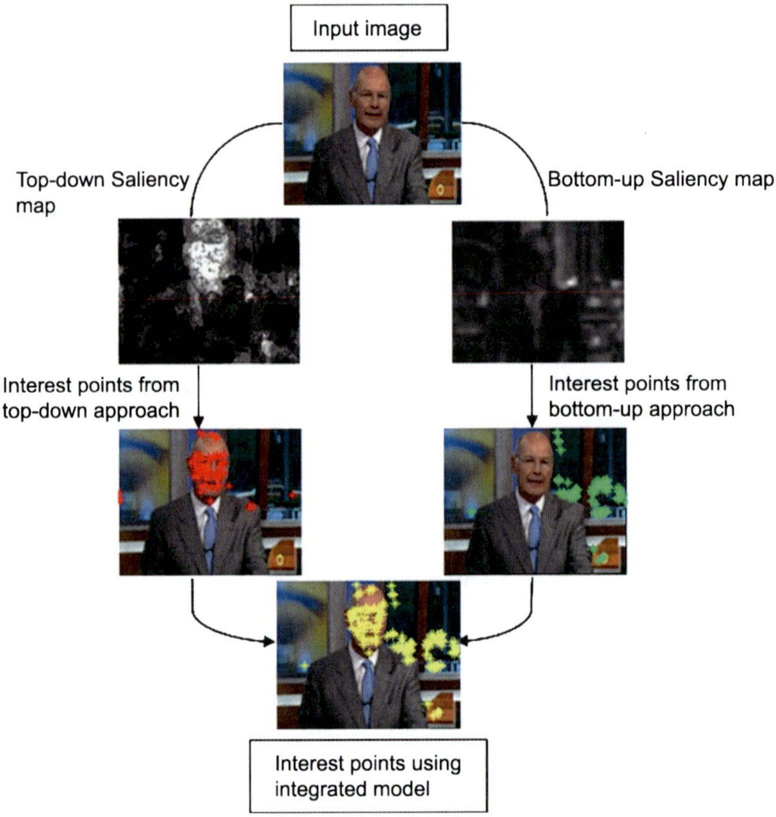

Fig. 20 An illustration of the prediction process integrating top–down and bottom–up frameworks

saliency map constructed using the learnt model highlights the region corresponding to the newsreader's face, since the faces of news readers were commonly looked at by the users in the training phase. On the other hand, the bottom–up approach using Itti's framework in [13] picks up regions having sharp movements, which consists of the yellow car speeding in the background. The final saliency map highlights the news reader's face as well as the speeding car in the background. In this case, it can be seen that the salient regions indicated by the two approaches are completely non-intersecting. This example was chosen to suitably illustrate the contrasting nature of the two approaches. This need not necessarily be the case. There could be some overlap between the regions indicated to be of top–down and bottom–up saliencies. This is illustrated in Fig. 21, where the face of the news reader is shown to be of salience by both the approaches.

Figure 22 illustrates a comparison of the performance of the proposed integrated approach with other popular spatiotemporal interest point detection approaches. The periodic detector and the approach of Kienzle et al. described in [19] are purely motion-based, and hence respond to all regions in the video frame that have movements. The 3-D Harris corner detector is known to have a drawback in that the

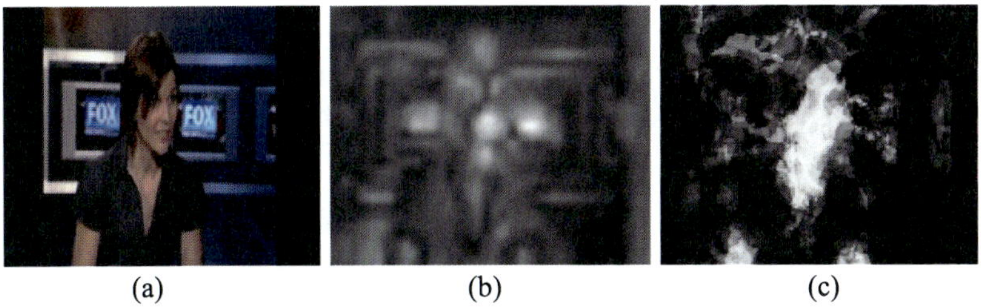

Fig. 21 Regions highlighted by the two saliency maps could have some overlap. (**a**) Is the original image. (**b**) Is saliency map from Itti's bottom–up approach. (**c**) Is the saliency map learnt from human eye movements (top–down or context-specific)

Fig. 22 Interest points detected using (**b**) presented approach, (**c**) Kienzle et al.'s detector, (**d**) periodic detector, (**e**) Itti's approach, and (**f**) 3-D Harris corner detector

selected points are sparse, as is indicated in Fig. 22(f). Results from Itti's approach are illustrated in Fig. 22(e). Itti's model detects the moving car as the most salient region in the frame. The integrated approach detects the face of the news reader as well as the moving car. Bottom–up approaches are preferred over interest point detectors for the integration process. The reason for this is that bottom–up approaches tend to detect regions that are significantly distinct from rest of an image, rather than detecting all regions having corners and motion. As a result, they highlight regions that 'pop out'.

The average fixations across the users on the frame is displayed in Fig. 23. It can be seen that a few of the users have viewed the cars in the background, while others continue to watch the newsreader. Events such as a moving car are not common in news videos. Hence, eye-gaze samples corresponding to regions from these unexpected events could be insignificant compared to those that are commonly viewed

Fig. 23 The averaged fixations of users on one of the frames while watching the news video

by users. As a result of this, a classifier trained on eye-gaze alone may be insufficient to detect all possible regions of interest. Though this may not be relevant in news videos, in high risk scenarios it is important to be able to detect all possible regions of interest. Hence, having a bottom–up approach integrated into the learnt detector is critical.

5 Conclusions and Future Work

Human visual attention modeling has always been of importance in computer vision research. While bottom–up frameworks have been relatively well studied over the years, models for top–down approaches have not been well defined. In this work, a novel approach to integrate both these saliencies is presented. Such an approach would be extremely useful in applications such as video surveillance or medical diagnosis to reduce the information overload. Saliency detection is also of relevance for problems involving recognition, tracking or compression. A seeming limitation of this approach is that the top–down component of the framework will need to be trained individually for every application. However, since eye tracking devices that can be integrated as part of a daily routine are commercially available, this limitation can be overcome by learning as the user goes about his daily activities (a radiologist or surveillance personnel, for example). Thus, the proposed framework can be used with ease in applications without any explicit procedure for data capture, or labeling methodology that may be needed every time the model needs to be learnt or updated.

In summary, this chapter shows that human eye movements can be used to determine saliency in videos. Such an approach is proven to be able to distinguish regions that are of saliency for a given context from those simply having distinctive features. This approach allows the model to be nonparametric since all the parameters are learnt by the classifier in the training phase. The framework also provides an intuitive interpretation of saliency in terms of probability, rather than having mere numbers corresponding to responses to predetermined filters. As a result of this, the thresholds used to predict saliency are also easier and intuitive to determine.

5.1 Possible Applications

The proposed learning process can be generalized to other applications as well. The presented model has been tested on news videos primarily as a proof of concept, since it was found that these news videos represent a typical real world scenario where there is a specific object of interest, along with background clutter and noise. Generic vision-based algorithms are not reliable in such scenarios since the regions that are relevant in a given scenario may not always have features that 'pop out'.

The presented framework is particularly useful in scenarios where professionals need to analyze huge number of videos belonging to a single application, such as surveillance video monitoring or medical video analysis. These scenarios demand that all relevant regions in the videos be analyzed. However, typically the duration of these videos are large, may have significant amount of clutter in the background, or there may be overload of simultaneous information (as in the case of a security personnel who may have to monitor multiple video feeds at the same time). In these cases, some regions that might have been salient could easily be overlooked. Introducing a model such as the presented learning framework in such scenarios will be of significant aid to the users in analyzing relevant regions. The classifier is trained on the regions commonly viewed by users for a specific application, and are is thereby capable of predicting regions that are of visual salience in similar videos.

All supervised learning algorithms require the training data to be labeled. Applications such as medical video analysis require extremely skilled professionals. As a result of this, obtaining labeled samples for training is a non-trivial and an expensive task in these scenarios. In these conditions, having an eye-tracker embedded into regular computer thereby enables an automated and inexpensive labeling process. State-of-the-art eye trackers can easily be integrated into regular desktop monitors to record the eye-gaze of the users as they are watching videos. This technology allows for unobtrusive and uninterrupted viewing, and hence allows the users to carry on their regular tasks while their eye movements get recorded simultaneously. In doing so, they implicitly label the regions that are potentially salient without having to go out of their way. This eliminates the need for an explicit labeling process.

5.2 Future Work

The work done in this chapter demonstrates the use of human eye movements in determining saliency for news videos. The results illustrate the performance of the presented framework in comparison with existing models. In future work, the learning process will also be validated with other applications. Available video datasets for computer vision have all been artificially created so as to have minimal background clutter. Hence, obtaining realistic datasets has been a challenge, and as mentioned earlier, the videos used to validate the framework were downloaded from popular news websites. Moreover, there has been no suitable eye tracking data that is readily available. These factors have proved to be limitations in validating the presented model with multiple applications.

Currently, the image patches that are used to extract the features are of constant size. A possible extension of work could involve employing a multi-scale approach in selecting the most optimal scale to improve the performance of the detector. One way to go about doing this would be to adopt a methodology similar to the Kadir and Brady detector [15], where entropy measures are used to optimize the spatial scale for detecting salient regions.

In this work, integrating bottom–up and top–down approaches has been presented as a joint distribution. It can also be modeled as a function of risk to vary its applicability in different kinds of applications. For example, in such an approach, the final saliency would take the following form:

$$P(S_{\text{td}}, S_{\text{bu}}|risk) = (1 - risk) \times \max\big(P(S_{\text{td}}), P(S_{\text{bu}})\big) + risk * \big(P(S_{\text{td}}) \times P(S_{\text{bu}})\big). \tag{5}$$

In a high-risk application, the final saliency is treated as an 'OR' function, and the greater of the two saliencies is chosen as the final saliency. On the other hand, when the risk is low, it is treated as an 'AND' function. In this case, the saliency is computed as a joint distribution, as presented in the current approach.

References

1. Dollar, P., Rabaud, V., Cottrell, G., Belongie, S.: Behavior recognition via sparse spatiotemporal features. In: 2nd Joint IEEE International Workshop on Visual Surveillance and Performance Evaluation of Tracking and Surveillance, pp. 65–72 (2005)
2. Duchowski, A.: A breadth-first survey of eye-tracking applications. Behav. Res. Methods Instrum. Comput. **34**(4), 455–470 (2002)
3. Findlay, J.M., Walker, R., Kentridge, R.W.: Eye Movement Research. Elsevier, Amsterdam (1995)
4. Gao, D., Vasconcelos, N.: Discriminant saliency for visual recognition from cluttered scenes. Adv. Neural Inf. Process. Syst. **17**, 481–488 (2005)
5. Gao, D., Vasconcelos, N.: Bottom–up saliency is a discriminant process. In: IEEE International Conference on Computer Vision (2007)
6. Gao, D., Han, S., Vasconcelos, N.: Discriminant saliency, the detection of suspicious coincidences, and applications to visual recognition. IEEE Trans. Pattern Anal. Mach. Intell. **31**(6), 989–1005 (2009)
7. Granka, L., Joachims, T., Gay, G.: Eye-tracking analysis of user behavior in WWW search. In: Proc. of the 27th Annual International ACM SIGIR Conference on Research and Development in Information Retrieval, pp. 478–479. ACM, New York (2004)
8. Guo, C., Ma, Q., Zhang, L.: Spatiotemporal saliency detection using phase spectrum of quaternion fourier transform. In: Proc. of IEEE Computer Society Conference on Computer Vision and Pattern Recognition (2008)
9. Harris, C., Stephens, M.: A combined corner and edge detector. In: Alvey Vision Conference, vol. 15, p. 50 (1988)
10. Hou, X., Zhang, L.: Saliency detection: a spectral residual approach. In: IEEE Conference on Computer Vision and Pattern Recognition, CVPR07. IEEE Computer Society, pp. 1–8. IEEE Comput. Soc., Los Alamitos (2007)
11. Itti, L.: Models of bottom–up attention and saliency. In: Neurobiology of Attention, vol. 582. Elsevier, Amsterdam (2005)
12. Itti, L., Koch, C.: Computational modelling of visual attention. Nat. Rev. Neurosci. **2**(3), 194–203 (2001)
13. Itti, L., Koch, C., Niebur, E.: A model of saliency-based visual attention for rapid scene analysis. IEEE Trans. Pattern Anal. Mach. Intell. **20**(11), 1254–1259 (1998)

14. Judd, T., Ehinger, K., Durand, F., Torralba, A.: Learning to predict where humans look. In: IEEE International Conference on Computer Vision, ICCV (2009)
15. Kadir, T., Brady, M.: Scale saliency: a novel approach to salient feature and scale selection. In: International Conference on Visual Information Engineering, VIE 2003, pp. 25–28 (2003)
16. Ke, Y., Sukthankar, R., Hebert, M.: Efficient visual event detection using volumetric features. In: 10th IEEE International Conference on Computer Vision, ICCV 2005, vol. 1 (2005)
17. Kienzle, W., Wichmann, F., Scholkopf, B., Franz, M.: Learning an interest operator from human eye movements. In: IEEE Conference on Computer Vision and Pattern Recognition, vol. 17, p. 22 (2006)
18. Kienzle, W., Wichmann, F., Scholkopf, B., Franz, M.: A nonparametric approach to bottom–up visual saliency. Adv. Neural Inf. Process. Syst. **19**, 689 (2007)
19. Kienzle, W., Scholkopf, B., Wichmann, F., Franz, M.: How to find interesting locations in video: a spatiotemporal interest point detector learned from human eye movements. Lect. Notes Comput. Sci. **4713**, 405 (2007)
20. Koch, C., Ullman, S.: Shifts in selective visual attention: towards the underlying neural circuitry. Hum. Neurobiol. **4**(4), 219–227 (1985)
21. Laptev, I.: On space-time interest points. Int. J. Comput. Vis. **64**(2), 107–123 (2005)
22. Laptev, I., Marszałek, M., Schmid, C., Rozenfeld, B.: Learning realistic human actions from movies. In: IEEE Conference on Computer Vision and Pattern Recognition (2008)
23. Liu, T., Sun, J., Zheng, N., Tang, X., Shum, H.: Learning to detect a salient object. In: Proc. IEEE Conference on Computer Vision and Pattern Recognition (2007)
24. Lowe, D.: Distinctive image features from scale-invariant keypoints. Int. J. Comput. Vis. **60**(2), 91–110 (2004)
25. Mohanna, F., Mokhtarian, F.: Performance evaluation of corner detection algorithms under similarity and affine transforms. In: British Machine Vision Conference, BMVC (2001)
26. Nabney, I.T.: Netlab, corrected edn. Springer, Berlin (2001)
27. Navalpakkam, V., Itti, L.: An integrated model of top–down and bottom–up attention for optimal object detection. In: Proc. IEEE Conference on Computer Vision and Pattern Recognition, CVPR, pp. 1–7 (2006)
28. Navalpakkam, V., Itti, L.: Optimal cue selection strategy. Adv. Neural Inf. Process. Syst. **18**, 987 (2006)
29. Niebles, J., Wang, H., Fei-Fei, L.: Unsupervised learning of human action categories using spatial-temporal words. Int. J. Comput. Vis. **79**(3), 299–318 (2008)
30. Oikonomopoulos, A., Patras, I., Pantic, M.: Human action recognition with spatiotemporal salient points. IEEE Trans. Syst. Man Cybern., Part B **36**(3), 710–719 (2006)
31. Oliva, A., Torralba, A., Castelhano, M., Henderson, J.: Top–down control of visual attention in object detection. In: IEEE Proc. of the International Conference on Image Processing, vol. 1, pp. 253–256 (2003)
32. Oyekoya, O., Stentiford, F.: An eye tracking interface for image search. In: ETRA '06: Proc. of the 2006 Symposium on Eye Tracking Research and Applications, New York, NY, USA, pp. 40–40. ACM, New York (2006)
33. Oyekoya, O., Stentiford, F.: Perceptual image retrieval using eye movements. Int. J. Comput. Math. **84**(9), 1379–1391 (2007)
34. Ruderman, D.: The statistics of natural images. Netw. Comput. Neural Syst. **5**(4), 517–548 (1994)
35. Salojarvi, J., Kojo, I., Simola, J., Kaski, S.: Can relevance be inferred from eye movements in information retrieval. In: Proc. of WSOM, vol. 3, pp. 261–266 (2003)
36. Schuldt, C., Laptev, I., Caputo, B.: Recognizing human actions: a local SVM approach. In: Proc. of the 17th International Conference on Pattern Recognition, ICPR 2004, vol. 3 (2004)
37. Scovanner, P., Ali, S., Shah, M.: A 3-dimensional sift descriptor and its application to action recognition. In: Proc. of the 15th International Conference on Multimedia, p. 360. ACM, New York (2007)
38. Srivastava, A., Lee, A., Simoncelli, E., Zhu, S.: On advances in statistical modeling of natural images. J. Math. Imaging Vis. **18**(1), 17–33 (2003)

39. Stentiford, F.: An estimator for visual attention through competitive novelty with application to image compression. In: Picture Coding Symposium, pp. 25–27 (2001)
40. Stentiford, F.: An attention based similarity measure with application to content based information retrieval. In: Storage and Retrieval for Media Databases, pp. 20–24 (2003)
41. Torralba, A.: Modeling global scene factors in attention. J. Opt. Soc. Am. A **20**(7), 1407–1418 (2003)
42. Treisman, A., Gelade, G.: A feature-integration theory of attention. Cogn. Psychol. **12**(1), 97–136 (1980)
43. Viola, P., Jones, M.: Rapid object detection using a boosted cascade of simple. In: Proc. of CVPR2001, vol. 1 (2001)
44. Wang, Z., Li, B.: A two-stage approach to saliency detection in images. In: IEEE International Conference on Acoustics, Speech and Signal Processing, ICASSP, pp. 965–968 (2008)
45. Ziou, D., Tabbone, S.: Edge detection techniques an overview. Int. J. Pattern Recognit. Image Anal. **8**(4), 537–559 (1998)

Video-Based Human Motion Estimation by Part-Whole Gait Manifold Learning

Guoliang Fan and Xin Zhang

Abstract This chapter presents a general gait representation framework for video-based human motion estimation that involves gait modeling at both the whole and part levels. Our goal is to estimate the kinematics of an unknown gait from image sequences taken by a single camera. This approach involves two generative models, called the kinematic gait generative model (KGGM) and the visual gait generative model (VGGM), which represent the kinematics and appearances of a gait by a few latent variables, respectively. Particularly, the concept of gait manifold is proposed to capture the gait variability among different individuals by which KGGM and VGGM can be integrated together for gait estimation, so that a new gait with unknown kinematics can be inferred from gait appearances via KGGM and VGGM. A key issue in generating a gait manifold is the definition of the distance function that reflects the dissimilarity between two individual gaits. Specifically, we investigate and compare three distance functions each of which leads to a specific gait manifold. Moreover, we extend our gait modeling framework from the whole level to the part level by decomposing a gait into two parts, an upper-body gait and a lower-body gait, each of which is associated with a specific gait manifold for part level gait modeling. Also, a two-stage inference algorithm is employed for whole-part gait estimation. The proposed algorithms were trained on the CMU Mocap data and tested on the HumanEva data, and the experiment results show promising results compared with the state-of-the-art algorithms with similar experimental settings.

G. Fan (✉) · X. Zhang
Oklahoma State University, Stillwater, OK 74078, USA
e-mail: guoliang.fan@okstate.edu

X. Zhang
e-mail: xin.zhang@okstate.edu

L. Wang et al. (eds.), *Machine Learning for Vision-Based Motion Analysis*,
Advances in Pattern Recognition,
DOI 10.1007/978-0-85729-057-1_9,

1 Introduction

Video-based human motion estimation has recently received great interest in the computer vision community due to its wide applications. On the one hand, it is a challenging research topic due to the variability and nonlinearity of human motion as well as the uncertainty and ambiguity of visual observations. On the other hand, this topic has been advanced by recent progress in fields of computer vision, artificial intelligence, machine learning and image processing. In this chapter, we are interested in the estimation of human body configurations from image sequences taken by a single camera. Specifically, we focus on the walking motion (i.e., gait) that is very useful for biometrics and many biomechanical modeling applications. Particularly, we define two terms about a gait: *gait kinematics* and *gait appearances*. The former one is represented by a sequence of Euler angles or 3D positions of body joints, and the latter one is a sequence of human silhouettes extracted from an image sequence captured by a single calibrated camera. Our goal is to estimate gait kinematics from gait appearances via explicit gait modeling in both kinematic and visual spaces. This research addresses several fundamental issues pertaining to the emerging markerless motion capture technology [40].

We have three important ideas in this research. The first one is that we could span a nonlinear low-dimensional space to represent a variety of human gait motions (in terms of kinematics or appearances) by learning from a set of representative (training) gaits. The second one is that a new gait with unknown kinematics or appearances can be *synthesized* (or interpolated) from the training gaits in this space. Particularly, we call this space the *gait manifold*. It is worth noting that the term of gait manifold used here has been *upgraded* from its original meaning in some previous research, for example, [15, 34], where the gait manifold is used to capture the low-dimensional intrinsic structure *among different poses* (either by their kinematics or appearances) *from a single gait*. Here, the gait manifold is used to represent the kinematic or visual variability *among different individuals*. The third proposed idea is that a gait can be modeled by two parts according to the hip joint, the upper-body (above the hip) and lower-body (below the hip) gaits, each of which is associated with a specific gait manifold for part-level gait modeling.

Correspondingly, our research involves three technical components. First, we develop a general gait representation framework that involves dual gait generative models, i.e., the *kinematic gait generative model* (KGGM) and *visual gait generative model* (VGGM). KGGM represents the kinematics of a gait by two variables, i.e., *gait* and *pose*, and VGGM characterizes the appearances of a gait by four variables, i.e., *view*, *shape*, *gait*, and *pose*. KGGM and VGGM are *temporally synchronized* by sharing the same pose variable. Second, we propose a new manifold learning technique to learn a low-dimensional gait manifold to capture the gait variability at either the whole-level or part-level among different individuals, by which KGGM and VGGM can be *semantically integrated*. This allows us to infer the kinematics of a new gait from its appearances. Third, considering the segmental variability of the gait variable in a long sequence, we develop an effective particle filtering-based inference algorithm that is embedded with a segmental jump diffusion Markov chain Monte Carlo (SJD-MCMC) scheme to support dynamic gait estimation. Also, we further develop a two-stage inference process for part-whole gait estimation. Two human motion databases were involved in our experiment, i.e., the CMU Mocap [11] and the Brown's HumanEva [56], which are used for algorithm training and testing, respectively. Both databases have been widely used by computer vision researchers who are interested in human motion analysis and pose estimation.

The remainder of this chapter is organized as follows. After reviewing some related works in Sect. 2, we present an overview of our research in Sect. 3. In Sect. 4, we discuss the learning of KGGM and VGGM that are the cornerstones of this research. The concept of gait manifold is introduced in Sect. 5. Section 6 presents two particle filtering-based inference algorithm for dynamic gait estimation at the whole and part levels. The experimental results are shown in Sect. 7. The conclusion and future research are given in Sect. 8.

2 Related Works

There have been a plethora of works on video-based human motion analysis. Previous surveys [36, 50] provide comprehensive reviews on this topic. Due to the nature of our research, we will present a brief review from the machine learning perspective where we mainly discuss *discriminative* and *generative* approaches, as shown by Fig. 1. Generally speaking, discriminative approaches are more efficient and can directly learn the image-to-pose (2D-to-3D) mapping relationship that is usually a one-to-multiple mapping, while generative ones involve explicit human motion/shape modeling and are more flexible to incorporate prior knowledge into the inference process.

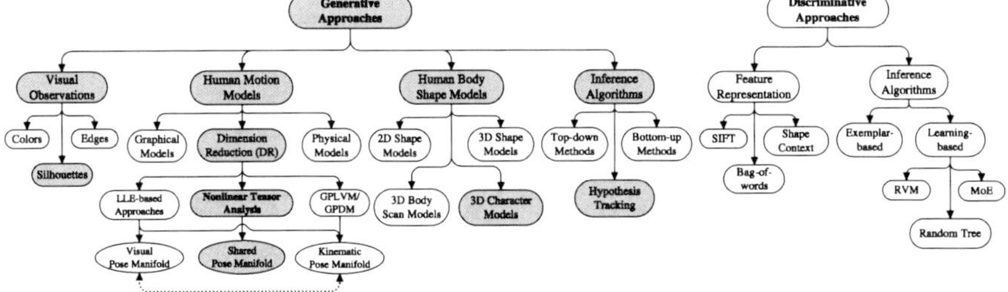

Fig. 1 The taxonomy of machine learning-based human motion analysis where shaded blocks indicate choices in our research. (Adapted from X. Zhang and G. Fan, Dual Generative Gait Models for Human Motion Estimation from a Single Camera, IEEE Trans. Systems, Man and Cybernetics, Part B: Cybernetics, Vol. 40, No. 4, 1034–1049, Aug. 2010. @ 2010 IEEE)

2.1 Discriminative Approaches

Discriminative approaches tend to directly recover the body configuration from visual inputs by learning an input–output mapping from training data. The major challenge is to handle the occlusion, ambiguity and uncertainty of visual data due to the variability of pose, view and shape. There are two major issues, the feature representation and the inference algorithms.

2.1.1 Feature Representation

The feature representation is required to be distinctive and informative enough. Previous methods have successfully demonstrated few powerful image descriptors, like shape contexts [2, 59], the histogram of gradient (HOG) [52], the scale invariant feature transform (SIFT) [35] and its variants like block of SIFT features and histogram of SIFT features [61], the bag-of-words [45] etc. These features can be obtained from block-wise statistical analysis from various visual observations, like colors, edges and silhouettes. Further, a set of hierarchical coarse-to-fine image descriptors are employed to characterize the variability of scale, depth and deformation. For example, the multilevel spatial blocks (MSB) method [26] is composed of various block sizes of SIFT descriptors. Above descriptors can be effective at estimating 3D human poses in discriminative methods but they may not be applicable to the generative approaches due to the difficulty of feature reconstruction.

2.1.2 Inference Algorithms

The inference algorithm can be loosely classified as *exemplar-based* or *learning-based* approaches [50]. The exemplar-based methods [22, 39, 47, 49] have a prelabeled database including both image descriptors and 3D poses. The goal of inference is to search for the most visually similar example or n closest examples for pose interpolation. These methods usually require a large set of training data to densely sample the pose space. The learning-based approaches aim at learning a model to bridge the gap between image descriptors and poses. Modeling image descriptors is difficult because it often produces multi-modal conditional distributions or one-to-multiple mappings. The regression model is an effective discriminate approach, especially the *nonlinear* regression models like neural network [53], the relevance vector machine (RVM) [1, 2], etc. Considering the one-to-multiple mapping problem, the random tree method [52] and mixture of experts (MoE) method [26, 59, 62] are employed and reached promising results. Specifically, the random tree approach is a statistical method based a learnt hierarchical tree structure. The MoE provides a multi-value regression model and its estimation is an intelligent combination of several individual experts. In [62], a discriminative graphical model with a temporal chain structure was proposed where density propagation rules were derived for Bayesian inference.

2.2 Generative Approaches

Generative approaches involve explicit models to explain the underlying visual and kinematic data via a few latent variables by which the motion/pose can be inferred. Our discussion below will focus on four major components in most generative approaches, as shown by Fig. 1.

2.2.1 Visual Observations

Visual observations are image descriptors extracted from image sequences for human body representation, including silhouettes, edges, colors and their combinations [50]. Silhouettes [9, 16, 19, 25, 41, 63] and edges [7, 44, 58] can be extracted efficiently from static background. Silhouettes are robust to color or texture variations of the human body but may be sensitive to the background noise or the shadow effect. Further analysis on extracted silhouette usually can provide richer representation, like level-sets method in [14] and radian distance function in [59]. Edge extraction may have some difficulties in cluttered background. Color [51, 57] can also be used to represent individual body parts, but self-occlusion may impose some problems. The combination of multiple visual cues [23, 48, 73] proves to be useful. In our research, we chose silhouette-based gait representation due to its robustness and simplicity.

2.2.2 Human Shape Models

Human shape models provide important shape priors to evaluate visual observations for pose/motion estimation, including 2D/3D shape models, 3D body scan models, and 3D computer models. Specifically, 2D shape models [27, 73] use rectangles or ellipses to approximate each body part that can be adjusted by a couple of parameters but may be limited to deal with complex shapes and self-occlusion. The parameters of each body part usually are predetermined or estimated off-line for better estimation results. 3D shape models [10, 23, 41, 48, 58] define the human body as an assembly of several rigid segments (i.e., cylinders or cuboid), each of which has maximum three degree-of-freedoms (DOFs). A 3D body scan model from a laser scanner can be subject specific [18, 54] or general enough to handle various body shapes [4, 42, 59]. For example, the SCAPE is a data-driven human shape model that can span variation in both shape and pose [3]. Usually, using this model requires multiple cameras for accurate and robust estimation. 3D computer models [2, 16, 19, 63] are very cost-effective and can be used for training data generation. Although each one is subject specific, multiple models can be used together to improve shape modeling even under a single camera [16]. Our research involves five computer models.

2.2.3 Inference Algorithms

The inference process aims to find the optimal solution (including both motion and shape) that best explains visual observations. The *top–down* approaches [10, 27, 28] match the 2D projection of a 3D body part with visual observations, or called "analysis-by-synthesis". The *bottom–up* ones [51, 57, 58] search for all body parts and assemble them into a human body for motion/pose estimation, and they may not need manual initialization. The *hypothesis tracking* can draw samples (or predictions) based on previous estimation and incorporate the temporal prior between poses or the spatial prior between parts for pose estimation. For example, the particle filter-based tracking algorithms [10, 48] are often used for inference which involve a dynamic model to predict a new pose [34] or the next part location [7]. Moreover, MCMC sampling can be embedded in the particle filter to further improve particle generation [31].

2.2.4 Human Motion Models

Motion models provide an important kinematic prior for generative models. *Graphic model-based approaches* represent the spatial and temporal priors of body parts by learning from a set of labeled images [27] or motion capture data [57, 58]. *Physical model-based approaches* [54, 71] incorporate various kinematic/dynamic constraints of body movements into the inference process. They may not need any training data, but a detailed physical model is hard to obtain which may also impose some challenges for inference due to the high-dimensional nature of the model. For example, the methods in [6–8] mainly focus on lower-body motion. *Dimension reduction (DR)—based methods* try to explore the low-dimensional intrinsic structure of human motion by learning from either kinematic or visual data that can be represented by a few latent variable and used for motion modeling.

Our research is focused on DR-based motion modeling. There are two major DR approaches, i.e., *deterministic* and *probabilistic* ones. The former one includes LLE [55] and Isomap [64] that can generate a latent space without providing a mapping function between the latent space and the data space. The latter one includes the Gaussian process latent variable model (GPLVM) [29] and its variants, such as Gaussian process dynamical models (GPDM) [38, 66, 72] and Back Constrained GPLVM [20, 30], which can learn not only the latent space but also the mapping function. In most DR methods, a pose manifold is often involved that captures the pose variability of a particular motion, and it is usually a 1D closed loop due to the cyclic nature of a gait.[1]

[1] There are several names in the literature for this concept, including the kinematic manifold [34], the gait manifold [15], or the pose latent space [19].

2.2.5 Single Pose Manifold

The pose manifold can be learnt either from visual observations (e.g., silhouettes) by LLE, [14, 32] or from kinematics data (i.e., motion capture data) by GPLVM or its variants, [13, 20, 65, 66, 68]. The kinematic pose manifold provides an accurate dynamic model for part-based body tracking where part-level shape modeling is needed and may have some difficulty due to the complexity of body parts. On the other hand, the visual pose manifold offers a direct way to generate visual hypotheses for likelihood computation, but it faces some challenges due to the one-to-many mapping problem as well as the view variability. For example, a view-dependent pose manifold was proposed in [14] where the view is treated as a discrete variable.

2.2.6 Dual Pose Manifolds

To take advantages of both the kinematic and visual pose manifolds, dual pose manifolds were proposed for video-based pose/motion estimation [19, 25, 63]. Different DR methods were used to learn the kinematic and visual pose manifolds from the kinematic and visual data, respectively. Especially, a mapping function is needed between two pose manifolds by which the visual data can be associated with the kinematic data for motion estimation.

2.2.7 Shared Pose Manifold

In [16, 33], a torus-shape manifold is designed for joint view-pose modeling that are shared in both the kinematic and visual spaces. Two mapping functions are needed to map both kinematic and visual data onto the same torus manifold via radial basis functions (RBFs). Although this approach does not involve manifold learning, it provides promising pose tracking along with continuous view estimation. In [34], a kinematic pose manifold is first learned via LLE that is also shared by the visual data using RBF-based mapping. Additionally, a continuous view manifold was proposed to supports smooth view estimation along with pose estimation.

2.3 Our Research

We focus on generative approaches, because we want to develop a general gait modeling framework. In most DR-based methods, the same subject is used for training and testing, and our research aims at estimating the kinematics of an unknown gait. Specifically, our research is inspired by the nonlinear tensor analysis approach proposed in [34] that combines manifold learning with multi-linear analysis and provides a compact generative model to represent a series of gait appearances by multiple factors, including pose, view, and shape.

3 Research Overview

3.1 Dual Gait Generative Models

We propose the *kinematic gait generative model* (KGGM) and the *visual gait generative model* (VGGM), for gait representation in the kinematic and visual spaces, respectively. Specifically, KGGM represents gait kinematics by two latent variables, *pose* and *gait*, where the pose variable defines a specific body configuration during a gait (or a gait phase) and the gait variable represents a specific *walking* motion. VGGM represents gait appearances by four latent variables, *pose*, *gait*, *view*, and *shape*, where the pose and gait variables are similar to the ones in KGGM and the view and shape variables reflect the view angle and the appearance of the subject, respectively. Both of KGGM and VGGM can be learnt from a set of training data by extending the nonlinear tensor decomposition method proposed in [34]. The learning of KGGM requires a set of gait motion data (or gait kinematics) acquired by a motion capture system. The same set of motion data is also used to generate a set of gait animations by commercial software that are used for learning VGGM. KGGM and VGGM are fully *temporally synchronized* by sharing the same pose variable during model learning, and they are the cornerstone of this research.

3.2 Gait Manifolds

In this research, we advocate the concept of *gait manifold* that captures the gait variability among different individuals and could span a low-dimensional latent space to represent all human gaits. Additionally, we also propose whole-part gait manifolds for gait modeling at both the whole and part levels. Correspondingly, KGGM and VGGM will involve three gait variables, one for the whole-level and two for the part-level (Fig. 3). A gait manifold can be learned from a set of representative training gaits, and it is defined at the gait variable of either KGGM or VGGM by which we can synthesize a new gait in terms of its kinematics or appearances.

We will use the whole-based gait manifold as an example for illustration. As shown in Fig. 2, given the motion data of five training gaits (Gaits 1–5), we can represent them in a low-dimensional (e.g., 2D) latent space where each gait is represented by a 2D vector (or gait vector). By treating the five gait vectors as the anchor points, we can span a 1D closed-loop gait manifold via curve fitting, where a new gait (Gaits A or B) can by synthesized by nonlinear interpolation. We assume a 1D nonlinear structure due to the simplicity of manifold generation, and the reason of using a closed-loop is to ease the inference process. In this work, we propose a new manifold learning technique that yields one whole-based and two part-based gait manifolds for part-whole gait modeling, by which KGGM and VGGM can be integrated via a non-linear manifold mapping function.

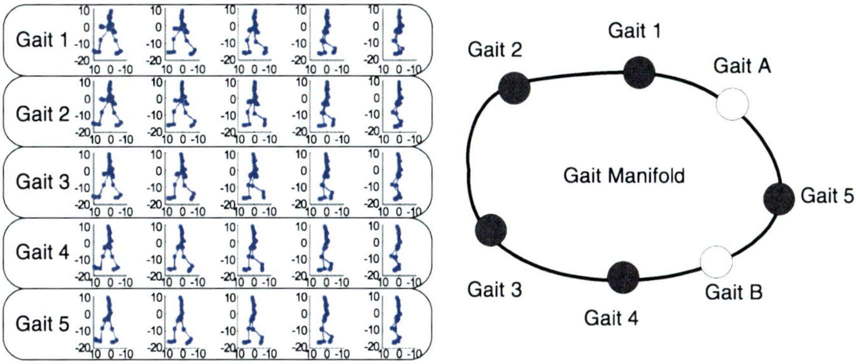

Fig. 2 The illustration of gait manifold used for gait synthesis or interpolation. (From X. Zhang and G. Fan, Dual Generative Gait Models for Human Motion Estimation from a Single Camera, IEEE Trans. Systems, Man and Cybernetics, Part B: Cybernetics, Vol. 40, No. 4, 1034–1049, Aug. 2010. @ 2010 IEEE)

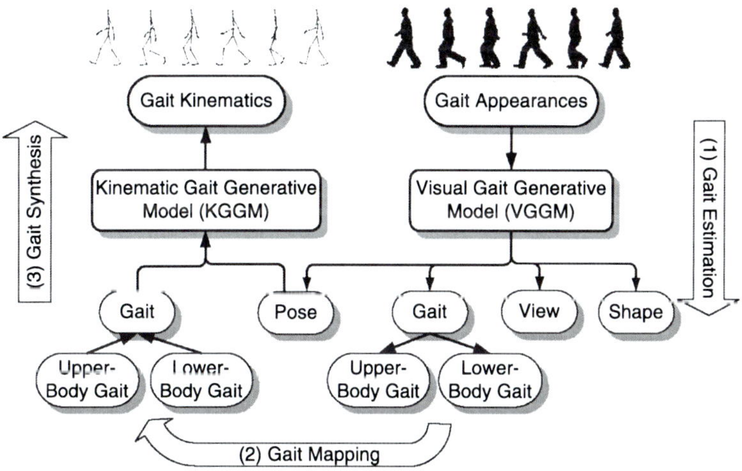

Fig. 3 The algorithm flow: (**1**) gait estimation via VGGM, (**2**) gait manifold mapping (VGGM→ KGGM), and (**3**) gait synthesis via KGGM

3.3 Inference for Gait Estimation

We need to estimate the underlying gait variable from observed gait appearances via VGGM that can be further mapped to KGGM for the synthesis of gait kinematics. The generative model fits well in the Bayesian approach, which attempts to construct the posterior probability density function (PDF) of the states based on all state and observation available. Here the states to be estimated are the four latent variables of VGGM, and the observation is a series of gait appearances. We develop a particle filter-based two-stage inference algorithm for part-whole gait estimation.

The algorithm flow is shown in Fig. 3, where three steps are involved. The first is *gait estimation* where the unknown gait is inferred from gait observations via VGGM by using a particle filtering-based inference algorithm where three gait variables (one whole-level and two part-level) are estimated. The second is *gait manifold mapping* that maps three gait variables from VGGM to KGGM via the non-linear manifold mapping. The third is *gait synthesis* by KGGM that yields the estimated gait kinematics either in whole or as two parts.

4 Dual Gait Generative Models

The learning of KGGM and VGGM is essentially a DR process, where we extend the nonlinear tensor decomposition method proposed in [34], as shown in Fig. 4.

4.1 Kinematic Gait Generative Model (KGGM)

The gait kinematics can be represented in different ways, like the 3-D positions of all joints or the angles between two adjacent joints. In order to reduce the effect of skeleton variability, gait kinematics are represented by a sequence of relative Euler angles between two adjacent joints. To learn KGGM, we need an universal pose manifold shared by different gaits based on which we can develop a unified gait

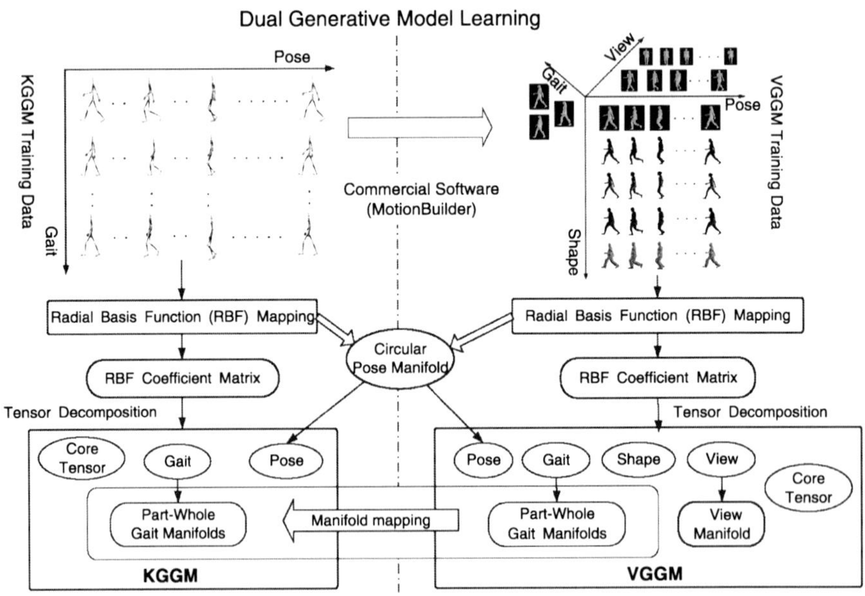

Fig. 4 The co-learning of KGGM and VGGM

representation. However, the pose manifold varies from gait to gait. Inspired by the conceptual torus manifold proposed in [33], we define a circular-shaped conceptual manifold in a 2-D space to represent a general pose variation in one gait cycle. The learning of KGGM is shown in Fig. 4(the left side).

Let $\mathcal{Z} = \{\mathbf{z}^{(i,q)} | i = 1, \ldots, N_g, q = 1, \ldots, N_p\}$ denote the set of N_g training gaits of N_p poses, where $\mathbf{z}^{(i,q)} \in \mathbb{R}^k$ encodes the k-dimensional kinematics for pose q of gait i. All poses are denoted by a set of 2-D coordinates, $\{\mathbf{p}_q \in \mathbb{R}^2, q = 1, \ldots, N_p\}$ uniformly sampled along the pose manifold. A nonlinear mapping function from $\mathbf{p}_q$ to $\mathbf{z}^{(i,q)}$ ($\mathbb{R}^2 \to \mathbb{R}^k$) can be learnt via a generalized radial basis function (RBF) as,

$$\mathbf{z}^{(i,q)} = \mathbf{B}^i \psi(\mathbf{p}_q), \tag{1}$$

where $\psi(\cdot)$ is a nonlinear kernel function defined as

$$\psi_L(\mathbf{p}_q) = \left[\phi\left(\mathbf{p}_q, \mathbf{c}_p^1\right), \ldots, \phi\left(p_q, \mathbf{c}_p^L\right)\right], \tag{2}$$

where $\phi(\cdot, \cdot)$ is a radial basis function (here we use Gaussian) and $\{\mathbf{c}_p^l | l = 1, \ldots, L\}$ are kernel centers along the pose manifold. $\mathbf{B}^i$ represents a $k \times L$ linear mapping matrix that encodes the individuality of training gait i.

All gait-dependent mapping matrices $\{\mathbf{B}^i | i = 1, \ldots, N_g\}$ can be stacked as a tensor and the high order singular value decomposition (HOSVD) [69] can be applied to decompose the tensor into two independent variables, i.e., the pose and gait. Then the generative model is defined as

$$\mathbf{z}^{(i,q)} = \mathcal{A} \times_1 \boldsymbol{\kappa}^i \times_2 \psi(\mathbf{p}_q), \tag{3}$$

where $\mathcal{A}$ is called *core tensor* ($k \times N_g \times L$) governing the interaction between two variables; $\boldsymbol{\kappa}^i$ ($N_g \times 1$) represents gait i and $\times_j$ denotes mode-j tensor product. Given a gait coefficient, this KGGM can synthesize the kinematics of an arbitrary pose.

4.2 Visual Gait Generative Model (VGGM)

We use commercial 3D animation software MotionBuilder to generate a set of gait animations, which involves multiple 3D human models and the same set of gait kinematic data used for learning KGGM. Each human model can be driven by N_g gait motions to produce different animations each of which can be recorded under different cameras views. Specifically, we use a global feature-free representation to represent gait appearances that can be obtained by the signed distance transform of the body silhouette extracted from an image [15]. The learning of VGGM is similar to that of KGGM but with more factors involved, as shown in Fig. 4 (the right side).

Let $\mathcal{Y} = \{\mathbf{y}^{(k,j,i,q)} | k = 1, \ldots, N_v, j = 1, \ldots, N_s, i = 1, \ldots, N_g, q = 1, \ldots, N_p\}$, represent the set of training gait appearances, where $\mathbf{y}^{(k,j,i,q)} \in \mathbb{R}^d$ is the d-dimensional appearance of gait i, pose q, shape j, and view k; and N_v, N_s, N_g, and N_p are the numbers of views, shapes, gaits, and poses, respectively. By sharing the same pose manifold with KGGM, we can represent a gait appearance by four factors, i.e., pose, gait, view and shape by assuming they are independent as follows:

$$\mathbf{y}^{(k,j,i,q)} = \mathcal{C} \times_1 \mathbf{v}^k \times_2 \mathbf{s}^j \times_3 \boldsymbol{\nu}^i \times_4 \varphi(\mathbf{p}_q), \tag{4}$$

where $\mathcal{C}$ is the core tensor $(d \times N_v \times N_s \times N_g \times N_p)$ governing the interaction between four variables; $\varphi(\cdot)$ is a nonlinear kernel function similar to the one in (2); $\mathbf{v}^k$, $\mathbf{s}^j$, $\boldsymbol{\nu}^i$ and $\mathbf{p}_q$ represent the view, shape, gait and pose, respectively. The first two are unique for VGGM, and the latter two have a close relationship with their counterparts in KGGM. Similar to [34], we can learn a *view manifold* from $\{\mathbf{v}^k|k = 1, \ldots, N_v\}$. Given a gait vector, a view coefficient along the view manifold, and a shape coefficient, this VGGM can synthesize the gait appearance of an arbitrary pose.

4.3 Two-Layer KGGM and VGGM

To support more detailed gait modeling, we can decompose one gait into two part-level gaits, a lower-body one and an upper-body one, according to the "hip" point. The hip point is well defined in both gait kinematics (the hip joint) and gait appearances (the image center) during the learning of KGGM and VGGM, making this part-based gait representation plausible and convenient. Therefore, both KGGM and VGGM can be added with two more part-level gait variables. It is worth mentioning all gait variables lead to a full gait synthesis by either KGGM or VGGM, while only part of it (either the lower-body or upper-body) is used for gait representation when a part-level gait variable is considered.

Since KGGM and VGGM share the same pose manifold during the learning process, the pose variable in (3) and that in (4) are equivalent and identical. However, three gait variables (one whole-based and two part-based) in KGGM and those in VGGM in (4), are quite different, since they represent the individuality of a specific training gait in the kinematic and visual spaces, respectively. The two generative models are not ready to be *integrated together* yet due to the different nature of their gait variables. Also, the gait variables defined in (3) and (4) corresponds to a set of gait vectors associated with the training gaits. In order to represent an unknown gait via valid gait interpolation, we propose the concept of gait manifold that is discussed in next section by which KGGM and VGGM can be integrated together. Thus the two-layer integrated VGGM-KGGM allows us to estimate the kinematics of an unknown gait at both the whole and part levels, as shown in Fig. 5, where gait kinematics and appearances are reconstructed by using KGGM (lower-half of the figure) and VGGM (upper-half of the figure) respectively.

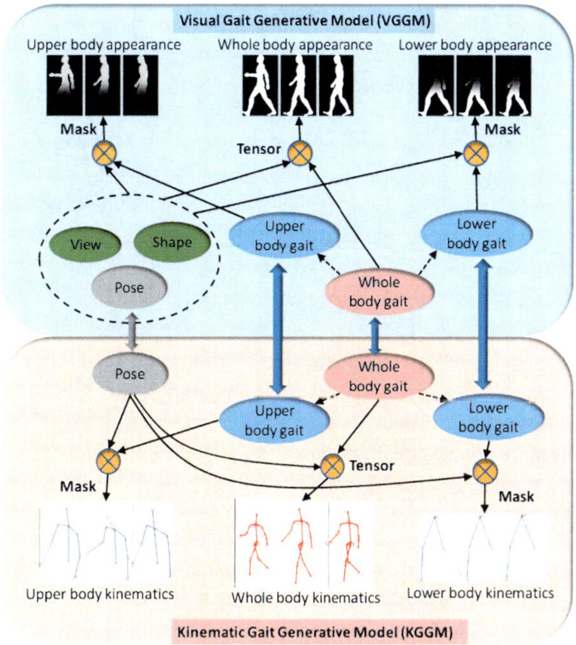

Fig. 5 Illustration of the connections between two-layer dual generative models. The pose manifold is shared between KGGM and VGGM to temporally synchronize two generative models. The connections between part-whole gait manifolds respectively are used for the physical gait synchronization. Hence, the part and whole visual silhouettes and kinematic skeletons can be reconstructed by (4) and (3) using corresponding latent variables from VGGM and KGGM respectively. (From X. Zhang, G. Fan, and L. Chou, Two-Layer Gait Generative Models for Estimating Unknown Human Gait Kinematics, in Proc. the 2nd International Workshop on Machine Learning for Vision-Based Motion Analysis (MLVMA'09), in conjunction with ICCV2009, Japan, Oct. 2009. @ 2009 IEEE)

5 Gait Manifolds

We first discuss the concept of gait manifold as well as a new manifold learning technique. Then we introduce the term of gait manifold topology where we compare three distance functions to compute the dissimilarity between two gaits. Each distance function leads to a specific manifold topology as well as a specific gait manifold. Third, we discuss part-whole gait modeling via one whole-based gait manifold and two part-based gait manifolds. Finally, two different manifold mappings are presented by which VGGM and KGGM can be integrated.

5.1 Gait Manifold Learning

The concept of *gait manifold* plays an important role in our research by which a new gait can be interpolated from training gaits via VGGM or KGGM. Given a

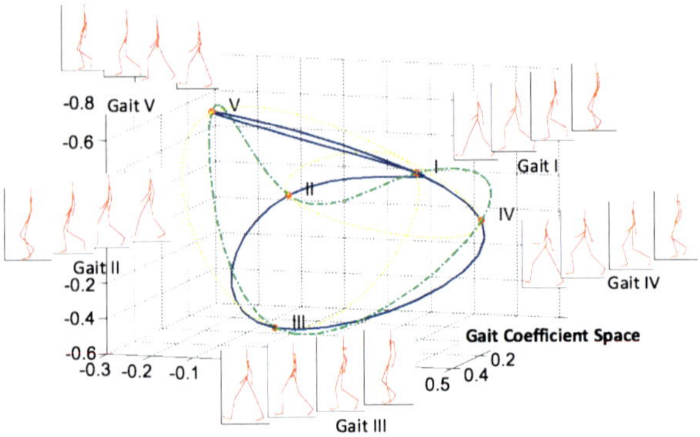

Fig. 6 Illustration of various gait manifolds with different topologies in the tensor coefficient space

set of training gaits represented in a latent space, e.g., the tensor coefficient spaces defined by (3) and (4), each training gait corresponds to a gait vector. We want to span a continuous gait manifold via curve fitting where all training gait vectors are treated as anchor points. We postulate two factors for the gait manifold. One is that it should have the best local smoothness to ensure effective and valid nonlinear gait interpolation, and the other is that it should be a 1D closed-loop to support efficient and continuous inference for gait estimation. In this work, we propose a two-step manifold learning technique to create a smooth closed-loop gait manifold. First, we will determine a *gait manifold topology* that defines a specific ordering relationship among all training gaits, and then accordingly, we use the spline-fitting method to connect all training gait vectors into a continuous closed-loop gait manifold in the tensor coefficient space. To ensure the best smoothness along the gait manifold, we define the optimal gait manifold topology to be the "shortest-closed-path" that travels through all training gaits under a certain distance metric. The key question is how to define an appropriate "*distance*" function between two training gaits. Under different distance functions, we will have different manifold topologies to connect all training gaits, leading to different gait manifolds, as shown in Fig. 6.

Moreover, we are interested in a *part-whole gait representation* that involves both *whole-based* and *part-based* gait manifolds. Rather than characterizing every limb, the "part" here is loosely defined as the *upper-body* and *lower-body*. These two parts can be easily divided along the *hip* that is well defined in gait kinematics and is always in the image center of gait appearances. Therefore, we need to define different gait manifold topologies that can reflect the dissimilarity between two gaits at both the whole and part levels. Consequentially, we can obtain multiple gait manifolds that reveals the intrinsic structures among training gaits with respect to the whole, upper and lower bodies. This hybrid part-whole gait representation can take advantage of the robustness of whole-level gait estimation and the accuracy of part-level gait refinement.

5.2 Gait Manifold Topology

Given a gait distance function $\mathcal{G}^{(w)}(,)$,[2] the gait manifold topology is defined as the shortest-closed-path to travel through all training gaits by this distance function as

$$\mathcal{T}^{w} = \arg\min_{Q} \sum_{i=1}^{N_g} \mathcal{G}^{(w)}(q_i, q_{i+1}), \tag{5}$$

where $Q = \{q_i \in [1, N_g] | i = 1, \ldots, N_g + 1, q_i \neq q_j \text{ for } i \neq j; q_1 = q_{N_g+1}\}$ (a specific order to connect N_g training gaits in a closed loop), and $\mathcal{T}^{w}$ specifies the optimal order from the shortest closed path according to the gait distance $\mathcal{G}^{w}(,)$. Similarly, we can obtain $\mathcal{T}^{u}$ and $\mathcal{T}^{l}$ which are the manifold topologies for the upper and lower bodies with respect to $\mathcal{G}^{(u)}(,)$ and $\mathcal{G}^{(l)}(,)$ respectively,

$$\mathcal{T}^{u} = \arg\min_{Q} \sum_{i=1}^{N_g} \mathcal{G}^{(u)}(q_i, q_{i+1}), \tag{6}$$

$$\mathcal{T}^{l} = \arg\min_{Q} \sum_{i=1}^{N_g} \mathcal{G}^{(l)}(q_i, q_{i+1}). \tag{7}$$

Based on three manifold topologies $\mathcal{T}_w$, $\mathcal{T}_u$, and $\mathcal{T}_l$, we can create three gait manifolds by connecting all training gait vectors in the tensor coefficient space associated with KGGM or VGGM, as shown in Fig. 6. It is obvious that the key issue of manifold topology is to define a distance function between two gaits. In the following, we will discuss three distance metrics, two of which can be extended to the part level comparison. All of them will be examined in the section of experiments in terms of their performance for gait synthesis by using KGGM.

5.2.1 Euclidean Distance Between Gait Vectors

In either KGGM or VGGM, we have N_g gait vectors ($\{\boldsymbol{\kappa}^i | i = 1, \ldots, N_g\}$ or $\{\boldsymbol{\nu}^i | i = 1, \ldots, N_g\}$) representing N_g training gaits in the tensor coefficient space. Since gait vectors in KGGM directly affect the performance of gait synthesis in terms of the capability of interpolating new gait kinematics. Hence, we can use the Euclidean distance between two gait vectors in KGGM to represent the dissimilarity between two training gaits,[3] written as:

$$\mathcal{G}^{(1)}(i, j) = \mathcal{D}\big(\boldsymbol{\kappa}^i, \boldsymbol{\kappa}^j\big), \tag{8}$$

[2]The superscripts w, l, and u denote the gait distances for the whole, lower and upper bodies.

[3]Since the decomposed vector is normalized and orthogonal, we discard the last two dimension data for the distance computation.

where $\mathcal{G}^{(1)}(i,j)$ is the distance between gait i and gait j according to their gait vectors $\boldsymbol{\kappa}^j$ and $\boldsymbol{\kappa}^i$ in the tensor coefficient space, and $\mathcal{D}$ is the Euclidean function. This distance function is robust to noise because it is defined on the gait variable in the latent space. However, it is unable to measure the gait similarity at the part-level.

5.2.2 Distance Between 3D Joint Positions

We can also define a distance function directly on the kinematic data of two gaits, which can reflect the gait dissimilarity at both whole and part levels. Given N_g training gaits, each gait has N_p poses and each pose has k body joints that can be divided into two subsets corresponding to the lower-body and upper-body (Fig. 7). As discussed in Sect. 4.1, the kinematics data are represented as the relative Euler rotation angle of each joint. We convert them into 3D joint positions by using a standard skeleton to minimize the effect of skeleton variability. The 3D kinematics of pose p in gait i of is represented by $\mathbf{z}^{(i,q)} = \{\mathbf{z}_u^{(i,q)}, \mathbf{z}_l^{(i,q)}\}$ where $\mathbf{z}_u^{(i,q)}$ and $\mathbf{z}_l^{(i,q)}$ are the 3D joint positions of the upper-body and lower-body, respectively. We use $\mathbf{Z}^i = \{\mathbf{z}^{(i,q)} | q = 1, \ldots, N_p\}$, $\mathbf{Z}_u^i = \{\mathbf{z}_u^{(i,q)} | q = 1, \ldots, N_p\}$, and $\mathbf{Z}_l^i = \{\mathbf{z}_l^{(i,q)} | q = 1, \ldots, N_p\}$ to present the kinematics of the whole, upper, and lower bodies, respectively. Correspondingly, three distance functions can be obtained to quantify the kinematic dissimilarity between gaits i and j at the part-whole levels as,

$$\mathcal{G}_w^{(2)}(i,j) = \mathcal{D}(\mathbf{Z}^i, \mathbf{Z}^j), \tag{9}$$

$$\mathcal{G}_u^{(2)}(i,j) = \mathcal{D}(\mathbf{Z}_u^i, \mathbf{Z}_u^j), \tag{10}$$

$$\mathcal{G}_l^{(2)}(i,j) = \mathcal{D}(\mathbf{Z}_l^i, \mathbf{Z}_l^j), \tag{11}$$

where $\mathcal{D}$ is the Euclidean distance function. Although this distance directly reveals the kinematic dissimilarity between two gaits in terms of the 3D positions of all joints, it does not capture the dynamics of each joint explicitly. Also, it may be prone to noise due to the high-dimensional nature of 3D joint positions.

5.2.3 Fourier Analysis of Joint Angles

As defined in Fig. 7, a joint in the skeleton has 1, 2, or 3 DOFs, and each DOF can be characterized by a sequence of rotation angles for a complete walking cycle. In [12] and [74], Fourier analysis is used to extract some dominant components from the rotation angle sequences of two lower-body joints, knees and ankles, for gait recognition, which are derived from the side-view of an image sequence of walking. This method was shown effective to capture the unique dynamics of a gait. In this work, we extend this idea to represent the gait kinematics of all joints in the 3D space. An illustration of a sequence of gait kinematics (top) and the plots of the rotation angle of four joints (one DOF only) (bottom) are shown in Fig. 8.

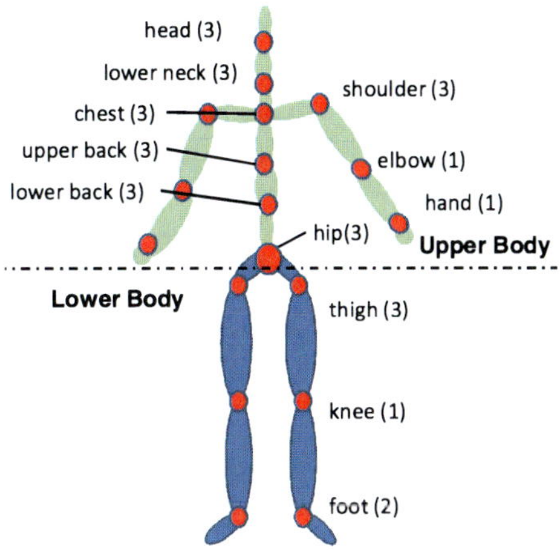

Fig. 7 The skeleton system with all 18 joints defined. The number in the parenthesis indicates the degree-of-freedom (DOF) of a joint. The *upper* and *lower bodies* are divided by the *hip* joint

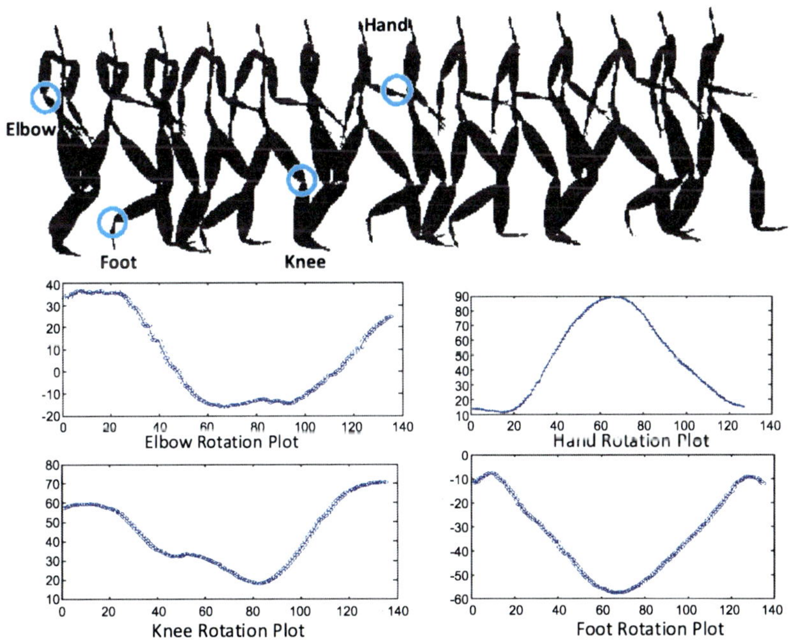

Fig. 8 Illustration of a gait sequence with the plots of the rotation angles of four joints

Given training gait i of $N_p^{'}$ poses, the nth DOF ($n = 1, 2$, or 3) of joint k is denoted by a sequence of rotation angles $\Theta_{k,n}^{i} = \{\theta_{k,n}^{(i,q)} | q = 1, \ldots, N_p^{'}\}$. Gait i can be represented by a matrix $\Phi^{i}_{N_r \times N_p^{'}}$, where N_r is the summation of DOFs of all joints and each row is a sequence of rotation angles of a joint along one DOF. Similar to [12] and [74], we use the discrete fourier transform (DFT) to extract top R DFT components for each row, and we adopt the phase-weighted magnitude to represent each rotation sequence in the frequency-domain. In the following, we omit the indexes of gait (i), joint (k) and DOF (n), for the sake of easiness, and we use Θ to represent a sequence of rotation angles of a joint along one DOF.

Let $\Theta \overset{\mathrm{DFT}}{\longleftrightarrow} \Psi(e^{j\omega})$ be the DFT pair of a sequence of rotation angles, then the R-order phase-weighted magnitude $\mathbf{x}$ is defined as

$$\mathbf{x} = \begin{bmatrix} x_1 \\ x_2 \\ \vdots \\ x_R \end{bmatrix}^T = \begin{bmatrix} |\Psi(e^{j\omega_1})| \cdot \arg \Psi(e^{j\omega_1}) \\ |\Psi(e^{j\omega_2})| \cdot \arg \Psi(e^{j\omega_2}) \\ \vdots \\ |\Psi(e^{j\omega_R})| \cdot \arg \Psi(e^{j\omega_R}) \end{bmatrix}^T, \tag{12}$$

where $|\Psi(e^{j\omega_r})|$ and $\arg \Psi(e^{j\omega_r})$ are the magnitude and phase of the rth DFT component. R is set to be 3 in this work. Then we can construct a compact frequency-domain representation of gait i as $\mathbf{X}^i_{N_r \times R} = [\mathbf{x}_1; \mathbf{x}_2; \ldots; \mathbf{x}_{N_r}]$ where each row is a 3-order phase-weighted magnitude defined in (12). Similarly, we can define the frequency-domain representations for the upper-body and lower-body as $\mathbf{X}^i_u$ and $\mathbf{X}^i_l$, which consider the joints in the upper-body and those in the lower-body, respectively. Hence, the gait distances between two training gaits for the whole, upper and lower bodies are defined as

$$\mathcal{G}^{(3)}_w(i,j) = \mathcal{D}(\mathbf{X}^i, \mathbf{X}^j), \tag{13}$$

$$\mathcal{G}^{(3)}_u(i,j) = \mathcal{D}(\mathbf{X}^i_u, \mathbf{X}^j_u), \tag{14}$$

$$\mathcal{G}^{(3)}_l(i,j) = \mathcal{D}(\mathbf{X}^i_l, \mathbf{X}^j_l). \tag{15}$$

5.3 Part-Whole Gait Manifolds

As shown in (3) and (4), each training gait can be represented by a gait vector in KGGM or VGGM. Given the gait topology defined as in Sect. 5.2, for N_g training gait vectors $\{\boldsymbol{\kappa}^i_* | i = 1, \ldots, N_g\}$ in the tensor coefficient space of KGGM defined in (3), we can generate one whole-based and two part-based gait manifolds by connecting all training gaits according to their individual topology, $\mathcal{T}^w$, $\mathcal{T}^u$, and $\mathcal{T}^l$, via the spline fitting technique, as defined below:

$$\begin{aligned} \mathcal{M}^w_\kappa &= \mathcal{S}\big(\boldsymbol{\kappa}^{\mathcal{T}_w(i)}_* | i = 1, \ldots, N_g + 1\big), \\ \mathcal{M}^u_\kappa &= \mathcal{S}\big(\boldsymbol{\kappa}^{\mathcal{T}_u(i)}_* | i = 1, \ldots, N_g + 1\big), \\ \mathcal{M}^l_\kappa &= \mathcal{S}\big(\boldsymbol{\kappa}^{\mathcal{T}_l(i)}_* | i = 1, \ldots, N_g + 1\big), \end{aligned} \tag{16}$$

where $\mathcal{S}$ is the spline fitting function that passes through N_g gait vectors by following a given topology. For simplicity, we use a 1D parameter $h \in [1, N_g + 1]$ to denote an arbitrary gait vector along a gait manifold. For example, $\mathcal{M}^w_\kappa(h)$ is

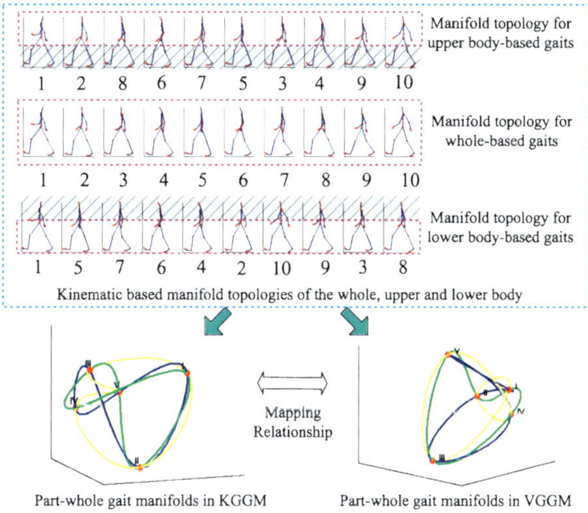

Fig. 9 Part-whole gait manifold topologies and corresponding gait manifolds associated with KGGM and VGGM. Each number (above) indicates a specific training gait with N_p poses. (From X. Zhang, G. Fan, and L. Chou, Two-Layer Gait Generative Models for Estimating Unknown Human Gait Kinematics, in Proc. the 2nd International Workshop on Machine Learning for Vision-Based Motion Analysis (MLVMA'09), in conjunction with ICCV2009, Japan, Oct. 2009. @ 2009 IEEE)

a specific gait vector along $\mathcal{M}_{\kappa}^{w}$. When $h =\in \{1, 2, \ldots, N_{g+1}\}$, it corresponds to a training gait. Otherwise, it represents a new gait vector that can be used to synthesize the *kinematics* of an unknown gait via KGGM.

Similarly, we can obtain three different gait manifolds from VGGM by connecting N_g training gait vectors, $\{\mathbf{v}_*^i | i = 1, \ldots, N_g\}$ defined in (4), according to $\mathcal{T}^w$, $\mathcal{T}^u$, and $\mathcal{T}^l$, which are represented by $\mathcal{M}_{\nu}^{w}$, $\mathcal{M}_{\nu}^{u}$, and $\mathcal{M}_{\nu}^{l}$. The same as before, $\mathcal{M}_{\nu}^{w}(h)$ is a specific gait vector along $\mathcal{M}_{\nu}^{w}$. When h is an integer, it corresponds to a given training gait. Otherwise, it denotes a new gait vector that can be used to synthesize the *appearances* of an unknown gait via VGGM. The part-whole gait manifolds are illustrated in Fig. 9, where the top box shows three different manifold topologies pertaining to the upper, whole and lower bodies, and the bottom presents three gait manifolds in each of KGGM and VGGM. In the following, we refer to those gait manifolds generated from KGGM as *kinematic gait manifolds*, and those from VGGM as *visual gait manifolds*.

5.4 Manifold Mapping Between KGGM and VGGM

Figure 10 demonstrates how KGGM and VGGM are used for gait estimation via manifold mapping. The gait variable is first estimated along $\mathcal{M}_{\nu}$ in VGGM (Step 1) where both the observed and synthesized gait appearances are involved in the particle filter-based inference process (Step 2). Then the estimated gait variable is mapped to the one along $\mathcal{M}_{\kappa}$ in KGGM via manifold mapping (Step 3), and the underlying gait kinematics can be synthesized by KGGM according to (3) (Step 4).

Basically, $\mathcal{M}_{\kappa}$ and $\mathcal{M}_{\nu}$ are defined on the gait variable in KGGM and that in VGGM, respectively. The manifold mapping between $\mathcal{M}_{\kappa}$ and $\mathcal{M}_{\nu}$ will naturally lead to the integration of KGGM and VGGM via their gait variables. Specifically, we

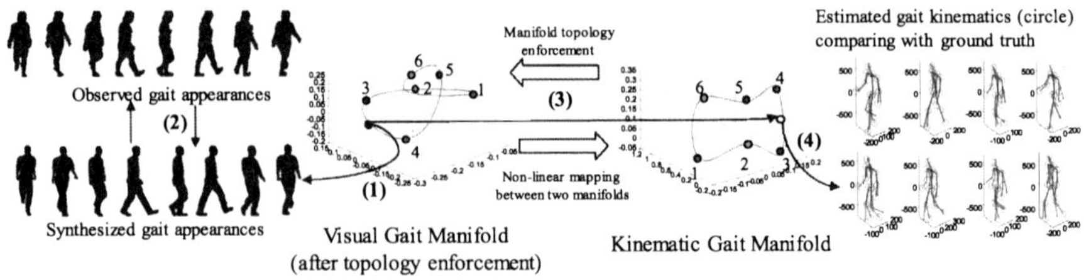

Fig. 10 Illustration of the integration of KGGM and VGGM by a mapping function between the two gait manifolds for gait estimation and synthesis

propose two manifold mapping functions, i.e., the *nonlinear kernel-based mapping* and the *similarity-preserving mapping*.

5.4.1 Nonlinear Mapping Functions (MAP-1)

We can develop a nonlinear RBF-based mapping function (MAP-1) from visual gait manifolds in VGGM to kinematic gait manifolds in KGGM as,

$$\boldsymbol{\kappa}^i = \mathcal{F}(\boldsymbol{\nu}^i) = \sum_{j=1}^{J} \omega_j \zeta(\boldsymbol{\nu}^i - \mathbf{c}_\nu^j), \tag{17}$$

where $\mathcal{F}(\cdot)$ maps $\boldsymbol{\nu}^i \in \mathcal{M}_\nu$ to $\boldsymbol{\kappa}^i \in \mathcal{M}_\kappa$ for each training gait; $\{\mathbf{c}_\nu^j | j = 1, \ldots, J\}$ are the kernel centers along $\mathcal{M}_\nu$; and $\zeta(\cdot)$ is a Gaussian function. Given a new gait vector that is between two training gaits along $\boldsymbol{\nu}' \in \mathcal{M}_\nu$, we can map it to $\mathcal{M}_\kappa$ as:

$$\boldsymbol{\kappa}' = \arg\min_{\boldsymbol{\kappa}} \{\mathcal{D}(\mathcal{F}(\boldsymbol{\nu}'), \boldsymbol{\kappa}) | \boldsymbol{\kappa} \in \mathcal{M}_\kappa\}, \tag{18}$$

where $\boldsymbol{\kappa}'$ is the corresponding gait variable in KGGM. After mapping $\boldsymbol{\nu}'$ into gait space of VGGM, $\arg\min_{\boldsymbol{\kappa}}$ finds the closest gait vector $\boldsymbol{\kappa}'$ along gait manifold $\mathcal{M}_\kappa$. We can use the same method to develop two part-level manifold mappings, one for lower-body gait manifolds and one for upper-body gait manifolds.

5.4.2 Similarity-Preserving Mapping Functions (MAP-2)

To improve the two-step mapping process of MAP-1, we also propose a *similarity-preserving mapping method* (MAP-2) that avoids any explicit mapping function. We assume the visual kinematic manifold in VGGM and the kinematic gait manifold in KGGM have the same nonlinear interpolation property. Then we would like to preserve the distance ratio between the new gait vector and its two nearest training gait vectors during manifold mapping, as shown in Fig. 11.

Given a new gait vector $\mathcal{M}_\nu^w(\tau)$ in VGGM, we can find its two nearest training gaits along $\mathcal{M}_\nu^w$, i.e., $\mathcal{M}_\nu^w(m)$ and $\mathcal{M}_\nu^w(n)$, which can be directly mapped to $\mathcal{M}_\kappa^w$,

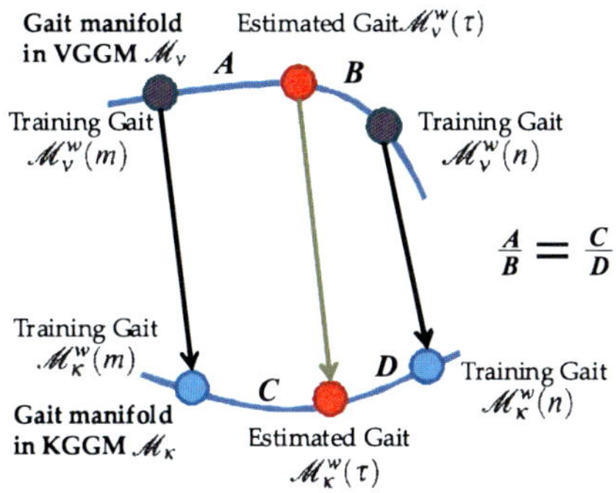

Fig. 11 The similarity-preserving mapping between KGGM and VGGM

i.e., $\mathcal{M}_\kappa^w(m)$ and $\mathcal{M}_\kappa^w(n)$. Then we can find the corresponding new gait vector in KGGM, i.e., $\mathcal{M}_\kappa^w(\tau)$, by ensuring that the ratio of its distances to the two nearest training gait vectors along $\mathcal{M}_\kappa^w$ are preserved, i.e.,

$$\frac{\mathcal{U}(\mathcal{M}_\nu^w(\tau), \mathcal{M}_\nu^w(m)|\mathcal{M}_\nu^w)}{\mathcal{U}(\mathcal{M}_\nu^w(\tau), \mathcal{M}_\nu^w(n)|\mathcal{M}_\nu^w)} = \frac{\mathcal{U}(\mathcal{M}_\kappa^w(\tau), \mathcal{M}_\kappa^w(m)|\mathcal{M}_\kappa^w)}{\mathcal{U}(\mathcal{M}_\kappa^w(\tau), \mathcal{M}_\kappa^w(n)|\mathcal{M}_\kappa^w)}, \tag{19}$$

where $\mathcal{U}(\cdot|\cdot)$ is a nonlinear distance function defined along a given manifold. Similarly, we can obtain the mapping relationships for $\mathcal{M}_\nu^u \leftrightarrow \mathcal{M}_\kappa^u$ and $\mathcal{M}_\nu^l \leftrightarrow \mathcal{M}_\kappa^l$ for upper-body and lower-body gait manifolds.

5.4.3 MAP-1 vs. MAP-2

Essentially, MAP-1 considers the *global* manifold mapping that associates all training gaits in VGGM with those in KGGM via one RBF-based mapping function, while MAP-2 deals with the *local* manifold mapping that only associates the two neighboring training gaits in VGGM with their counterparts in KGGM. No explicit mapping function is involved in MAP-2. Also, MAP-2 was found to be slightly better than MAP-1 in practice. Therefore, we chose MAP-2 in our implementation.

6 Inference Algorithms

This section discusses three inference issues. (1) How to formulate gait estimation as a sequential inference problem via graphical models? (2) How to cope with the dynamic nature of the gait variable in a long sequence for whole-based gait estimation? (3) How to conduct part-basedpart-based gait estimation via two-stage inference?

6.1 Graphical Models

We employ a graphical model to integrate all related variables along with their conditional dependencies, as shown in Fig. 12. The key parameters to be estimated are

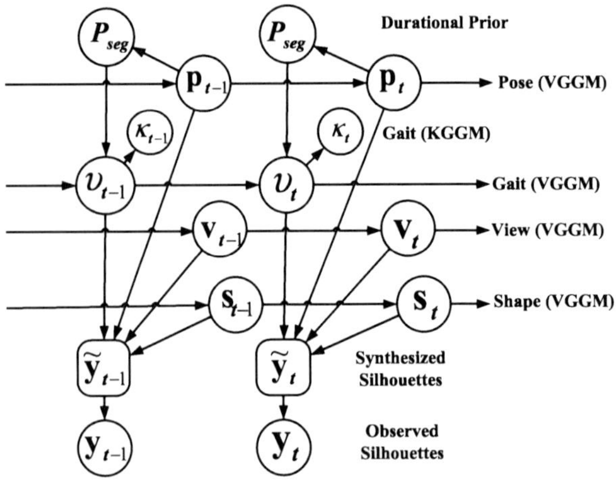

Fig. 12 The graphic model for gait tracking and estimation. (From X. Zhang, G. Fan, and L. Chou, Two-Layer Gait Generative Models for Estimating Unknown Human Gait Kinematics, in Proc. the 2nd International Workshop on Machine Learning for Vision-Based Motion Analysis (MLVMA'09), in conjunction with ICCV2009, Japan, Oct. 2009. @ 2009 IEEE)

two gait variables i.e., $\boldsymbol{\kappa}_i$ in KGGM (3) and $\boldsymbol{\nu}_i$ in VGGM (4). VGGM specifies four latent variables, the pose $\mathbf{p}_t$, view $\mathbf{v}_t$, shape $\mathbf{s}_t$, and gait $\boldsymbol{\nu}_t$, all of which have to be estimated. According to Bayes' rule, we can recursively estimate the posterior for time step t and latent variables are estimated via maximum a posterior probability (MAP) as

$$\hat{\mathbf{x}}_t = \arg\max_{\mathbf{x}_t} p(\mathbf{x}_t|\mathbf{y}_{t-1})p(\mathbf{y}_t|\mathbf{x}_t), \tag{20}$$

where $\mathbf{x}_t = [\mathbf{p}_t, \mathbf{v}_t, \mathbf{s}_t, \boldsymbol{\nu}_t]$ encapsulates the four latent variables; $p(\mathbf{x}_t|\mathbf{y}_{t-1})$ specifies the prediction based on previous observation $\mathbf{y}_{t-1}$, and $p(\mathbf{y}_t|\mathbf{x}_t)$ defines the observation model that involves the comparison between the observed gait appearance $\mathbf{y}_t$ and the one synthesized by VGGM given four hypothesized latent variables, as defined below,

$$p(\mathbf{y}_t|\mathbf{x}_t) = p(\mathbf{y}_t|\mathbf{p}_t, \mathbf{v}_t, \mathbf{s}_t, \boldsymbol{\nu}_t) \propto \exp -\frac{\|\mathbf{y}_t - \tilde{\mathbf{y}}_t\|^2}{2\sigma^2}, \tag{21}$$

where $\mathbf{y}_t$ is the observed gait appearance represented by a signed distance transform defined in [15]; $\tilde{\mathbf{y}}_t$ is synthesized by VGGM according to (4); $\|\cdot\|^2$ denotes the mean square error, and σ^2 controls the sensitivity of observation evaluation.

Since the four variables are independent, we can approximate $p(\mathbf{x}_t|\mathbf{y}_{t-1})$ in (20) by the product of the prior distribution of each latent variable given previous state estimation, that is,

$$p(\mathbf{x}_t|\mathbf{y}_{t-1}) \approx p(\mathbf{p}_t|\mathbf{x}_{t-1})p(\mathbf{v}_t|\mathbf{x}_{t-1})p(\mathbf{s}_t|\mathbf{x}_{t-1})p(\boldsymbol{\nu}_t|\mathbf{x}_{t-1}), \tag{22}$$

where four dynamic models are involved for four latent variables. Specifically, three of them are constrained by their 1D nonlinear manifold, i.e., the pose manifold for $\mathbf{p}_t$, the view manifold for $\mathbf{v}_t$, and the visual gait manifold (i.e., $\mathcal{M}_\nu^*$) for $\boldsymbol{\nu}_t$. The shape variable $\mathbf{s}_t = [w_t^1, \ldots, w_t^{N_s} | \sum_{j=1}^{N_s} w_t^j = 1]$ represents the linear combinations coefficients of N_s prototype shapes specified by $\{\mathbf{s}^j | j = 1, \ldots, N_s\}$ given in (4).

Therefore, we use a constant speed dynamic model for $\mathbf{p}_t$ defined along the circular-shaped pose manifold, and a random walk to propagate the view samples along the view manifold. We also use a random walk to sample the shape variable in the tensor coefficient space. Regarding $\boldsymbol{\nu}$, we need a special dynamics to sample it along the gait manifold due to its segmental variability that will be discussed shortly.

For simplicity, we can sequentially estimate four latent variables one by one in the order of their robustness, i.e., pose $\rightarrow$ view $\rightarrow$ shape $\rightarrow$ gait $\rightarrow$ pose $\rightarrow$ view $\rightarrow$ shape. In other words, (20) is decomposed into four steps where the four latent variables are estimated individually and sequentially. The pseudo code of the inference algorithm is listed in Algorithm 1 below. For each variable, we resort to the MCMC sampling approach to rejuvenate the sample distribution for state estimation in each time step. Basically, the inference algorithm is not sensitive to the parameter initialization. The pose and view variables need a rough initial estimation to reduce the ambiguity introduced by a single camera settings. The shape and gait initial parameters can be randomly chosen.

Algorithm 1 Inference algorithm

1: Given observations $\mathbf{y}_t$ and previous state estimation $\mathbf{x}_{t-1}$ (for the first frame, $\mathbf{p}_1$ and $\mathbf{v}_1$ are determined by a rough estimation, and $\boldsymbol{\nu}_1$ and $\mathbf{s}_1$ are randomly chosen)
2: Predict pose $\mathbf{p}'_t$, view $\mathbf{v}'_t$, shape $\mathbf{s}'_t$ and gait $\boldsymbol{\nu}'_t$ according to their own dynamic models
3: Update pose $\mathbf{p}''_t$ using MCMC sampling given $\mathbf{v}'_t$, $\mathbf{s}'_t$, $\boldsymbol{\nu}'_t$
4: Update view $\mathbf{v}''_t$ using MCMC sampling given $\mathbf{p}''_t$, $\mathbf{s}'_t$, $\boldsymbol{\nu}'_t$
5: Update shape $\mathbf{s}''_t$ using MCMC sampling given $\mathbf{p}''_t$, $\mathbf{v}''_t$, $\boldsymbol{\nu}'_t$
6: Estimate gait $\hat{\boldsymbol{\nu}}_t$ as discussed in Sect. 6.2 and Algorithm 2
7: Refine pose $\hat{\mathbf{p}}_t$ using MCMC sampling given $\mathbf{v}''_t$, $\mathbf{s}''_t$, $\hat{\boldsymbol{\nu}}_t$
8: Refine view $\hat{\mathbf{v}}_t$ using MCMC sampling given $\hat{\mathbf{p}}_t$, $\mathbf{s}''_t$ and $\hat{\boldsymbol{\nu}}_t$
9: Refine shape $\hat{\mathbf{s}}_t$ using MCMC sampling given $\hat{\mathbf{p}}_t$, $\hat{\mathbf{v}}_t$ and $\hat{\boldsymbol{\nu}}_t$

(Note: $\mathbf{p}'_t$, $\mathbf{p}''_t$, $\hat{\mathbf{p}}_t$ represent the *predicted*, *updated*, and *refined* pose variables, respectively. The same notation applies to other variables.)

6.2 Whole-Based Gait Estimation

In practice, we observed that in a long sequence, the gait variable may exhibit significant variations, especially when the subject is not walking straight. Usually, the gait variable is dominated by one value within each half-cycle, and it may *jump* to another value (along the gait manifold) in the next half-cycle. We believe it is because that the 1D gait manifold used here is a simplified representation of the unknown gait space that is very likely to be a higher dimension one, and the continuity in that space is not well preserved in this 1D space. Moreover, we also observed that the jump usually occurs around the *contact* pose when both feet are on the ground and a new half-cycle starts. To accommodate these two factors, we propose a new segmental jump-diffusion MCMC inference scheme (SJD-MCMC) that is embedded in the particle filter for dynamic gait estimation. Specifically, *jump* enables the sample generation to traverse along the gait manifold and to explore globally while *diffusion* draws samples intensively in a local area of the gait manifold.

6.2.1 Segmental Modeling

We represent a gait manifold by a 2D circle in order to facilitate the inference process. Then the gait is denoted by an angular variable $\nu \in [0, 2\pi)$ that can be mapped to the original one defined in the tensor coefficient space, i.e., $\boldsymbol{\nu} = \mathfrak{f}(\nu)$. For simplicity, we will use them exchangeably in the following discussion. The idea of segmental modeling allows us to control the switch between two kinds of dynamics, jump and diffusion, in a probabilistic way. Especially, we define a probabilistic model of duration prior (as shown in Fig. 13) that is a function of pose, as follows:

$$P_{\text{seg}} = \beta \exp \frac{-\tau_{\mathbf{p}}^2}{2\sigma_{\mathbf{p}}^2}, \tag{23}$$

where $\tau_{\mathbf{p}}$ is the arc distance between the current pose and the contact pose along the pose manifold; $\sigma_{\mathbf{p}}^2$ controls the relative frequency of *jump*; β is a normalization constant to let $P_{\text{seg}} \in (0, 1]$. This model indicates how likely a *jump* should be triggered given current pose estimation.

6.2.2 Mode-Based Gait Estimation

Due to the hybrid nature of the gait variable, we will use the jump-diffusion MCMC to generate gait samples. The challenge is that the sampling space is continuous, which is different from the traditional *jump-diffusion* applications where a mixed

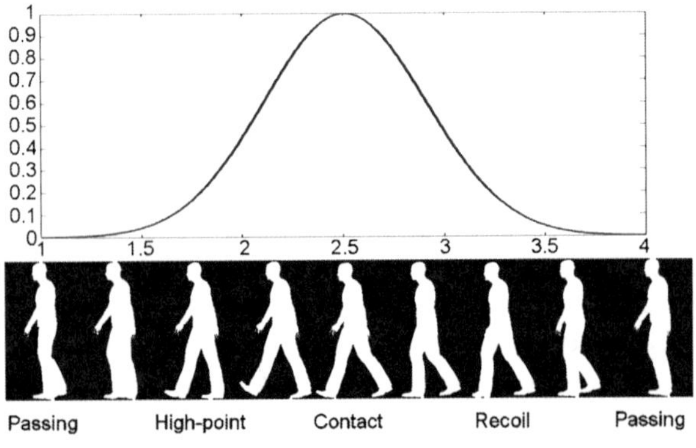

Fig. 13 The segmental prior model, where the pose is defined according [24]. (From X. Zhang and G. Fan, Dual Generative Gait Models for Human Motion Estimation from a Single Camera, IEEE Trans. Systems, Man and Cybernetics, Part B: Cybernetics, Vol. 40, No. 4, 1034–1049, Aug. 2010. @ 2010 IEEE)

discrete-continuous state space is involved [75]. Therefore, we propose the concept of *mode*. A mode is a local continuous model defined along the angular gait manifold, while a set of modes cover all possible gait values. Then, *jump* indicates switches among modes and *diffusion* exploits within a mode. Gait estimation will involve several modes defined as

$$\mathcal{M}_{(1,\ldots,R)} = \{(\mu_1, \delta_1^2), \ldots, (\mu_R, \delta_R^2)\}, \tag{24}$$

where $\mathcal{M}_{(1,\ldots,R)}$ represents R modes and each mode $\mathcal{M}_r$ is a Gaussian function with mean μ_r and variance δ_r^2 defined along the angular gait manifold. For an unknown test subject, we will estimate on-line these modes from which we can draw samples along the angular gait manifold that can be further mapped to the hypothesized gait coefficients in the tensor space via $\boldsymbol{v} = \mathfrak{f}(\nu)$ for VGGM-based gait synthesis defined in (4). Initially, we can define a fixed number of modes which are uniformly distributed along the angular gait manifold with equally large variances to cover all possible gait values. During inference, the mean and variance of some modes will be updated, as discussed in the following.

6.2.3 Segmental Jump-Diffusion MCMC Inference

We use the Metropolis–Hasting algorithm as the inference framework that incorporates two kinds of dynamics, *jump* and *diffusion*.

- *Jump* At the $i$the MCMC iteration, the state vector is denoted $\mathbf{x}^{(i)}$ that includes the gait sample $(\nu^{(i)})$, and the gait mode is $(m^{(i)})$. We randomly choose a mode $m^* \in [1, R]$ with a probability $1/R$ and generate a new sample ν^* according to $\mathcal{M}_{m^*}$. The proposal distribution defined is independent with $\nu^{(i)}$ and $m^{(i)}$:

$$q(\nu^*, m^* | \nu^{(i)}, m^{(i)}) = N(\nu^*; \mu_{m^*}, \delta_{m^*}^2). \tag{25}$$

  Hence, the acceptance ratio is

$$\alpha = \min\left\{1, \frac{p(\mathbf{x}^*|\mathbf{y}) N(\nu^{(i)}; \mu_{m^{(i)}}, \delta_{m^{(i)}}^2)}{p(\mathbf{x}^{(i)}|\mathbf{y}) N(\nu^*; \mu_{m^*}, \delta_{m^*}^2)}\right\}, \tag{26}$$

  where $\mathbf{x}^*$ is a new version of $\mathbf{x}^{(i)}$ by incorporating ν^* as the gait sample. $p(\mathbf{x}^*|\mathbf{y})$ is the posterior probability computed by the product of (21) and (22).
- *Diffusion* We randomly sample the gait variable ν^* with a proposal distribution as

$$q(\nu^* | \nu^{(i)}) = N(\nu^*; \nu^{(i)}, \delta_d^2), \tag{27}$$

  where δ_d^2 is the diffusion variance, that is decreased in a simulated annealing way. The acceptance ratio is

$$\alpha = \min\left\{1, \frac{p(\mathbf{x}^*|\mathbf{y}) N(\nu^{(i)}; \nu^*, \delta_d^2)}{p(\mathbf{x}^{(i)}|\mathbf{y}) N(\nu^*; \nu^{(i)}, \delta_d^2)}\right\} = \min\left\{1, \frac{p(\mathbf{x}^*|\mathbf{y})}{p(\mathbf{x}^{(i)}|\mathbf{y})}\right\}. \tag{28}$$

The pseudo code of SJD-MCMC is presented in Algorithm 2 that is the core of our inference algorithm and embedded in each time step of the particle filter.

Algorithm 2 Segmental jump-diffusion MCMC for whole-based gait estimation

Initialization: Let the initial MCMC sample $\nu_t^{(0)}$ be the MAP estimated gait variable from previous time step and keep the previous gait mode, i.e., $\nu_t^{(0)} = \hat{\nu}_{t-1}$ and $m_t^{(0)} = m_{t-1}$
Compute the segmental probability P_{seg} using (23)
for $i = 1, \ldots, (B + MN)$ (N is the number of samples, B is the length of the burn-in period and M is the length of the thinning interval) **do**
    Randomly sample $\gamma \sim U[0, 1]$
    if $P_{\text{seg}} \geq \gamma$ **then**
        Jump Sample the mode variable $m^* \in [1, R]$, and the gait variable ν^* according to (25). Compute the acceptance ration α using (26)
    else
        Diffusion Sample ν^* according to (27). Compute the acceptance ratio α using (28)
    end if
    Randomly sample $\eta \sim U[0, 1]$
    if $\alpha \geq \eta$ **then**
        Accept ν^* as $\nu_t^{(i+1)} = \nu^*$
        if ν^* is generated by *diffusion* **then**
            Decrease diffusion variance δ_d^2
        end if
    else
        Reject ν^* and let $\nu_t^{(i+1)} = \nu_t^{(i)}$
    end if
end for
Return the new sample set $\{\nu_t^{(B+kM)} | k = 1, \ldots, N\}$, and the estimated gait variable is $\hat{\nu}_t = \frac{1}{N}\sum_{k=1}^{N} \nu_t^{(B+kM)}$
Estimate the current gait mode m_t with respect to $\hat{\nu}_t$ by using the maximum likelihood estimation as $m_t = \arg\max_{i=1,\ldots,R}\{N(\hat{\nu}_t; \mu_i, \delta_i^2)\}$
Update the mean and variance for the current mode m_t by $\mu_{m_t} \leftarrow \frac{1}{2}(\mu_{m_t} + \hat{\nu}_t)$ and $\delta_{m_t} \leftarrow \delta_{m_t}^{\frac{1}{ct}}$, where c controls the annealing speed
The diffusion variance is also updated by $\delta_d = \delta_{m_t}$

6.3 Part-Based Gait Estimation

We propose a two-stage inference algorithm for part-whole gait estimation, as shown in Fig. 14, where the first stage is for whole-based gait estimation discussed

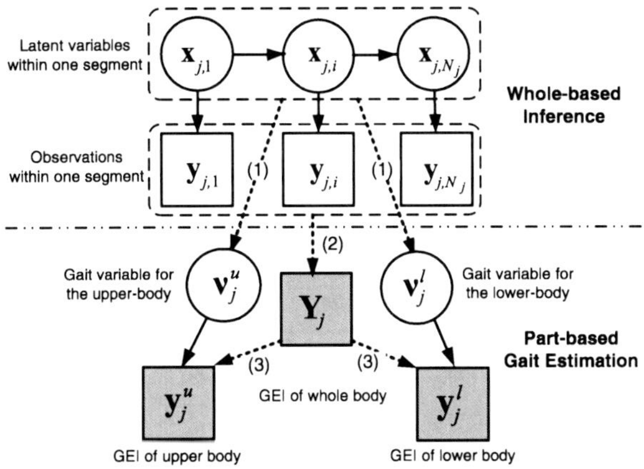

Fig. 14 The two-stage inference algorithm. *Dash lines* indicate the data association between whole-based (*top*) and part-based (*bottom*) gait estimation. Specifically, (**1**) represents priors for part-based gait estimation, and (**2**) and (**3**) are GEI-based observations generated from one segment at the whole-level and part-level respectively. (From X. Zhang, G. Fan, and L. Chou, Two-Layer Gait Generative Models for Estimating Unknown Human Gait Kinematics, in Proc. the 2nd International Workshop on Machine Learning for Vision-Based Motion Analysis (MLVMA'09), in conjunction with ICCV2009, Japan, Oct. 2009. @ 2009 IEEE)

in Sect. 6.2, and the second stage is for part-based gait estimation to be discussed in the following. Unlike whole-based gait estimation that is for each frame, part-based gait estimation is performed for each segment where we use the gait energy image (GEI) [21] as the part-level observation. Also whole based gait estimation provides some useful priors for part-based gait estimation.

6.3.1 Part-Level Gait Priors

The results of whole-based gait estimation can be used to speed-up part-based gait estimation by defining a search region along part-based gait manifolds. For segment j, the estimated whole-based gait variables for N_j frames are represented by $\boldsymbol{\nu}_j^w = \{\boldsymbol{\nu}_{j,i}^w \in \mathcal{M}_\nu^w | i = 1, \ldots, N_j, \}$ from which we can find the two closest training gait vectors $\mathcal{M}_\nu^w(w_1)$ and $\mathcal{M}_\nu^w(w_2)$ that can embrace $\boldsymbol{\nu}_j^w$ along $\mathcal{M}_\nu^w$ (as shown in Fig. 15). Now we need to pass this prior knowledge into part-based gait manifolds, $\mathcal{M}_\nu^l$. In other words, we need to find the corresponding gait vectors in $\mathcal{M}_\nu^l$ with respect to $\mathcal{M}_\nu^w(w_1)$ and $\mathcal{M}_\nu^w(w_2)$. By using the given manifold topologies, we can find l_1 and l_2 that make $\mathcal{M}^w(w_1) = \mathcal{M}^l(l_1)$ and $\mathcal{M}^w(w_2) = \mathcal{M}^l(l_2)$, where l_1 and l_2 are the 1D parameters corresponding to two training gaits in $\mathcal{M}_\nu^l$. Basically, l_1 and l_2 decide the search region for lower-body gait estimation along $\mathcal{M}_\nu^l$.

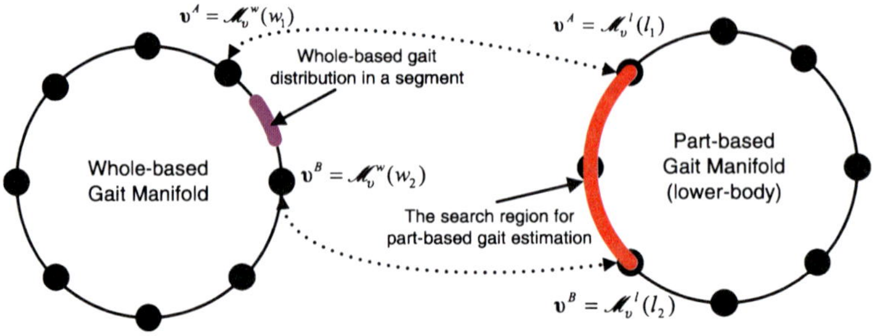

Fig. 15 Part-level gait priors from whole-based gait estimation, where $\boldsymbol{\upsilon}^A$ and $\boldsymbol{\upsilon}^B$ are the two training gait vectors defined in VGGM by (4). Each *solid circle* represents a training gait vector

6.3.2 Part-Level Likelihood Functions

Given a set of gait appearances in segment j, $\mathbf{Y}_j = \{\mathbf{y}_{j,i} | i = 1, \ldots, N_j\}$, we use the gait energy image (GEI) [21] to compute the part-level observations for a segment where two *mask* images are involved to weight the whole-based observation. For example, the GEI of lower body is obtained by:

$$\mathcal{Y}_j^l = \mathcal{I}^l \cdot \frac{1}{N_j} \sum_{i=1}^{N_j} \mathbf{y}_{j,i}, \tag{29}$$

where $\mathcal{I}^l$ is the *mask* image for the lower-body. We also need $\mathcal{I}^u$ for the upper-body. They are defined as:

$$\begin{aligned} \mathcal{I}^u(x, y) &= f\left(x | b, d, w, c^u\right), \\ \mathcal{I}^l(x, y) &= 1 - f\left(x | b, d, w, c^l\right), \end{aligned} \tag{30}$$

where c^u and c^l control the boundary between the lower and upper bodies. In this work, we use the "hip" to define the lower-body and upper-body that is always in the center of gait appearances. Hence, c^u and c^l are equal to the half of the image height. $f(\cdot)$ is a 1D edge model [70] that defines a smooth transition between two parts. Figure 16 shows how the part-level observations are computed.

Given $\mathbf{X}_j = \{\mathbf{x}_{j,i} | i = 1, \ldots, N_j\}$ obtained from whole-based gait estimation for segment j as well as the hypothesized part-level gait variable $\boldsymbol{\upsilon}_j^l$, the corresponding hypothesized lower-body GEI in segment j is defined as:

$$\tilde{\mathcal{Y}}_j^l = \mathcal{I}^l \cdot \tfrac{1}{N_j} \textstyle\sum_{i=1}^{N_j} \mathcal{V}\left(\mathbf{v}_{j,i}, \mathbf{s}_{j,i}, \boldsymbol{\upsilon}_j^l, \mathbf{p}_{j,i}\right), \tag{31}$$

where $\mathcal{V}(\cdot)$ is VGGM defined in (4). Given a hypothesized gait vector $\boldsymbol{\upsilon}_j^l$ for lower-body in segment j, the likelihood function for part-level gait estimation is defined as

$$p(\mathbf{Y}_j | \boldsymbol{\upsilon}_j^l, \mathbf{X}_j) \propto \exp\left(-\frac{\|\mathcal{Y}_j^l - \tilde{\mathcal{Y}}_j^l\|}{2\sigma^2}\right), \tag{32}$$

where σ is a predefined observation variance that controls the algorithm sensitivity, and $\|\cdot\|$ is the mean square error.

The complete inference algorithm is listed below.

Algorithm 3 Two-stage inference algorithm for part-based gait estimation

---

Given an observation $\mathbf{y}_t$, estimate its pose $\mathbf{p}_t$, view $\mathbf{v}_t$, shape $\mathbf{s}_t$ and gait $\boldsymbol{\nu}_t$ as Algorithm 1

for Whole-based gait estimation in Algorithm 2 **do**

  if estimated pose $\mathbf{p}_t$ is the *contact pose* **then**

    Define segment j with N_j observations $\mathbf{Y}_j$ (from previous *contact pose* to the current one) along with the estimated state variables $\mathbf{X}_j$

    Obtain the two bounds along $\mathcal{M}_\nu^w$ defined by two training gaits

    for Gait estimation for the lower-body **do**

      Compute the prior for lower body gait estimation as in Sect. 6.3.1

      Compute lower-part GEI $\mathcal{Y}_j^l$ according to (29)

      According to the prior, draw hypotheses of the lower-body gait variable along $\mathcal{M}_\nu^l$

      Compute the synthesized lower-body GEI via VGGM using (31)

      Evaluate synthesized GEIs with $\mathcal{Y}_j^l$ according to (32)

    end for

    Obtain the lower-body gait $\boldsymbol{\nu}_j^l$ for segment j by MAP estimation and map it into the kinematic gait manifold $\boldsymbol{\kappa}_j^l$ associated with KGGM

    Similarly, do part-level gait estimation for the upper-body and get $\boldsymbol{\nu}_j^u$ and $\boldsymbol{\kappa}_j^u$

    Generate the gait kinematics for frames $1, \ldots, N_j$ in segment j, where the upper-body and lower-body gait kinematics are reconstructed by using $\boldsymbol{\kappa}_j^u$ and $\boldsymbol{\kappa}_j^l$ along with estimated poses $\{\mathbf{p}_{j,i}\}_{i=1}^{i=N_j}$ in $\mathbf{X}_j$ via KGGM (3)

  end if

end for

---

7 Experimental Results and Discussions

We first discuss the experimental setup for training and testing as well as two error analysis methods for algorithm evaluation. Secondly, we examine and compare proposed five gait distance functions on KGGM in terms of their capability of gait synthesis. Thirdly, several inference algorithms are tested to show the advantage of SJD-MCMC for part-whole gait estimation. Fourth, our algorithm is compared with

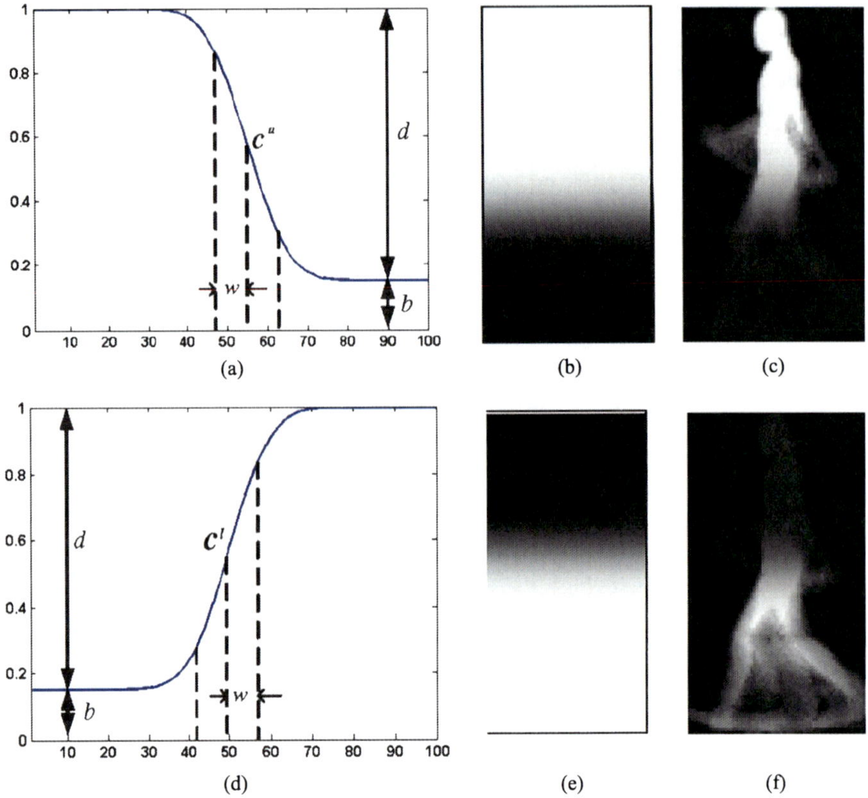

Fig. 16 Two 1D edge models (**a**) and (**d**) used for generating two mask images; (**b**) Upper-body mask; (**c**) Upper-body GEI; (**e**) Lower-body mask; (**f**) Lower-body GEI. (From X. Zhang, G. Fan, and L. Chou, Two-Layer Gait Generative Models for Estimating Unknown Human Gait Kinematics, in Proc. the 2nd International Workshop on Machine Learning for Vision-Based Motion Analysis (MLVMA'09), in conjunction with ICCV2009, Japan, Oct. 2009. @ 2009 IEEE)

a set of state-of-the-art algorithms in details. We also discuss some limitations of our algorithms.

7.1 Experimental Setups

7.1.1 Training Data Collection

The CMU Mocap library [11] provides a wealth of various human motion data, where we selected 20 walking motions (i.e., gait kinematics), represented by a series of Euler angles of joints, to learn KGGM. The gait appearances used for learning VGGM are generated by AutoDesk MotionBuilder. We rendered 100 3D gait animations by using the 20 gaits (that are used for KGGM training) and five human

Fig. 17 Five 3D human models. The first and last one are from the MotionBuilder Clip of Art and the others are from aXYZ design 3D

Fig. 18 Some gait animations generated by MotionBuilder that are about one gait (*the first one*) on five shapes under one view

models (Fig. 17). Each 3D gait animation was recorded under 12 camera views (30° apart).[4] Totally, we created 1200 30-frame gait animations (100 × 80). Figure 18 shows a set of gait motion data followed by five gait animations that are created by MotionBuilder using the same motion data under human five shape models. Moreover, we extracted the binary silhouettes that were further "softened" by the signed distance transform used in [34] to create the gait appearances for VGGM training. Given a binary image containing one object, the signed distance transform assigns to

[4]We need a skeleton model to convert the motion data represented by Euler angles into joint positions (mm) for animation generation. We chose one from the 20 CMU training gaits that best matches the five human models to generate all gait animations. All animations were created by setting the hip position as the image center.

Fig. 19 Silhouette sequences extracted for three HumanEva-I subjects. (From X. Zhang and G. Fan, Dual Generative Gait Models for Human Motion Estimation from a Single Camera, IEEE Trans. Systems, Man and Cybernetics, Part B: Cybernetics, Vol. 40, No. 4, 1034–1049, Aug. 2010. @ 2010 IEEE)

each pixel, both inside (positive) and outside (negative) of the object, the minimum distance from that pixel to the nearest pixel on the border of the object. Such representation imposes smoothness of the distance between different gait appearances.

7.1.2 Testing Data Collection

We tested our algorithm on Subjects 1, 2 and 3 in the HumanEva-I dataset [56]. We used the background subtraction technique in [17] to extract the foreground object.[5] We also developed two specific schemes to improve foreground extraction results. (1) We divided each frame into two regions vertically according to the overall intensity value (roughly along the boundary of the carpet), and background subtraction is applied in each region independently. (2) We also employed some simple morphological operators to clean up the isolated foreground pixels and to fill the holes (in the upper body). To extract the silhouettes, we need the 3D/2D hip positions in all frames (to be discussed shortly), the subject height and the camera calibration information. Then, for each frame, the silhouette size is determined by the distance between the 3D hip position and the camera as well as the subject height, and the silhouette center is the 2D hip position computed from the 3D hip position and the camera model. All extracted silhouettes that have different sizes have to be normalized to the size of training data (100×80). Some examples of normalized silhouettes of three subjects are shown in Fig. 19. Like training data collection, we need to apply the signed distance transform to convert binary silhouettes into gray-scale gait appearances.

7.1.3 Local Error Analysis

Essentially, our algorithm computes the local motion that records the relative joint position with respect to the hip. The local error (ERR-I, mm) measures the 3D distance between the estimated and ground-truth joint positions that excludes the global

[5] We used the C++ code from http://cvlab.epfl.ch/~tola/open_source.html.

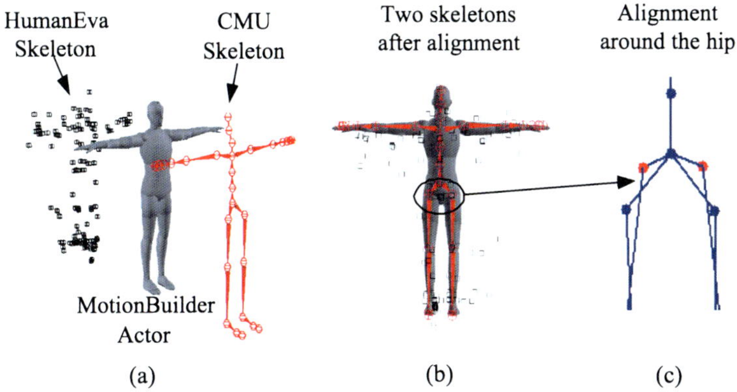

Fig. 20 The illustration of scaling/skeleton mappings. (**a**) Alignment of two skeletons after scale mapping. (**b**) Two skeletons after alignment. (**c**) The *red* and *blue dots* indicate different thigh configurations around the hip. (From X. Zhang and G. Fan, Dual Generative Gait Models for Human Motion Estimation from a Single Camera, IEEE Trans. Systems, Man and Cybernetics, Part B: Cybernetics, Vol. 40, No. 4, 1034–1049, Aug. 2010. @ 2010 IEEE)

position of the subject. Because the skeleton (or the marker system) in HumanEva-I (15 joints) and the one in CMU (18 joints excluding fingers, thumbs, toes) have partially different joint configurations, we need to establish a mapping relationship between them under the T-pose (Fig. 17) in order to compute a valid ERR-I value. Specifically, two mapping operations are needed, as shown in Fig. 20.

- *Scale mapping* accounts for the height difference between the training and testing subjects, and it resizes the training skeleton to match the testing one according to their height ratio. This operation implies that the lengths of all body parts are proportional to the height [42]. After scale mapping, the eight major joints (two elbows, hands, knees and feet) are usually matched very well because they are commonly shared by most skeleton systems. However, the rest of six joints (head, chest, two shoulders, and two thighs) are defined quite differently between the training and testing skeletons that need to be aligned further.
- *Skeleton mapping* is inspired by the motion retargeting technique in computer graphics [37] that can adapt one motion data captured from one figure to another one. This operation can be implemented efficiently via MotionBuilder. We first define a reference model by which we can align two skeletons to share the same hip point. Then, for each of six joints, we compute the translational displacement between two skeletons by which we can associate the same joint in two skeletons. This operation assumes that the relative position between each joints to the hip is fixed that may not very accurate if some local deformation occurs. This step can be avoided if we can use the same skeleton for training and testing.

Estimated gait kinematics are initially represented by a sequence of Euler angles of joints that can be converted into relative joint positions by using the training skeleton. Then via scale/skeleton mappings, the joint positions under the training skeleton can be mapped to the ones under the testing skeleton. Then ERR-I for each joint can be computed by aligning the estimated hip and the ground-truth one. ERR-I is used to validate the usefulness of KGGM and VGGM.

7.1.4 Global Error Analysis

The global error (ERR-II) computes the distance between the estimated joint positions and the ground-truth ones in the global 3D space. ERR-II is used for overall performance evaluation in terms of the accuracy of video-based human motion estimation. Since our algorithm directly outputs the local motion for each frame, we need to convert it into the global motion by following three steps.

- *Global hip localization* is implemented by a particle filter based on a series of silhouettes. The particle dynamics is a 3D motion model that has a constant angular velocity and random walks in other two dimensions. The tracker is initialized by giving the ground-truth 3D hip position in the first frame with zero velocity. For a new frame, the hypotheses of 3D hip position are generated by the motion model that are mapped to the 2D image plane via the camera model. Then we apply the previous silhouette to find the best 2D/3D hip hypothesis. We only use the upper part of silhouettes for matching where the quality of foreground extraction is relatively stable. In our experiment, the errors (mm) of global hip localization are 21.75, 16.67, and 18.40, for Subjects 1, 2, and 3, respectively.
- *Local motion estimation* is accomplished by using the proposed algorithm for a series of gait appearances extracted according to the estimated 2D/3D hip positions, as discussed in Sect. 7.1.2. Local motion records the relative position between each joint and the hip.
- *Gait motion direction* is computed by comparing the present and previous 3D hip positions. It is needed to impose the local motion estimation result (after scale/skeleton mappings) onto the 3D hip position to reconstruct the complete 3D motion estimation that records the global joint positions.

7.2 Experiments on KGGM

The experiments on KGGM have two purposes. One is to test its capability of gait synthesis where a gait manifold is involved to approximate an unknown gait, and the other is to examine three distance functions each of which can lead to a specific gait manifold. Specifically, two out of three distance functions support part-whole gait modeling with one whole-based gait manifold and two part-based ones. According to (3), given a gait vector along the kinematic gait manifold $\boldsymbol{\kappa}' \in \mathcal{M}_\kappa$ defined in (16), KGGM can synthesize the corresponding gait kinematics of an arbitrary pose by (3). Given an unknown gait, we use KGGM to approximate it by an exhaustive search along $\boldsymbol{\kappa}' \in \mathcal{M}_\kappa$ and find the optimal gait vector that yields the best synthesized gait with the smallest ERR-I. We call the smallest ERR-I provided by KGGM the *lower error bound* (LEB) that indicates the best performance we can achieve for

Table 1 The LEB results (mm) of using KGGM for gait synthesis under three distance functions

	Whole/Part	Sub1	Sub2	Sub3
Single best matched gait	Whole	63.26	79.91	79.32
Gait tensor coefficient	Whole	32.20	46.25	44.87
3D joint position	Whole	32.38	49.09	48.31
3D joint position	Part	26.53	39.51	36.85
Fourier analysis	Whole	**31.37**	**44.59**	**43.72**
Fourier analysis	Part	**23.24**	**33.05**	**33.91**

this specific gait. The key to KGGM-based gait synthesis is the gait manifold that is used for gait interpolation. Since different distance functions lead to different gait manifolds, the optimal distance function should be the one that provides the smallest LEB and will be used in our following experiments.

Given the gaits of three HumanEva subjects, we tried to approximate each of them by using KGGM (learned from Mocap) under different gait manifolds at both whole and part levels. We also included the result of using the best matched training gait without gait synthesis. The numerical results of ERR-I are shown in Table 1. It is obvious that KGGM provides much more accurate results comparing with the one without gait synthesis. For the three distance functions, the best result is achieved by the *Fourier analysis method* that compares two gaits in terms of their dynamics at each joint and can reduce the noise effect by only using the top DFT components for gait representation. Also, the distance based on gait tensor coefficients is quite robust and effective due to the dimension reduction nature of tensor decomposition, but it does not support part-level gait modeling. The distance based on 3D joint positions has a slightly larger error than other two metrics due to two reasons. One is that its high-dimensional nature makes the Euclidean distance less accurate, and the other is that it does not directly reflect the dynamic dissimilarity between two gaits. Moreover, part-level gait modeling always improve the gait synthesis results by over 20%, showing the usefulness and potential of part-whole gait modeling.

7.3 Evaluation of Two-Stage Inference

7.3.1 Segmental Gait Modeling

We used Subject I as the example for this experiment. We first derived the optimal gait value (with the least ERR-I) for each frame using an exhaustive search along the gait/pose manifolds in KGGM that serves as the ground-truth value for algorithm evaluation. Then we tested SJD-MCMC regarding its performance of dynamic gait estimation based on observed gait appearances. Results are illustrated in Fig. 21 where the interpolated gait along the gait manifold is shown in the vertical axes as

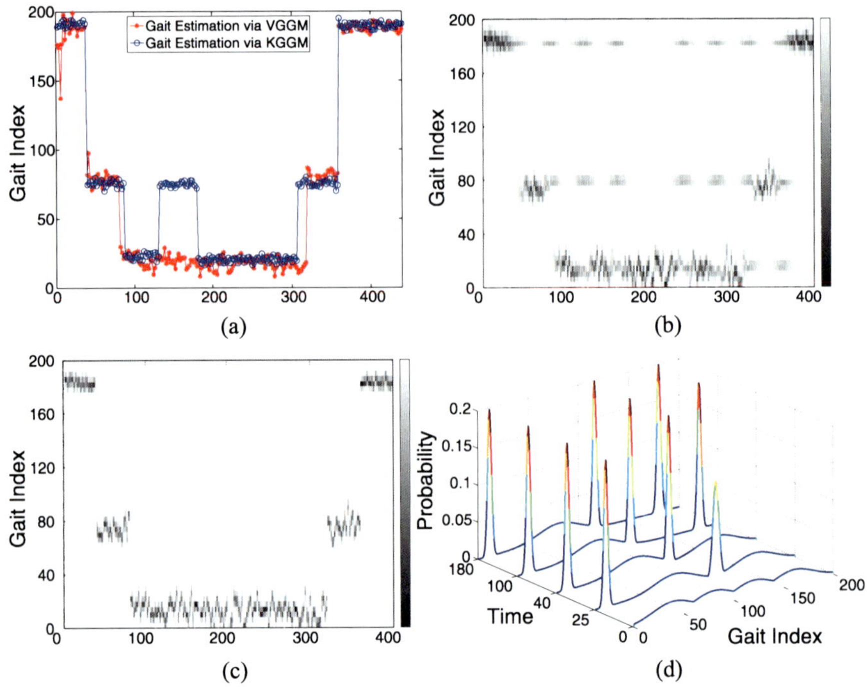

Fig. 21 Gait estimation results for Subject 1 where the *horizontal/vertical axes* show the frame index and gait index (indicating interpolated gaits along the gait manifold), respectively: (**a**) the comparison between the estimated and ground-truth gait values, (**b**) the distribution of generated gait samples, (**c**) the distribution of accepted samples, and (**d**) the online learning results of the four gait modes defined in (24)

gait index. In Fig. 21(a), the segmental variability of the gait variable is evident, and gait estimation results are mostly accurate except few gait mode jumps are missed. Specifically, Figs. 21(b) and 21(c) present the distribution of the gait samples generated before and after evaluation during SJD-MCMC. The mixed jump-diffusion dynamics in conjunction with segmental modeling can capture the dynamic nature of the gait variable very well. Figure 21(d) displays the learning of four modes in SJD-MCMC among which only three modes are updated over time.

7.3.2 Local Motion Estimation

We implemented four algorithms (Alg-1, Alg-2, Alg-3, Alg-4) for whole-based gait estimation and two algorithms (Alg-5, Alg-6) for part-whole gait estimation. The numerical results are shown in Table 2. Specifically, Alg-1 is the basic particle filter without segmental gait modeling. Alg-2 is the off-line version of Alg-1 with the gait variable fixed to be the one estimated by Alg-1. In other words, Alg-1 and Alg-2 do not consider dynamic gait estimation. Alg-3 is the particle filter embedded with the

Table 2 Comparison of six inference algorithms (ERR-I, mm). The first four use whole-based gait modeling, while the last two involve part-whole gait modeling

ERR-I (mm)	LEB	Alg-1	Alg-2	Alg-3	Alg-4	Alg-5	Alg-6
Subject 1	32.20	95.54	89.84	83.46	**78.79**	79.11	**74.93**
Subject 2	46.25	100.26	92.86	85.01	**82.11**	77.25	**74.53**
Subject 3	44.87	103.32	94.31	91.72	**87.37**	84.33	**81.64**

SJD-MCMC with online mode learning, while Alg-4 is the off-line version of Alg-3 and uses the four modes pre-learnt by Alg-3 for dynamic gait estimation. Alg-5 involves two-stage inference for part-whole gait estimation, while Alg-6 is the off-line version of Alg-5 with all part-whole gait variables pre-learned. The ERR-I results improve from Alg-1 to Alg-4 progressively, showing that segmental gait modeling clearly improves the accuracy of motion estimation. Moreover, part-whole gait modeling further reduces ERR-I by 5%–10% (from Alg-3 to Alg-5 or from Alg-4 to Alg-6). The improvement for Subject 1 is relatively small (near 5%). This is due to the low quality of the silhouettes.

7.3.3 Whole-Based Gait Estimation

Figure 22 depicts the estimated whole-based gait, pose, view and ERR-II for three subjects, where we have four important observations. (1) The gait variable of three subjects exhibit obvious segmental variability, and the estimated values (by VGGM) are close to the ground-truth ones (estimated by KGGM with LEB). In some frames in the video sequence of Subjects 1 and 2, there exist differences of the gait variable estimated from VGGM comparing to the one from KGGM. Usually, the gait estimation discrepancy occurs in the front or back view where VGGM-based gait estimation is often hindered by the lack of motion information from gait appearances. The gait estimation discrepancy contributes the ERR-I and ERR-II but it is not the dominant factor. The reason is that the gait estimation is constraint by the gait manifold mode model, which has provided distribution models along the gait manifold according to the similarity to the testing subject. (2) The estimated pose and view well reflect the circular and periodic walking patterns. (3) Three ERR-II plots roughly exhibit a periodic pattern that is consistent with the nature of a gait motion. (4) The result Subject 3 is the worst one in terms of both ERR-I (Table 2) and ERR-II (Fig. 22). It is mainly because that Subject 3 is leaning inwards during walking (more than other two subjects) while all training gaits follow a straight-line motion with an upright posture. We show more visual results in Fig. 23, where the ground-truth (in red) and estimated (in green) joint positions are plotted. In most frames, the motion estimation results are fairly accurate.

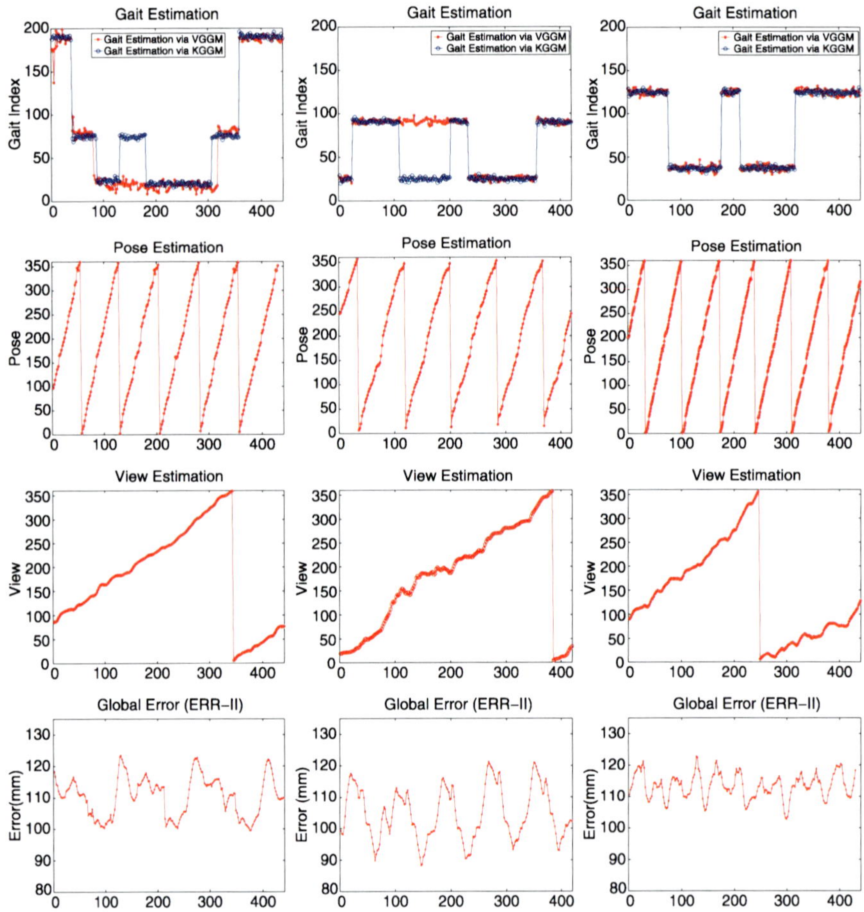

Fig. 22 The whole-based results of gait, pose, view and ERR-II from Subjects 1, 2, and 3 (*from left to right*)

7.3.4 Part-Whole Gait Estimation

We show detailed results of part-whole gait estimation of Subject 3 in Fig. 24. Figure 24(a) compares whole-based gait estimation and the ground-truth ones that yield the LEB in KGGM. In most frames, the estimated gait is close to the ground-truth. Figures 24(b) and 24(c) show the part-based gait estimation results, where dashed boxes are the prior from whole-based gait estimation and the solid line are the estimated part-based gait variable for each segment. Figures 24(d), 24(e), and 24(f) show the estimated view, pose and averaged ERR-II for each frame. As shown in Fig. 24(f), part-based gait estimation outperforms the whole-based one significantly in terms of ERR-II.

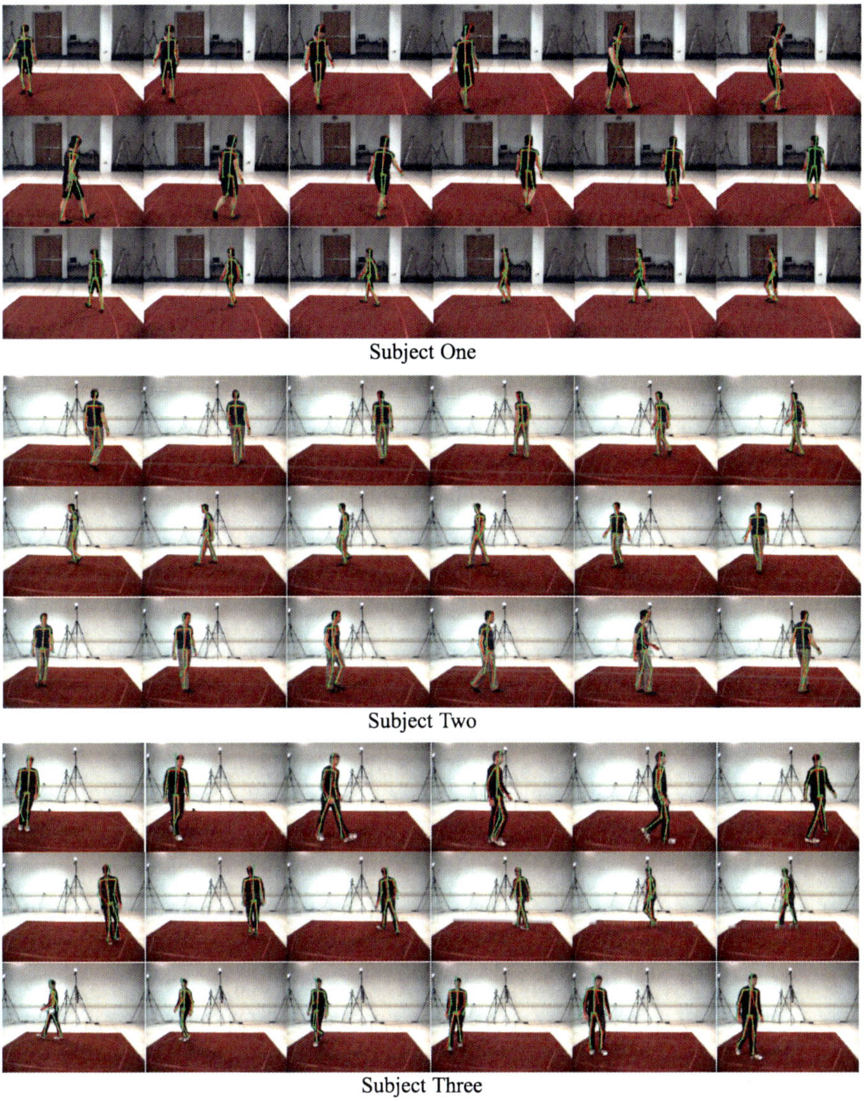

Fig. 23 Illustration of experimental results on three HumanEva-I subjects (Alg-6). The estimated (as *green* '*') and ground-truth joint positions (as *red* 'o') are drawn on each image

7.4 Overall Performance Evaluation

We compare our algorithms (Alg-4: the whole-based approach, and Alg-6: the part-based approach) with a series of recent algorithms in Table 3, where most methods were tested on HumanEva-I, some [10, 18] on HumanEva-II, and one [23] on both datasets (only the HumanEva-I result is reported here). HumanEva-II is more chal-

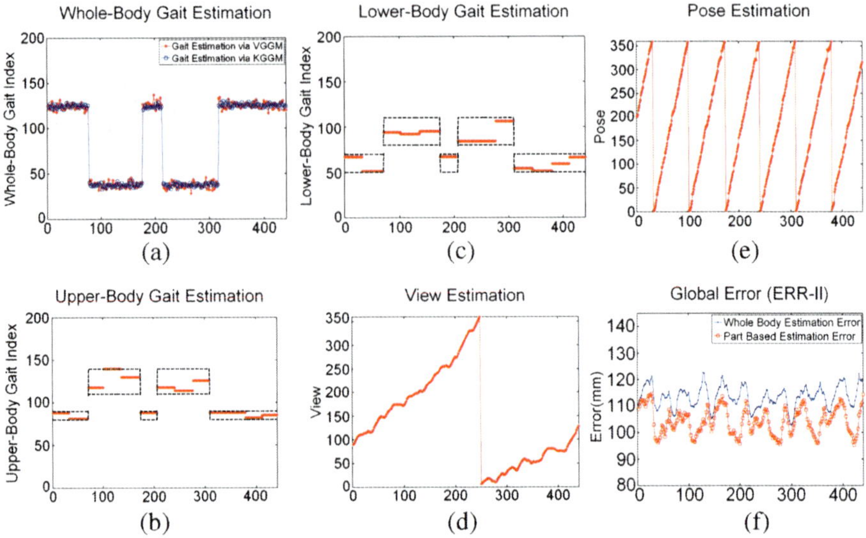

Fig. 24 (**a**) Whole-based gait estimation; (**b**) part-based gait estimation (*upper-body*) and (**c**) part-based gait estimation (*lower-body*), where the boxed areas with *dashed lines* show the local search regions (the priors from whole-based gait estimation) in each segment; (**d**) view estimation; (**e**) pose estimation; (**f**) ERR-II of whole-based gait estimation ($*$) and part-based gait estimation (○)

lenging due to the mixed motion types (walking and jogging) in one sequence, but the results from [10, 18] that are listed here are only for the walking portion from Subject 2 who is also included in HumanEva-I. Therefore, it is still reasonable to include them for a comprehensive comparison. Furthermore, all algorithms are discussed in three groups according to how motion data are used for training. The Group-I algorithms requires motion data for training, and the same subject is used for training and testing. The Group-II algorithms do not require any motion data for training. The Group-III algorithms requires motion data for training, and the subjects used for training do not include the ones used for testing. Both of our algorithms belong to Group-III. Additionally, we evaluate all algorithms in each group by considering their experimental settings, including the number of cameras; global or local motion estimation (G/L), motion data for training, new testing subjects, discriminative or generative approaches, visual observations. In the following, we discuss each group in details.

7.4.1 Group-I

Essentially, the algorithms in this group aim at pose estimation instead of motion estimation due to the fact that the same subject is involved for training and testing. Most algorithms involve certain temporal prior in inference to ensure smooth and continuous pose estimation. Specifically, the discriminative approaches, e.g.,

Table 3 Comparison of recent algorithms tested on the HumanEva-I or Human-II dataset (*), including estimation errors of three testing subjects (Subjects 1, 2, 3) and experimental settings. (Adapted from X. Zhang and G. Fan, Dual Generative Gait Models for Human Motion Estimation from a Single Camera, IEEE Trans. Systems, Man and Cybernetics, Part B: Cybernetics, Vol. 40, No. 4, 1034–1049, Aug. 2010. @ 2010 IEEE)

	Sub. one	Sub. two	Sub. three	Camera number	Global/ local	Motion training	New sub.	Discrimi- native/ generative	Visual observation
Elgammal [16]	24.71	31.16	38.21	1	L	Y	N	G	silhouette
Poppe [49]	37.54	40.09	55.25	3	L	Y	N	D	HOG
Howe [22]	99			1	G	Y	N	D	silhouette + optical flow
Urtasun [67]	31.4	19.3	47.4	1	G	Y	N	D	HOG + edge
Sigal [60]	64.63			3	G	Y	N	D	shape context
Okada [46]	41.19	35.03	37.69	1	L	Y	N	D	HOG
Ni [43]	8.57	8.57	8.57	7	G	Y	N	G	silhouette
Bo [5]	23	13.7	40.3	1	L	Y	N	D	shape context
Gall [18]		32.23*		4	G	N	N	G	silhouette
Brubaker [7]	104			1	G	N	Y	G	edge
Mundermann [41]	53.1			7	G	N	Y	G	silhouette
Husz [23]	105.7			3	G	N	Y	G	silhouette
Canton-Ferrer [9]		115.21	115.21	4	G	N	Y	G	silhouette
Vondrak [71]	93.4	93.4	93.4	3	G	N	Y	G	silhouette
Xu [73]	140.35	149.37	156.3	4/7	G	Y	N	G	silhouette + edge
Cheng [10]		125*		4	G	Y	Y	G	silhouette
Peurum [48]	85.5	116.9	84.7	3	G	Y	Y	G	silhouette + edge
Whole-based	112.26	104.37	114.53	1	G	Y	Y	G	silhouette
Part-whole	108.79	98.34	106.29	1	G	Y	Y	G	silhouette
Whole-based	88.09	98.56	105.29	1	G	Y	Y	G	silhouette (cleaned)
Part-whole	80.12	91.63	97.84	1	G	Y	Y	G	silhouette (cleaned)

[5, 22, 46, 49, 60, 67], involve a direct mapping between visual observations to body configurations without explicitly pose modeling. The key issue of this mapping is how to handle the multiple-to-one problem due to the ambiguity of visual observations. On the other hand, the generative approaches require explicit pose modeling for visual or kinematic data where the key issue is how to deal with the view variability of visual observations. For example, the method in [16] involves a torus-shaped manifold for multi-view pose modeling. Two mappings are learned to map both visual and kinematic data onto the torus. Then, the pose/view can be jointly estimated via maximum a priori estimation along the torus manifold. The best pose estimation result was reported in [43] where a hybrid sample-and-refine framework was proposed by combining both stochastic sampling and deterministic optimization for pose estimation and which also requires seven cameras.

7.4.2 Group-II

Most algorithms in this group are generative approaches, since they usually need an explicit human model based on which motion estimation is accomplished. They can deal with unknown testing subjects, except the one in [18] that reports the best result in Group-II and requires the 3D body scan data of the testing subject. The human shape model plays a key role for Group-II algorithms that is expected to be flexible and general enough to handle different subjects. For example, a physics-based biomechanical model was used in [7, 71]. In [41], a large set of body scan data was collected and used to learn a parametric 3D human model that is general enough to handle various body shapes. Additionally, physical constraints can be incorporated into the human model that provide useful priors for estimation and inference [9, 23]. The anthropometric measurements are also crucial in some algorithms in this group that can be learned online by using different shape models, such as cylinders [71] or visual hulls [9, 23, 41].

7.4.3 Group-III

Similar to Group II, most algorithms in this group are generative approaches, which need explicit shape and motion modeling. For example, the body shape can be modeled by bounding boxes [48], visual hulls [10], cylinders [73], or 3D character models (like ours). Motion modeling can be implemented by using a graphical model [10, 48] or a DR method (like ours) that can be trained from a set of kinematic data. In [73], the symmetric property of gait kinematics was used for motion modeling. Most algorithms model the motion and shape separately, and the two models are only used together during inference. However, one highlight of our research is that the two models are integrated together via VGGM and KGGM during both learning and inference. We also studied both whole-level (Alg-4) and part-level (Alg-6) gait modeling. Our algorithms were tested on both actual observations and ones with manual clean-up, showing both the real performance and potential of our approaches. Since the actual observations of Subject 1 have strong shadows around feet, part-whole gait estimation shows less improvement compared with that of Subjects 2 and 3. On the other hand, the improvement is consistent 7–8 mm for three subjects when cleaned observations are used. Overall, our algorithm provides very promising results compared with the peers in the same group, considering only a single camera is used.

7.5 Limitations and Discussion

There are some limitations in the proposed algorithm that will guide our future research.

- The proposed algorithms are mainly designed for a specific type of motion, such as walking. The framework may be applied on other motion types but may not handle different motion types together (such as walking, running, and jogging) due to the fact that the strong dissimilarity between them may not be generalized well by one generative model.
- The algorithm's accuracy heavily relies on the quality and richness of training gaits. Although increasing training gaits and shape models would improve the algorithm performance, it will also drastically increase the computational complexity. The algorithm is moderately sensitive to the quality of the silhouette detection and extraction. The noisy silhouette decreases the algorithm performance as shown in Table 3.
- There are a couple of systematic errors in this algorithm. One is that we ignore the local variability of a gait by assuming that any gait can be approximated locally by a straight motion with an upright posture. This assumption may not be accurate when the subject exhibits some nonstraight or inclined motion patterns. The other is about the two assumptions made for the scale/skeleton mappings that is needed to compute ERR-II for performance evaluation. This error could be reduced if the same marker system is used for training and testing.
- The global hip-position estimation assumes that the ground-truth position is available in the first frame. In practice, we can initialize the hip position by estimating the height and 3D position of the subject using camera calibration information and ground plane constraint (one foot is always on the ground plane).
- The computational load of the proposed algorithms is quite high due to the complexity of the generative models. Particularly, VGGM involves four or six parameters that have to be estimated during inference where each Monte Carlo run involves a tensor product for the synthesis of gait appearance. Our algorithm was developed in Matlab 2009, and current implementation (without program optimization) is about 35 seconds per frame (including both global hip localization and local motion estimation for part-whole gait estimation) on a PC computer (2G memory, dual-core, 2.2 GHz).

8 Conclusion and Future Research

In this chapter, we have presented a new approach to video-based human motion estimation that involves two gait generative models, KGGM and VGGM, to represent the kinematics and appearances by a few latent variables, respectively. The main idea is to learn part-whole gait manifolds by which two generative models can be integrated for gait estimation at both whole and part levels. A key issue for gait manifold learning is to define an appropriate distance function that can reflect the similarity between two gaits and determines the topology of gait manifold. Specifically, a whole-based and two part-based gait manifolds are introduced for gait modeling at both the whole and part levels, respectively. A two-stage particle filtering-based inference algorithm was proposed to support sequential whole-based and part-based gait estimation. The experimental results demonstrate the usefulness of part-whole gait modeling via two generative models.

Our future research will focus on three issues: (1) how to create an adaptive yet simple human model to accommodate more shape variability; (2) how to span a more informative gait manifold that can support more accurate gait estimation; (3) how to extend the proposed framework to multiple other human activities. Our current implementation only involves five fixed character models that have limited capability for complex human shape modeling. We plan to use a more advanced human shape modeling method in future work. Moreover, we only consider a simple 1D structure for gait manifold learning that offers limited interpolation capability, and a higher dimension (e.g., 2D) may be more helpful. However, due to the sparseness of training gaits, it will be challenging to span a 2D gait manifold where the inference algorithm has to be carefully designed to ensure efficient gait estimation. The proposed dual generative model framework can be extended to other human motions by using corresponding training visual and kinematic data. To cope with a sequence with multiple motions together like HumanEva-II testing sequences, a human behavior recognition method is needed to control and switch among different generative models for various motions.

Acknowledgements This work is supported by the National Science Foundation (NSF) under Grant IIS-0347613 and an OHRS award (HR09-030) from the Oklahoma Center for the Advancement of Science and Technology (OCAST).

References

1. Agarwal, A., Triggs, B.: 3D human pose from silhouettes by relevance vector regression. In: Proc. IEEE Conference on Computer Vision and Pattern Recognition (2004)
2. Agarwal, A., Triggs, B.: Recovering 3D human pose from monocular images. IEEE Trans. Pattern Anal. Mach. Intell. **28**, 44–58 (2006)
3. Anguelov, D., Srinivasan, P., Koller, D., Thrun, S., Rodgers, J.: SCAPE: Shape completion and animation of people. ACM Trans. Graph. **24**, 408–416 (2005)
4. Balan, A.O., Sigal, L., Black, M.J., Davis, J.E., Haussecker, H.W.: Detailed human shape and pose from images. In: Proc. IEEE Conference on Computer Vision and Pattern Recognition (2007)

5. Bo, L., Sminchisescu, C., Kanaujia, A., Metaxas, D.: Fast algorithms for large scale conditional 3D prediction. In: Proc. IEEE Conference on Computer Vision and Pattern Recognition (2008)
6. Brubaker, M., Fleet, D.: The kneed walker for human pose tracking. In: Proc. IEEE Conference Computer Vision and Pattern Recognition (2008)
7. Brubaker, M., Fleet, D., Hertzmann, A.: Physics-based human pose tracking. In: Proc. NIPS Workshop on Evaluation of Articulated Human Motion and Pose Estimation (2006)
8. Brubaker, M., Fleet, D., Hertzmann, A.: Physics-based person tracking using simplified lower-body dynamics. In: Proc. IEEE Conference on Computer Vision and Pattern Recognition (2007)
9. Canton-Ferrer, C., Casas, J., Pardas, M.: Exploiting structural hierarchy in articulated objects towards robust motion capture. In: Conference on Articulated Motion and Deformable Objects (2008)
10. Cheng, S.Y., Trivedi, M.M.: Articulated human body pose inference from voxel data using a kinematically constrained Gaussian mixture model. In: Proc. CVPR 2nd Workshop on Evaluation of Articulated Human Motion and Pose Estimation (2007)
11. CMU Human Motion Capture Database. Available at http://mocap.cs.cmu.edu
12. Cunado, D., Nixon, M.S., Carter, J.N.: Automatic extraction and description of human gait models for recognition purposes. Comput. Vis. Image Underst. **90**, 1–41 (2003)
13. Ek, C.H., Torr, P., Lawrence, N.: Gaussian process latent variable models for human pose estimation. In: Proc. Machine Learning and Multimodal Interaction (2007)
14. Elgammal, A., Lee, C.S.: Inferring 3D body pose from silhouettes using activity manifold learning. In: Proc. IEEE Conference on Computer Vision and Pattern Recognition, vol. 2, pp. 681–688 (2004)
15. Elgammal, A., Lee, C.S.: Separating style and content on a nonlinear manifold. In: Proc. IEEE Conference on Computer Vision and Pattern Recognition (2004)
16. Elgammal, A., Lee, C.S.: Tracking people on torus. IEEE Trans. Pattern Anal. Mach. Intell. **31**, 520–538 (2009)
17. Elgammal, A., Harwood, D., Davis, L.: Non-parametric model for background subtraction. In: Proc. European Conference on Computer Vision (2000)
18. Gall, J., Rosenhahn, B., Brox, T., Seidel, H.P.: Optimization and filtering for human motion capture—a multi-layer framework. Int. J. Comput. Vis. **87**(1–2), 75–92 (2010). doi:10.1007/s11263-008-0173-1
19. Guo, F., Qian, G.: Monocular 3D tracking of articulated human motion in silhouette and pose manifolds. EURASIP J. Image Video Process. **2008**, 1–18 (2008)
20. Gupta, A., Chen, T., Chen, F., Kimber, D., Daivs, L.: Context and observation driven latent variable model for human pose estimation. In: Proc. IEEE Conference on Computer Vision and Pattern Recognition (2008)
21. Han, J., Bhanu, B.: Individual recognition using gait energy image. IEEE Trans. Pattern Anal. Mach. Intell. **28**(2), 316–322 (2006)
22. Howe, N.R.: Recognition-based motion capture and the HumanEva II test data. In: Proc. CVPR 2nd Workshop on Evaluation of Articulated Human Motion and Pose Estimation (2007)
23. Husz, Z.L., Wallace, A., Green, P.: Evaluation of a hierarchical partitioned particle filter with action primitives. In: Proc. CVPR 2nd Workshop on Evaluation of Articulated Human Motion and Pose Estimation (2007)
24. http://www.idleworm.com/how/anm/02w/walk1.shtml
25. Jaeggli, T., Koller-Meier, E., Gool, L.V.: Multi-activity tracking in LLE body pose space. In: Proc. International Conference on Computer Vision 2nd Workshop on Human Motion (2007)
26. Kanaujia, A., Sminchisescu, C., Metaxas, D.: Semi-supervised hierarchical models for 3D human pose reconstruction. In: Proc. IEEE Conference on Computer Vision and Pattern Recognition (2007)
27. Lan, X., Huttenlocher, D.: A unified spatio-temporal articulated model for tracking. In: Proc. IEEE Conference on Computer Vision and Pattern Recognition (2004)

28. Lan, X., Huttenlocher, D.: Beyond trees: common-factor models for 2D human pose recovery. In: Proc. IEEE International Conference on Computer Vision (2005)
29. Lawrence, N.: Gaussian process latent variable models for visualization of high dimensional data. In: Advances in Neural Information Processing. MIT Press, Cambridge (2003)
30. Lawrence, N., Candela, J.: Local distance preservation in the GPLVM through back constraints. In: International Conference on Machine Learning (2006)
31. Lee, M.W., Cohen, I.: Proposal maps driven MCMC for estimating human body pose in static images. In: Proc. IEEE Conference on Computer Vision and Pattern Recognition (2004)
32. Lee, C.S., Elgammal, A.: Body pose tracking from uncalibrated camera using supervised manifold learning. In: Proc. of NIPS Workshop on Evaluation of Articulated Human Motion and Pose Estimation (2006)
33. Lee, C.S., Elgammal, A.: Simultaneous inference of view and body pose using torus manifolds. In: Proc. International Conference on Pattern Recognition (2006)
34. Lee, C.S., Elgammal, A.: Modeling view and posture manifolds for tracking. In: Proc. IEEE International Conference on Computer Vision (2007)
35. Lowe, D.G.: Distinctive image features from scale-invariant keypoints. Int. J. Comput. Vis. **60**, 91–110 (2004)
36. Moeslund, T.B., Hilton, A., Kruger, V.: A survey of advances in vision-based human motion capture and analysis. Comput. Vis. Image Underst. **104**, 90–126 (2006)
37. Monzani, J.S., Baerlocher, P., Boulic, R., Thalmann, D.: Using an intermediate skeleton and inverse kinematics for motion retargeting. Comput. Graph. Forum **19**, 11–19 (2000)
38. Moon, K., Pavlovic, V.: Impact of dynamics on subspace embedding and tracking of sequences. In: Proc. IEEE Conference on Computer Vision and Pattern Recognition (2006)
39. Mori, G., Malik, J.: Recovering 3D human body configurations using shape contexts. IEEE Trans. Pattern Anal. Mach. Intell. **28**, 1052–1062 (2006)
40. Mudermann, L., Corazza, S., Andriacchi, T.P.: The evolution of methods for the capture of human movement leading to markerless motion capture for biomechanical applications. J. Neuroeng. Rehabil. **3** (2006). doi:10.1186/1743-0003-3-6
41. Mudermann, L., Corazza, S., Andriacchi, T.P.: Markerless human motion capture through visual hull and articulated ICP. In: Proc. NIPS Workshop on Evaluation of Articulated Human Motion and Pose Estimation (2006)
42. Mundermann, L., Corazza, S., Andriacchi, T.P.: Accurately measuring human movement using articulated ICP with soft-joint constraints and a repository of articulated models. In: Proc. IEEE Conference on Computer Vision and Pattern Recognition (2007)
43. Ni, B., Kassim, A.A., Winkler, S.: A hybrid framework for 3D human motion tracking. IEEE Trans. Circuits Syst. Video Technol. **18**, 1075–1084 (2008)
44. Ning, H., Tan, T., Wang, L., Hu, W.: Kinematics-based tracking of human walking in monocular video sequences. Image Vis. Comput. **22**, 429–441 (2004)
45. Ning, H., Xu, W., Gong, Y., Huang, T.: Discriminative learning of visual words for 3D human pose estimation. In: Proc. IEEE Conference on Computer Vision and Pattern Recognition (2008)
46. Okada, R., Soatto, S.: Relevant feature selection for human pose estimation and localization in cluttered images. In: Proc. European Conference on Computer Vision (2008)
47. Ong, E.J., Micilotta, A., Bowden, R., Hilton, A.: Viewpoint invariant exemplar-based 3D human tracking. Comput. Vis. Image Underst. **104**, 178–189 (2006)
48. Peurum, P., Venkatesh, S., West, G.: A study on smoothing for particle-filtered 3D human body tracking. Int. J. Comput. Vis. **87**, 53–74 (2010). doi:10.1007/s11263-009-0205-5
49. Poppe, R.: Evaluating example-based pose estimation: experiments on the HumanEva set. In: Proc. CVPR 2nd Workshop on Evaluation of Articulated Human Motion and Pose Estimation (2007)
50. Poppe, R.: Vision-based human motion analysis: an overview. Comput. Vis. Image Underst. **108**, 4–18 (2007)
51. Ramanan, D., Forsyth, D.A., Zisserman, A.: Strike a pose: tracking people by finding stylized poses. In: Proc. IEEE Conference on Computer Vision and Pattern Recognition (2005)

52. Rogez, G., Rihan, J., Ramalingam, S., Orrite, C., Torr, P.H.: Randomized trees for human pose detection. In: Proc. IEEE Conference on Computer Vision and Pattern Recognition (2008)
53. Rosales, R., Sclaroff, S.: Estimating 3D body pose using uncalibrated cameras. In: Proc. IEEE Conference on Computer Vision and Pattern Recognition (2001)
54. Rosenhahn, B., Schmaltz, C., Brox, T.: Markerless motion capture of man-machine interaction. In: Proc. IEEE Conference on Computer Vision and Pattern Recognition (2008)
55. Roweis, S., Saul, L.: Nonlinear dimensionality reduction by locally linear embedding. Science **290**, 2323–2326 (2000)
56. Sigal, L., Black, M.: HumanEva: Synchronized video and motion capture dataset for evaluation of articulated human motion. Technical Report CS-06-08, Brown University (2006)
57. Sigal, L., Black, M.J.: Measure locally, reason globally: occlusion-sensitive articulated pose estimation. In: Proc. IEEE Conference on Computer Vision and Pattern Recognition (2006)
58. Sigal, L., Bhatia, S., Roth, S., Black, M., Isard, M.: Tracking loose-limbed people. In: Proc. IEEE Conference on Computer Vision and Pattern Recognition (2004)
59. Sigal, L., Balan, A., Black, M.J.: Combined discriminative and generative articulated pose and non-rigid shape estimation. In: Advances in Neural Information Processing Systems, pp. 1337–1344. MIT Press, Cambridge (2007)
60. Sigal, L., Memisevic, R., Fleet, D.J.: Shared kernel information embedding for discriminative inference. In: Proc. IEEE Conference on Computer Vision and Pattern Recognition (2009)
61. Sminchisescu, C., Kanaujia, A., Metaxas, D.: Learning joint top–down and bottom–up processes for 3D visual inference. In: Proc. IEEE Conference on Computer Vision and Pattern Recognition (2004)
62. Sminchisescu, C., Kanaujia, A., Metaxas, D.N.: BM3E: Discriminative density propagation for visual tracking. IEEE Trans. Pattern Anal. Mach. Intell. **29**(11), 2030–2044 (2007)
63. Tangkuampien, T., Suter, D.: Real-time human pose inference using kernel principal component pre-image approximations. In: Proc. British Machine Vision Conference (2006)
64. Tenenbaum, J.B., de Silva, V., Langford, J.C.: A global geometric framework for nonlinear dimensionality reduction. Science **290**, 2319–2323 (2000)
65. Tian, T.P., Li, R., Sclaroff, S.: Articulated pose estimation in a learned smooth space of feasible solutions. In: Proc. IEEE Conference on Computer Vision and Pattern Recognition (2005)
66. Urtasun, R.: Motion model for robust 3D human body tracking. Ph.D. Thesis, EPFL (2006)
67. Urtasun, R., Darrell, T.: Sparse probabilistic regression for activity-independent human pose inference. In: Proc. IEEE Conference on Computer Vision and Pattern Recognition (2008)
68. Urtasun, R., Fleet, D., Hertzmann, A., Fua, P.: Priors for people tracking from small training sets. In: Proc. IEEE International Conference on Computer Vision (2005)
69. Vasilescu, M.A.O., Terzopoulos, D.: Multilinear subspace analysis of image ensembles. In: Proc. IEEE Conference on Computer Vision and Pattern Recognition, pp. 93–99 (2003)
70. vanBeck, P.J.L.: Edge-based image representation and coding. Ph.D. Thesis, Delft University of Technology, the Netherlands (1995)
71. Vondrak, M., Sigal, L., Jenkins, O.C.: Physical simulation for probabilistic motion tracking. In: Proc. IEEE Conference on Computer Vision and Pattern Recognition (2008)
72. Wang, J., Fleet, D., Hertzmann, A.: Gaussian process dynamic models. In: Advances in Neural Information Processing Systems. MIT Press, Cambridge (2006)
73. Xu, X., Li, B.: Learning motion correlation for tracking articulated human body with a Rao-Blackwellised particle filter. In: Proc. IEEE International Conference on Computer Vision (2007)
74. Yam, C., Nixon, M.S., Carter, J.N.: Automated person recognition by walking and running via model-based approaches. Pattern Recognit. **37**, 1057–1072 (2004)
75. Zhao, T., Nevatia, R.: Tracking multiple humans in crowded environment. In: Proc. IEEE Conference on Computer Vision and Pattern Recognition (2004)

Spatio-Temporal Motion Pattern Models of Extremely Crowded Scenes

Louis Kratz and Ko Nishino

Abstract Extremely crowded scenes present unique challenges to motion-based video analysis due to the large quantity of pedestrians within the scene and the frequent occlusions they produce. The movement of pedestrians, however, collectively form a spatially and temporally structured pattern in the motion of the crowd. In this work, we present a novel statistical framework for modeling this structured pattern, or steady-state, of the motion in extremely crowded scenes. Our key insight is to model the motion of the crowd by the spatial and temporal variations of local spatio-temporal motion patterns exhibited by pedestrians within the scene. We divide the video into local spatio-temporal sub-volumes and represent the movement through each sub-volume with a local spatio-temporal motion pattern. We then derive a novel, distribution-based hidden Markov model to encode the temporal variations of local spatio-temporal motion patterns. We demonstrate that by capturing the steady-state of the motion within the scene, we can naturally detect unusual activities as statistical deviations in videos with complex activities that are hard for even human observers to analyze.

1 Introduction

The large number of surveillance cameras currently deployed has created a dire need for computational methods that can assist or ultimately replace human operators. Surveillance cameras record large amounts of video that is typically only viewed after an incident occurs. Live monitoring of video feeds requires human personnel who are frequently tasked with observing multiple cameras simultaneously. Vision-based video analysis systems attempt to augment human security personnel by an-

L. Kratz (✉) · K. Nishino
Drexel University, Philadelphia, PA, USA
e-mail: lak24@drexel.edu

K. Nishino
e-mail: kon@drexel.edu

L. Wang et al. (eds.), *Machine Learning for Vision-Based Motion Analysis*,
Advances in Pattern Recognition,
DOI 10.1007/978-0-85729-057-1_10,

alyzing surveillance videos computationally, enabling the automatic monitoring of large numbers of video cameras.

Despite the general interest and active research in video analysis, an important area of video surveillance has been overlooked. Extremely crowded scenes, such as those shown in Fig. 2, are perhaps in the most need of computational methods for analysis. Crowded, public areas require monitoring of a large number of individuals and their activities, a significant challenge for even a human observer. Videos with this level of activity have yet to be analyzed via a computational method. Common scenes for video analysis, such as the PETS 2007 database [15], contain less than one hundred individuals in even the most crowded videos. Extremely crowded scenes, however, contain hundreds of people in any given frame and possibly thousands throughout the duration of the video.

The people and objects that compose extremely crowded scenes present an entirely different level of challenges due to their large quantity and complex activities. The sheer number of pedestrians results in frequent occlusions that make modeling the movement of each individual extremely difficult, if not impossible. The movement of each pedestrian, however, contributes to the overall motion of the crowd. In extremely crowded scenes, the crowd's motion may vary spatially across the frame, as different areas contain different degrees of traffic, and temporally throughout the video due to natural crowd variations. As a result, analysis of extremely crowded scenes must account for the spatial and temporal variations in the local movements of pedestrians that comprise the crowd.

In this work, we present a novel statistical framework to model the steady-state motion of extremely crowded scenes. Our key insight is that the crowd's motion is a spatially and temporally structured pattern of the local motion patterns exhibited by pedestrians. We model the spatial and temporal variations of the motion in local space-time sub-volumes to capture the latent motion structure (i.e., the steady-state) of the scene. We use this model to describe the typical motion of the crowd, that is, usual events within the scene, and demonstrate its effectiveness by identifying unusual events in videos of the same scene.

First, we divide the video volume into local spatio-temporal sub-volumes, or cuboids, defined by a regular grid. Next, we use a local spatio-temporal motion pattern to describe the possible complex movements of pedestrians within each cuboid. We then identify prototypical local spatio-temporal motion patterns that describe the typical movements within the scene, and estimate a distribution of local motion patterns to compactly represent the entire video. Finally, we capture the spatial and temporal variations in the motion of the crowd by training a novel, distribution-based hidden Markov model (HMM) on the local spatio-temporal motion patterns at each spatial location of the video volume. In other words, we model the steady-state motion of the crowd by a spatially and temporally varying structured pattern of the movements of pedestrians in local areas.

2 Related Work

Motion-based video analysis characterizes the scene by the movement of the scene's constituents within a video sequence. Trajectory-based approaches [5, 8, 9], for example, track the objects within the scene and describe their motion by their changing spatial location in the frame. These techniques are suitable for scenes with few moving objects that can easily be tracked, such as infrequent pedestrian or automobile traffic. Trajectory-based approaches focus on each subject individually, but the behavior of extremely crowded scenes depends on the concurrent movement of multiple pedestrians. In addition, the frequent occlusions in extremely crowded scenes makes tracking significantly difficult.

Other approaches estimate the optical flow [2, 3] or the motion within spatio-temporal volumes [4, 6, 7, 10]. Flow-based approaches have used HMMs [2] or Bayes' classifiers [3] to represent the overall motion within the scene. The large number of people in extremely crowded scenes, however, make modeling specific activity by the collective optical flow difficult. Extremely crowded scenes may contain any number of concurrent, independent activities taking place in different areas of the same frame. This makes global approaches [2, 20] that model the motion over the entire frame unsuitable since the local events of interest would not be discernible from the rest of the scene.

Spatio-temporal approaches [4, 6, 7, 10] directly represent the motion in local sub-volumes of the video. Though these representations are well suited to extremely crowded scenes, their use has been limited to volume distance [4, 10] or interest points [7], requiring the explicit modeling of each event to be detected. In other words, they have not been used to represent the overall motion of a scene, just specific events. In addition, most spatio-temporal representations assume that the sub-volume contains motion in a single direction. In extremely crowded scenes, however, the motion within each cuboid may be complex and consist of movement in multiple directions.

Other work have modeled the motion of the crowd by flow fields [1], topical models [17], or dynamic textures [12]. Ali and Shah [1] model the expected motion of pedestrians in order to track pedestrians at a distance. More recently, Rodriguez et al. [17] model a fixed number of possible motions at each spatial location using a topical model. These approaches, however, fix the number of possible motions at each location within the frame. In extremely crowded scenes the pedestrian motions vary depending on the spatial and temporal location within the video. Pedestrian movements may be severely limited in one area of the frame and highly variable in others. As such, models that fix the number of possible movements may not represent the rich variations in motion that can occur in extremely crowded scenes.

3 Local Spatio-Temporal Motion Patterns

The motion of the crowd in extremely crowded scenes is formed by the collective movement of the pedestrians. The movement of each pedestrian depends on the

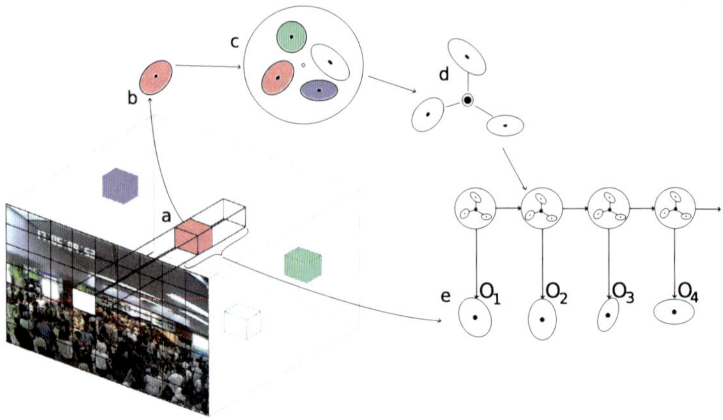

Fig. 1 A schematic overview of the framework. (**a**) The video sequence is divided into local spatio-temporal volumes (*cuboids*). (**b**) The local spatio-temporal motion pattern within each cuboid is represented by a 3D Gaussian of spatio-temporal gradients. (**c**) Prototypical local motion patterns are identified by grouping similar local motion patterns that may lie at disjoint space-time locations. (**d**) Prototypes model the motion pattern variation in the video with a 1D normal distribution of 3D Gaussians, that is, a distribution of distributions. (**e**) A distribution-based HMM is trained for each tube using the prototypical motion patterns for hidden states and 3D Gaussians as observations

physical structure of the scene, surrounding pedestrians, and the individual's goals. As a result, the motion of the crowd changes naturally spatially across the frame and temporally over the video. It is exactly these spatial and temporal variations in the local movements of pedestrians that we model to characterize the motion of the crowd.

In order to model the movement of pedestrians in local areas, we subdivide the video into a set of local spatio-temporal sub-volumes, or cuboids, of a fixed size as shown in Fig. 1(a). A finer approach, such as the pixel-level optical flow, would only provide the motion at a single pixel, not the collective movements of possibly multiple pedestrians. Conversely, the motion of the entire frame would not provide the level of detail necessary for differentiating between independent, concurrent activities within the same frame. By dividing the video into spatio-temporal sub-volumes, we can extract rich local motion patterns while discerning different movements in local space-time areas.

We represent the local movement of pedestrians by the collection of spatio-temporal gradients within the cuboid. For each pixel i in the cuboid, we compute the spatio-temporal gradient ∇I_i, a 3-dimensional vector representing the gradient in the horizontal, vertical, and temporal dimensions. Previous work on analyzing persistent motion patterns [19] and correlating video sequences [18] have used the collection of spatio-temporal gradients in the form of the structure tensor matrix

$$\mathbf{G} = \frac{1}{N}\sum_{i=1}^{N} \nabla I_i \nabla I_i^T, \tag{1}$$

where N is the number of pixels in the cuboid. These methods assume that the spatio-temporal gradients lie on a plane in 3D gradient space, and thus the cuboid contains a single motion vector [18]. In extremely crowded scenes, however, a cuboid may contain complex motion caused by movement in multiple directions such as a pedestrian changing direction or two pedestrians moving past one another.

As illustrated in Fig. 1(b), we represent the possibly complex motion within each cuboid by estimating the distribution of spatio-temporal gradients as a 3D Gaussian $\mathcal{N}(\boldsymbol{\mu}, \boldsymbol{\Sigma})$ where

$$\boldsymbol{\mu} = \frac{1}{N}\sum_{i}^{N} \nabla I_i, \qquad \boldsymbol{\Sigma} = \frac{1}{N}\sum_{i}^{N} (\nabla I_i - \boldsymbol{\mu})(\nabla I_i - \boldsymbol{\mu})^T . \tag{2}$$

Thus for each spatial location n and temporal location t, the local spatio-temporal motion pattern is represented by $\boldsymbol{\mu}_t^n$ and $\boldsymbol{\Sigma}_t^n$. Intuitively, by modeling the distribution of spatio-temporal gradients, we are representing the possibly multiple motion vectors that may occur within the cuboid by estimating the shape of the 3D gradients that may not lie on a single plane.

The amount of pedestrian motion represented depends on the size of the cuboid. Since the camera recording the scene is fixed, we set the cuboid size manually. We consider this an acceptable cost of our approach since the size of pedestrians remains similar over the duration of the scene.

4 Prototypical Motion Patterns

By directly modeling the video as a collection of local spatio-temporal motion patterns, we reduce the size of the video representation from a set of raw pixels to a collection of Gaussian parameters. This representation, however, is still quite large. For example, a one minute video with resolution 720×480 will have $19,440$ cuboids of size $40 \times 40 \times 20$, resulting in $233,280$ parameters. As shown in Fig. 1(c), we further reduce the size of this representation by identifying common local spatio-temporal motion patterns. We extract prototypical local spatio-temporal motion patterns (prototypes) that represent the characteristic movements of pedestrians within the scene. Note that similar local spatio-temporal motion patterns can occur at disjoint space-time locations in the video, and it is this recurrence that forms the underlying steady-state of the motion of the crowd.

In order to identify similar local spatio-temporal motion patterns, we directly compare the 3D Gaussian distributions of spatio-temporal gradients. Previous approaches have used algebraic metrics which are sensitive to noise [14], or have assumed that the cuboid volume only consists of a single motion vector [10, 18]. We use the symmetric Kullback–Leibler (KL) divergence [11] to measure the difference between local spatio-temporal motion patterns. Since the local spatio-temporal motion patterns are Gaussian distributions, the divergence has a closed analytical form [13]. To reduce numeric instabilities, we take the log of the divergence, compare the norm differences between the distributions, and limit the condition numbers of the covariance matrices. We also add a 1 to the divergence prior to taking the log to ensure that the measure is positive. Thus we measure the difference between two local spatio-temporal motion patterns A and B by

$$\tilde{d}(A,B) = \begin{cases} 0 & \mathcal{K}(\boldsymbol{\Sigma}_a) > d_{\mathcal{K}},\ \mathcal{K}(\boldsymbol{\Sigma}_b) > d_{\mathcal{K}}, \\ & \|\boldsymbol{\Sigma}_a - \boldsymbol{\Sigma}_b\|_F < d_{\Sigma},\ \|\boldsymbol{\mu}_a - \boldsymbol{\mu}_b\| < d_{\mu}, \\ \log(d(A,B)+1) & \text{otherwise,} \end{cases} \tag{3}$$

where $d(A,B)$ is the KL divergence, $\mathcal{K}(\cdot)$ is the condition number of the matrix, and d_{Σ} and d_{μ} are limits on the norms to ensure the distributions are reasonably similar. We refer to this measure as the KL distance for the remainder of this paper. By using the KL divergence, we distinguish the possibly complex movements of pedestrians from different cuboids directly from the 3D Gaussian distributions.

Using the KL distance, we identify the prototypical local spatio-temporal motion patterns, or prototypes, that characterize the local movements of pedestrians within the scene. Off-line clustering techniques such as K-means, however, require that all of the local spatio-temporal motion patterns be available. We use an online method that computes the KL distance between each local spatio-temporal motion pattern O_t^n and the prototypes as we parse the video. If the KL distance is greater than a specified threshold, d_{KL}, for all prototypes $\{P_s | s = 1, \ldots, S\}$, then the cuboid is considered a new prototype. Otherwise, the prototype P_s is updated with the new observation O_t^n by

$$P_s = \frac{1}{N_s+1} O_t^n + \left(\frac{N_s}{N_s+1}\right) \widetilde{P}_s, \tag{4}$$

where N_s is the total number of observations associated with the prototype P_s at time t and $\widetilde{P}_s$ is the previous value of P_s. The set of prototypes is initially empty.

Since the motion patterns O_t^n and P_s are multi-variate Gaussian distributions, (4) should reflect the spatio-temporal gradients that the local spatio-temporal motion patterns represent. To solve (4) with respect to the KL divergence, we use the expected centroid presented by Myrvoll and Soong [13]

$$\boldsymbol{\mu}_s = \frac{1}{N_s+1} \boldsymbol{\mu}_t^n + \frac{N_s}{N_s+1} \boldsymbol{\mu}_s, \tag{5}$$

$$\boldsymbol{\Sigma}_s = \frac{1}{N_s+1} \left(\boldsymbol{\Sigma}_t^n + \boldsymbol{\mu}_t^n (\boldsymbol{\mu}_t^n)^T\right) + \frac{N_s}{N_s+1} \left(\widetilde{\boldsymbol{\Sigma}}_s + \widetilde{\boldsymbol{\mu}}_s \widetilde{\boldsymbol{\mu}}_s^T\right) - \boldsymbol{\mu}_s \boldsymbol{\mu}_s^T, \tag{6}$$

where $\boldsymbol{\mu}_t^n$, $\boldsymbol{\Sigma}_t^n$, $\boldsymbol{\mu}_s$, and $\boldsymbol{\Sigma}_s$ are the mean and covariance matrices of O_t^n and P_s, respectively. Thus, we compute prototypes that represent the collection of spatio-temporal gradients for each typical movement within the scene. By extracting prototypical local spatio-temporal motion patterns, we construct a canonical representation of the video as a collection of the characteristic movements of pedestrians in local areas.

5 Distribution-Based Hidden Markov Models

While the set of prototypes provides a picture of similar movements of pedestrians within the scene, it does not capture the relationship between their occurrences. By modeling the temporal relationship between sequential local spatio-temporal motion patterns, we characterize a given video by its temporal variations. We assume that cuboids in the same spatial location exhibit the Markov property in the temporal domain since the scene is comprised of physically moving objects. We create a single HMM for each tube of observations in the video as shown in Fig. 1(e) to model the temporal evolution of local spatio-temporal motion patterns in each local region. Since each local spatio-temporal motion pattern is a 3D Gaussian of spatio-temporal gradients, we derive an HMM that can handle observations that are distributions themselves.

Ordinary HMMs are defined by the parameters $M = \{H, \mathbf{o}, \mathbf{b}, \mathbf{A}, \boldsymbol{\pi}\}$, where H is the number of hidden states, $\mathbf{o}$ the possible values of observations, $\mathbf{b}$ a set of H emission probability density functions, $\mathbf{A}$ a transition probability matrix, and $\boldsymbol{\pi}$ an initial probability vector. We model a single HMM $M^n = \{H^n, \mathbf{O}^n, \mathbf{b}^n, \mathbf{A}^n, \boldsymbol{\pi}^n\}$ for each spatial location $n = 1, \ldots, N$ and associate the hidden states H^n with the number of prototypes S^n in the tube. The set of possible observations $\mathbf{O}^n$ is the range of 3D Gaussian distributions of spatio-temporal gradients. Complex observations for HMMs are often quantized for use in a discrete HMM. Such quantization, however, would significantly reduce the rich motion information that the local spatio-temporal motion patterns represent. Using a distribution-based HMM allows the observations to remain 3D Gaussian distributions. Therefore, the emission probability density function for each prototype must be a distribution of distributions.

We model each emission distribution as a Gaussian centered around the prototype and using the KL distance to the observed local spatio-temporal motion pattern. The probability of a local spatio-temporal motion pattern O_t^n given the hidden state s is computed by

$$p\left(O_t^n|s\right) = \frac{1}{\sqrt{\pi\sigma_s^2}} \exp\left[\frac{-\tilde{d}(O_t^n, P_s)^2}{2\sigma_s^2}\right], \tag{7}$$

where P_s is the prototype given in (4), and $\tilde{d}(\cdot)$ is the KL distance. This retains the rich motion information represented by each local spatio-temporal motion pattern and provides a probability calculation consistent with our distance measure. We compute the standard deviation by the maximum likelihood estimator

$$\sigma_s^2 = \frac{1}{N_s} \sum_j^{N_s} \tilde{d}(O_j, P_s)^2, \tag{8}$$

where N_s is the number of local spatio-temporal motion patterns associated with the prototype P_s. In practice, however, there may be too few cuboids in a specific group to estimate σ_s. On such occasions, we use a 99.7 percent confidence window around d_{KL}, letting $\sigma_s = 3d_{\text{KL}}$.

The distribution-based HMM represents the temporal variations of local spatio-temporal motion patterns in a sound statistical framework. The emission probability distributions are created using (4) and (8) for each prototype in the scene. Note that, while a single HMM is trained on the local spatio-temporal motion patterns in each tube, the prototypes are created using samples from the entire video volume. The parameters $\mathbf{A}^n$ and $\boldsymbol{\pi}^n$ are estimated by expectation maximization.

Modeling the temporal variations of the local spatio-temporal motion patterns in a statistical framework provides a natural way to define unusual activities as statistical deviations from the learned model of the motion of the crowd. Given a query video, the likelihood of a sequence of local spatial-temporal motion patterns can be calculated using the forwards–backwards algorithm [16]. Let $\mathcal{T}_t^n$ be the likelihood of the tth temporal sequence of observations within a given video tube n. Thus

$$\mathcal{T}_t^n = \mathrm{p}\big(O_t^n, \ldots, O_{t+w}^n\big), \tag{9}$$

where w is the sequence length, and $O_t^n, \ldots, O_{t+w}^n$ is a subsequence of observed local spatio-temporal motion patterns at location n. Ideally, we would like to measure the likelihood of each individual cuboid. Since $\mathcal{T}_t^n$ is calculated for every sequence within the tube of length w, each observation can be associated with w likelihood measures by sliding a window of size w over the entire video. We define an ensemble function that selects a measure from the set of likelihoods associated with the observation. We use a window size of 2 and let the ensemble function maximize over the likelihoods. This correctly classifies the cuboids with the exception of one case when a usual cuboid is temporally surrounded between two unusual cuboids, which is rare and errs on the side of caution (a false positive). Since extremely crowded scenes may contain larger variations in one location than another, we normalize the likelihood of each observation by the minimum likelihood value of the training data in each spatial location n.

6 Experimental Results

We demonstrate the effectiveness of our model of the crowd's motion by detecting unusual events in videos of three scenes: one simulated crowded scene and two real-world extremely crowded scenes. For each scene, we train a collection of distribution-based HMMs on a finite length training video that represents the typical motion of the crowd. We then use these HMMs to detect unusual activity on query videos of the same scene. To quantitatively evaluate our results, we manually annotate cuboids containing pedestrians moving in unusual directions. We fix the cuboid size to $40 \times 40 \times 20$ for all experiments, as such a size captures the distinguishing characteristics of the movement of the pedestrians. The threshold values are selected empirically since they directly depend on the variations in the motion within the specific scene.

We generate a synthetic crowded scene by translating a texture of a crowd across the frame, resulting in large motion variations and nonuniform motion along border areas. The image sequence consists of 216 tubes and 9,072 total cuboids. We

Fig. 2 We evaluate our approach by detecting unusual events in videos from a subway station during rush hour courtesy of Nippon Telegraph and Telephone Corporation. The *first*, from the station's concourse (*top*), contains a large number of pedestrians moving in different directions. The *second*, from the station's ticket gate (*bottom*), has pedestrians moving in similar directions as they pass through the gate but still contains significant variations

then insert several smaller images moving in arbitrary directions to simulate unusual pedestrian movements. The thresholds used in this experiment are $d_{\mathrm{KL}} = 0.02$, $d_{\Sigma} = 5$, $d_{\mu} = 1$, and $d_{\mathcal{K}} = 400$. The Receiver Operating Characteristic (ROC) curve for this example is shown in Fig. 3, and is produced by varying values of the classification threshold. Our approach achieve a false positive rate of 0.009 and a true positive rate of 1.0. The false positives occur in cuboids lacking motion and texture.

Figure 2 shows frames[1] from two real-world extremely crowded scenes that we use to evaluate our method. For both scenes, we set $d_{\Sigma} = 1$, $d_{\mu} = 1$, and $d_{\mathcal{K}} = 1000$. The first scene, shown on the top in Fig. 2, is from an extremely crowded concourse of a subway station. The scene contains a large number of moving and loitering pedestrians and employees directing traffic flow. The query video contains station employees walking against the flow of traffic. The training video contains 3,020 frames and the query video contains 380 frames. The threshold d_{KL} is 0.06. The second real-world extremely crowded scene, shown on the bottom in Fig. 2, is a wide-angle view of pedestrian traffic at the station's ticket gate. The motion of the crowd occurs in a more constant direction than in the concourse, but still contains excessive occlusions. The query video contains instances of people reversing direction or stopping in the ticket gate. The training video contains 2,960 frames and the query video contains 300 frames. The threshold d_{KL} is 0.05.

[1]The original videos courtesy of Nippon Telegraph and Telephone Corporation.

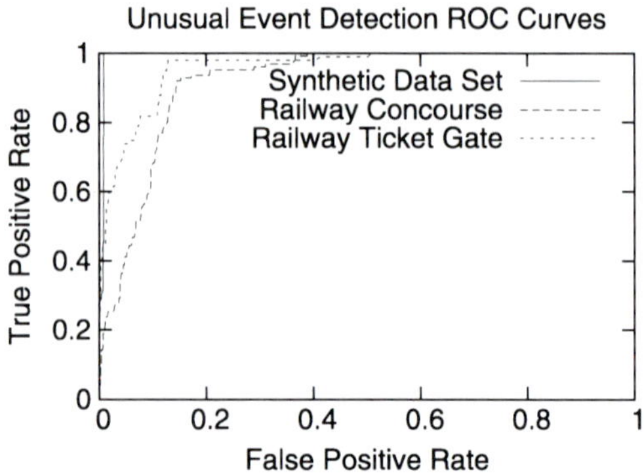

Fig. 3 ROC curves for the videos of the synthetic and real-world extremely crowded scenes. Our approach achieves significant accuracy on videos of both scenes, and almost perfect detection on our synthetic video

Fig. 4 Detection of unusual events in the concourse video (*top*) and the ticket gate (*bottom*). Correctly classified unusual cuboids are highlighted in *blue*, usual cuboids in *green*, and false positives in *magenta*. The intensity of magenta blocks indicates the severity of false positives. Individuals reversing direction, loitering, and moving in irregular directions are correctly classified as unusual. False positives indicate a sensitivity to cuboids with little motion

As shown in Fig. 3, our approach achieves a false positive rate of 0.15 and a true positive rate of 0.92 for the video from the concourse scene, and a false positive rate of 0.13 and a true positive rate of 0.97 for the video of the ticket gate scene. Figure 4 shows a visualization of the detection results. Employees moving against the flow of traffic are successfully detected in the video of the concourse scene, and pedestrians loitering and reversing direction are successfully detected in the video of the ticket gate scene.

False positives occur in both experiments for slightly irregular motion patterns such as after pedestrians exit the gate, and areas of little motion such as where the floor of the station is visible. The few false negatives in both real-world examples occur adjacent to true positives, which suggests they are harmless in practical scenar-

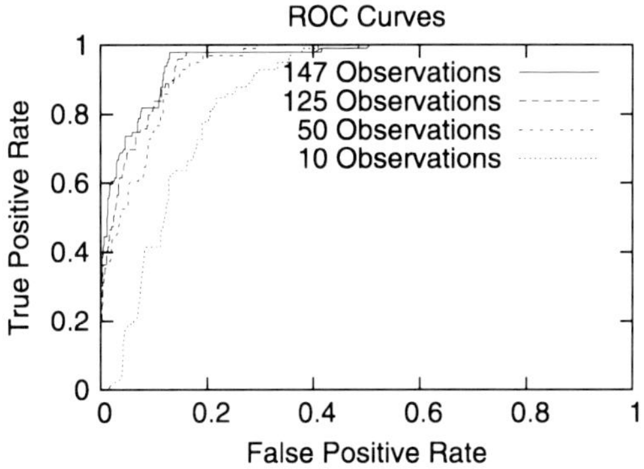

Fig. 5 Effects of increased training data on the video of the ticket gate scene. Our approach achieves significant results with as few as 10 observations, and good performance with only 50 observations (about 30 seconds of video)

ios. Intuitively, most unusual behavior that warrants personnel intervention would be less subtle than those detected here, and as such would result in a smaller number of errors. The ability to detect subtle unusual events, however, is made possible by training the HMMs on the local spatio-temporal motion patterns themselves.

Figure 5 shows the effect of increasing the amount of training data for the ticket gate video. As expected, the performance increases with the size of the training data, and approaches a steady-state with about 100 observations per tube. The performance with only 10 observations per tube achieves a false positive rate of 0.22 and true positive rate of 0.84. This is largely due to the window around d_{KL} for σ_s in (8) when there is insufficient observations for specific prototypes and the inclusion of disjoint local motion patterns in the prototypes. The performance with such small training data reflects the rich descriptive power of the distribution-based HMMs.

7 Conclusion

In this work, we introduced a novel statistical framework using local spatio-temporal motion patterns to represent the motion of the crowd in videos of extremely crowded scenes. We represented the movement of pedestrians in local areas by a local spatio-temporal motion pattern in the form of a multivariate Gaussian. We then identified prototypical local spatio-temporal motion patterns to canonically represent the characteristic movements within the video. Using a novel distribution-based hidden Markov model, we learned the statistical temporal variations of local motion patterns from a training video of an extremely crowded scene. Finally, we used this model of the motion of the crowd to identify unusual events as statistical anomalies. Our results indicate that local spatio-temporal motion patterns are a suitable representation for analyzing extremely crowded scenes. We evaluated our approach on videos of real-world extremely crowded scenes and successfully detected unusual local spatio-temporal motion patterns including movement against the normal flow of traffic, loitering, and traffic congestion. We believe our proposed framework plays an important role in the analysis of dynamic video sequences with spatially and temporally varying local motion patterns.

Acknowledgements This work was supported in part by Nippon Telegraph and Telephone Corporation and the National Science Foundation grants IIS-0746717 and IIS-0803670.

References

1. Ali, S., Shah, M.: Floor fields for tracking in high density crowd scenes. In: Proc. of European Conference on Computer Vision (2008)
2. Andrade, E., Blunsden, S., Fisher, R.: Modelling crowd scenes for event detection. In: Proc. of International Conference on Pattern Recognition, pp. 175–178 (2006)
3. Black, M.: Explaining optical flow events with parameterized spatio-temporal models. In: Proc. of IEEE International Conference on Computer Vision and Pattern Recognition, pp. 326–332 (1999)
4. Boiman, O., Irani, M.: Detecting irregularities in images and in video. In: Proc. of IEEE International Conference on Computer Vision, pp. 462–469 (2005)
5. Dee, H., Hogg, D.: Detecting inexplicable behaviour. In: Proc. of British Machine Vision Conference, pp. 477–486 (2004)
6. DeMenthon, D., Doermann, D.: Video retrieval using spatio-temporal descriptors. In: Proc. of the 11th ACM International Conference on Multimedia, pp. 508–517 (2003)
7. Dollar, P., Rabaud, V., Cottrell, G., Belongie, S.: Behavior recognition via sparse spatio-temporal features. In: International Workshop on Performance Evaluation of Tracking and Surveillance, pp. 65–72 (2005)
8. Hu, W., Xiao, X., Fu, Z., Xie, D., Tan, T., Maybank, S.: A system for learning statistical motion patterns. IEEE Trans. Pattern Anal. Mach. Intell. **28**(9), 1450–1464 (2006)
9. Johnson, N., Hogg, D.: Learning the distribution of object trajectories for event recognition. In: Proc. of British Machine Vision Conference, pp. 583–592 (1995)
10. Ke, Y., Sukthankar, R., Hebert, M.: Event detection in crowded videos. In: Proc. of IEEE International Conference on Computer Vision, pp. 1–8 (2007)
11. Kullback, S., Leibler, R.A.: On information and sufficiency. Ann. Math. Stat. **22**(1), 79–86 (1951)
12. Ma, Y., Cisar, P.: Activity representation in crowd. In: Proc. of the Joint IAPR International Workshop on Structural, Syntactic, and Statistical Pattern Recognition, pp. 107–116 (2008)
13. Myrvoll, T., Soong, F.: On divergence based clustering of normal distributions and its application to HMM adaptation. In: Proc. of European Conference Speech Communication and Technology, pp. 1517–1520 (2003)
14. Nishino, K., Nayar, S.K., Jebara, T.: Clustered blockwise PCA for representing visual data. IEEE Trans. Pattern Anal. Mach. Intell. **27**(10), 1675–1679 (2005)
15. PETS: 10th IEEE International Workshop on Performance Evaluation of Tracking and Surveillance. http://www.pets2007.net/ (2007)
16. Rabiner, L.: A tutorial on hidden Markov models and selected applications in speech recognition. Proc. IEEE **77**(2), 257–286 (1989)
17. Rodriguez, M., Ali, S., Kanade, T.: Tracking in unstructured crowded scenes. In: Proc. of IEEE International Conference on Computer Vision (2009)
18. Shechtman, E., Irani, M.: Space-time behavior based correlation. In: Proc. of IEEE International Conference on Computer Vision and Pattern Recognition, pp. 405–412 (2005)
19. Wright, J., Pless, R.: Analysis of persistent motion patterns using the 3D structure tensor. In: IEEE Workshop on Motion and Video Computing, pp. 14–19 (2005)
20. Zhong, H., Shi, J., Visontai, M.: Detecting unusual activity in video. In: Proc. of IEEE International Conference on Computer Vision and Pattern Recognition, pp. 819–826 (2004)

Learning Behavioral Patterns of Time Series for Video-Surveillance

Nicoletta Noceti, Matteo Santoro, and Francesca Odone

Abstract This chapter deals with the problem of learning behaviors of people activities from (possibly big) sets of visual dynamic data, with a specific reference to video-surveillance applications. The study focuses mainly on devising meaningful data abstractions able to capture the intrinsic nature of the available data, and applying similarity measures appropriate to the specific representations. The methods are selected among the most promising techniques available in the literature and include classical curve fitting, string-based approaches, and hidden Markov models. The analysis considers both supervised and unsupervised settings and is based on a set of loosely labeled data acquired by a real video-surveillance system. The experiments highlight different peculiarities of the methods taken into consideration, and the final discussion guides the reader towards the most appropriate choice for a given scenario.

1 Introduction

The study and the understanding of human activities from video has been widely addressed in the last decades, particularly with the availability of an enormous amount of installed video-cameras, and the consequently increased interest for video-based applications. In this chapter, we refer in particular to video-surveillance systems, assuming they observe possibly complex scenarios from a distance where the "action of interest" is the trajectory of a moving object as a whole, and no information is available or needed on the motion of object's parts. We thus explicitly refer to data

N. Noceti (✉) · M. Santoro · F. Odone
DISI, Università degli Studi di Genova, Genova, Italy
e-mail: noceti@disi.unige.it

M. Santoro
e-mail: santoro@disi.unige.it

F. Odone
e-mail: odone@disi.unige.it

L. Wang et al. (eds.), *Machine Learning for Vision-Based Motion Analysis*,
Advances in Pattern Recognition,
DOI 10.1007/978-0-85729-057-1_11,

that may be modeled as time-series of instantaneous observations. In such settings, depending on the specific application, we rely on some sort of supervision, i.e., we may be able to manually label a data set and use it to learn common behaviors. In this case, since the learning process follows a subjective view of what is "normal" in the scene, we may build models that are either too simplistic or too structured for the actual amount of information that may be extracted from the data. An example of this, in our setting, occurs if the labeling process is inspired by the dynamics of the 3D world with no consideration for the information loss due to image formation. In other more common cases, labels may not be available in any way, and therefore one has to rely on an unsupervised approach where the main problem is to discover common behaviors from unlabeled examples.

Even if the research on video analysis and scene understanding is greatly contributing to these problems, in this general setting a number of questions are still open—how to represent data limiting the amount of ambiguities, when there is no or little knowledge on the underlying data acquisition process? And, more specifically, how to exploit the data internal structure induced by time-coherence of observations? Also, how to learn from large amounts of noisy data, particularly when they belong to high dimensional spaces? Moreover, the specific domain we refer to opens various *practical* issues, related to the scene complexity, the specifications of the video-surveillance system (e.g., the number and the position of the cameras), and the amount of prior information available on the observed phenomena.

This chapter deals with some of such general issues, trying to adopt data driven strategies for representing data in flexible ways, studying approaches specifically designed for time-series analysis, and adopting mathematically sound machine learning methods proved able to learn from large sets of high dimensional data. Some practical issues related to the application domain are addressed in the experimental phase since we deal with data acquired from a real single-camera video-surveillance system installed in a rather complex environment.

The general pipeline we refer to in this work is rather standard and may be organized in three phases: a *low-level video processing* phase to extract information on dynamic events occurring in the scene; a *data abstraction* phase that performs a mapping into an appropriate feature space where data may be better observed and understood; finally, a *learning from examples* phase, that exploits available knowledge to build models able to understand common behaviors and, if possible, associate yet to be seen data with known behaviors. One of the aims of the data abstraction phase is to build general models that could be applied to different input representations. Once an effective representation is found, well established machine learning methods (adequately equipped with ad hoc similarity functions) are applied to deal with the supervised or unsupervised problem under consideration.

For what concerns the first phase in this chapter, we simply report the specific model adopted in our work, referring the interested researcher to more appropriate readings. Instead, we focus more specifically to the second and third stage. The goals of our study are (i) to analyse the pertinent state-of-the-art with a reference to time series data; (ii) to devise promising data representations and similarity measures in both supervised and unsupervised settings; (iii) to validate them on a relatively big and complex set of data acquired by a video-surveillance system and loosely labeled by means of a fast calibration procedure. All along the work, we exploit the peculiarity of video-surveillance domain, where it is usual that a system observes the same scene for very long periods and it is thus reasonable to imagine systems automatically learning what is common in the observed scene.

More in detail, we analyze three different data abstraction approaches and their associated similarity measures: a parametric fitting approach to trajectories that maps temporal series in the parameters space of the adopted fitting curves (B-Splines in our case); a string-based approach that learns an appropriate alphabet from the training data available and then maps all data in a string based representation over the alphabet; finally, a method based on hidden Markov models (HMMs) that builds a probabilistic model of the present observation given all past observations. With each representation, we associate an appropriate similarity measure to adopt it in the learning phase. As for the learning algorithms in the supervised case, we adopt a regularized least squares (RLS) scheme for B-Splines and string-based models, while with HMMs we estimate the highest likelihood with respect to the available models; instead, in the unsupervised case we adopt Spectral Clustering. A thorough experimental analysis will highlight the peculiarities of the various models, suggesting what is more appropriate for each circumstance.

The structure of the chapter is as follows. Section 2 discusses the relevant state of the art, with a specific reference to learning from examples approaches and to time-series analysis. Section 3 is a brief account of the low-level video processing module we adopt to extract dynamic events from video streams. Section 4 deals with the second phase of the above mentioned pipeline, discussing a selection of methods for representing temporal series data available from the literature; Sect. 5 focuses on the third phase and reports possible approaches to the problem of learning behavioral patterns, both from labeled and unlabeled data. In this case, specific reference to the choice of appropriate similarity measures is made. Section 6 reports in details an experimental analysis that compares the different studied approaches in the context of a real and rather complex scenario. Finally, Sect. 7 is left to discussions and open issues.

2 Related Works

Event analysis and recognition have been addressed by many authors in the last decades, mainly in the context of visual surveillance, domotics, and video annotation. The number of related subproblems is large, as well as the number of possible different applications. Consequently, the amount of literature on these topics is huge and becomes more and more unmanageable. Therefore, this section does not aim at providing an exhaustive overview of current state of research, which can be found—for instance—in the recent special issue [28] and in the references therein. Instead, we focus mainly on the crucial subproblem of analyzing and modeling motion trajectories, and more specifically on the key issues of representation and matching of the trajectories.

From the viewpoint of possible applications, two different scenarios have been considered in the literature. First, if labeled data are available then a supervised approach to trajectory analysis is the best option and the ultimate goal is to build suitable classifiers representing different types of behavioral patterns. Second, if there is no prior knowledge about the behavior classes, then one is forced to adopt unsupervised learning methods (specifically, clustering) in order to extract information from sets of unlabeled data. In this latter case, however, the goal is necessarily different since the best one can do is to model automatically the notion of "common behavior" and try to detect anomalies.

For what concerns the representation of the trajectories, several authors adopted an approach in which the temporal component of the observed behaviors is kept explicit [14], therefore relying on methods for time-series analysis. In what follows, instead of considering representations that explicitly refer to 2D or 3D trajectories, we focus on methods that do not make any assumption on the size of each instantaneous measurement.

Both with the supervised and the unsupervised approaches, an important issue in knowledge discovery from sequential data is what similarity measure to use to extract meaningful information: [25] offers a complete survey to such a topic. From the point of view of data abstraction, the contributions proposed in the literature may be roughly divided between the ones relying on an explicit modeling of the events of interest (in the learning from examples framework it is assumed that an exhaustive set of labeled data is available) and the ones aiming at an automatic modeling of "normal behaviors" from a (possibly) big set of unlabeled data.

A priori knowledge on the analyzed environment and its dynamics may be exploited in different ways. A first example, widely explored in the literature, refers to stochastic grammars (see, for instance, [10]). In certain application domains (e.g., traffic control) the amount of structures contained in the expected events may be profitably used to model the dynamics of the scene. For instance, in [17], the results obtained by a robust low-level processing module are associated with context information to identify possible abnormal behaviors in aerial images.

In general, however, the most appropriate way to exploit prior knowledge is to rely on learning from examples. If the available data are labeled, i.e., with each one of them we may associate a label of a known behavior, the use of state-of-the-art machine learning algorithms lead to effective behavior categorization methods. On this respect, we mention the work by Pittore et al. [23], where the available data are mapped in the parameter space induced by the fit of each trajectory with B splines, then support vector machines (SVMs) are applied. Otherwise, the sequential structure of trajectories may be captured by training a well designed hidden Markov models (HMMs) [24] or other similar dynamical systems (see, for instance, [2] and references therein).

For what concerns the literature of unsupervised approaches, the general goal is to model normal behaviors from (possibly big) sets of unlabeled observations. Many successful methods have been proposed to discover classes of similar pattern in a data set. Unfortunately, most of them are not tailored for clustering sequential data, such as temporal series. In fact, while analyzing temporal data, the *temporal dimension* usually induces specific, inherent structure to the sequence of feature vectors that is likely to be disregarded by traditional clustering methods. Furthermore, sequences of different lengths cannot be compared in a unique way.

Among the first contributions to this topic is the influential work by Stauffer and his coworkers (see, for instance, [29]), based on learning from data a codebook derived from data quantization. Co-occurrence of motion patterns in the analyzed trajectories is evaluated, through the use of a hierarchical classification module. More recently, [9] proposes a pipeline based on k-means to track objects, represent trajectories with respect to an estimated codebook, and cluster them. Finally, a Bayes method is applied to detect anomalies. Experimental analysis is based on both synthetic and real traffic sequences. In [22], normal behaviors are associated with one class only. Then a one-class SVM is used to model the class of normal trajectories, against which identify abnormal behaviors. The representation is a sequence of 2D coordinates and the fixed length property is restored by means of trajectory subsampling. Again, the reference application is outdoor traffic control analysis. Anjum and Cavallaro [1] adopts a multi-feature representation of each trajectory based on a cumulative approach: each measurement at time t refers to all previous observations. Clustering is performed with mean-shift. A trajectory is labeled as normal or anomalous, according to the trajectories density among the clusters.

More specific to our problem is previous work on temporal series clustering. In this context, the literature addressed the problem from two distinct viewpoints (see, for instance, [11, 13, 14] and references therein). The first approach relies on modifying suitable existing algorithms for clustering static data in such a way that time series data can be handled. The rationale of the second approach is to convert temporal series data into the form of static data so that the existing algorithms for clustering static data can be directly used. An interesting alternative to these two mainstream approaches, is to tackle directly the problem of building suitable similarity/kernel functions that encompass dynamical properties of the events. In [31], the authors provide a unifying theoretical framework to study and design different kernels based on dynamical systems, which can be used for behavioral analysis (through the so called ARMA models [5]), diffusion processes, graphs-based systems. A number of efficient methods for computing such kernels are also discussed, making the paper a valuable starting point for a more detailed study of this approach.

A rather complete account of the open issues related to events classification in an unsupervised setting is reported in [18].

Before concluding the section, we believe it is worth spending a few words about the problem of assessing the results and comparing the different methods. Since we consider both the supervised and unsupervised scenario, it is not possible to use a unique evaluation measure. On the one hand, when the *true* labels are available the quality of the classifiers can be easily measured by computing the number of misclassification errors on a separate test set. On the other hand, in the unsupervised learning scenario, evaluating the quality of the results on real world data is considered a challenging aspect. Indeed, if one relies on clustering methods, then there is no standard way to assess the obtained clusters both with respect of the initial set of data and with respect to possible future observations. As reported in [3, 7, 8], most of the work is devoted to estimate an appropriate model (usually related to the choice of the number of clusters to consider) by comparing the estimated membership against a ground-truth, when available. Otherwise, when several instances have been computed (e.g., for different values of the parameters) an analysis based on quality measures, or quality *indices*, is used to select the most promising computation.

3 Low Level Processing and Initial Representation

The starting point of our behavior analysis module is a low-level processing phase extracting dynamic information from video streams. Once the occurred dynamic events have been identified, it is important to find an appropriate initial representation for them, with a potential to discriminate among different behavioral models. This choice should embed most of the prior information available on a specific scenario.

3.1 Video Processing

The low-level processing phase is based on an efficient object tracking algorithm [21] which combines motion and appearance information.

A *motion tracker* based on Kalman and considering position, size and velocity, is associated with an *appearance tracker* based on weighted color histograms. The two modules are combined according to a modular pipeline (see Fig. 1) so to cope with different levels of complexity of the monitored environment while keeping the computational cost low. To this purpose, appearance information is introduced only when appropriate.

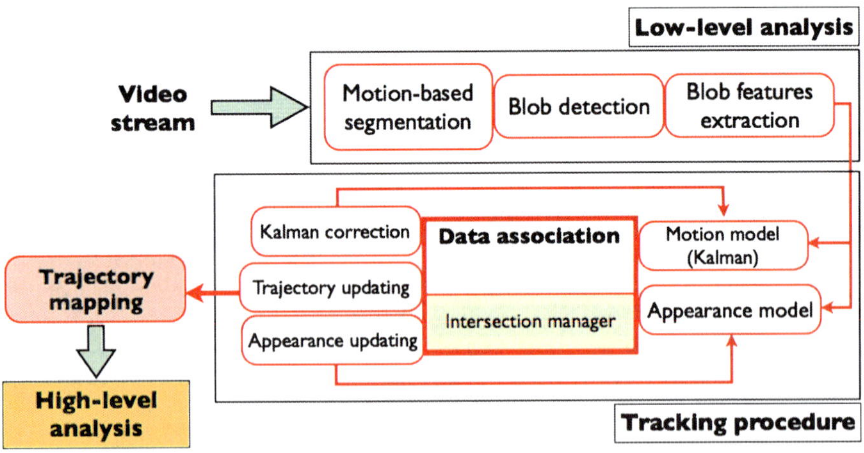

Fig. 1 Architecture of the motion and appearance tracker [21]

Motion and appearance information are measured, stored, and updated independently, but jointly contribute to represent a new observation into the data association procedure. This procedure detects the similarity between new observations and previously seen targets, to correct and update the state of the observed objects. To cope with temporary occlusions among targets, typical causes of failure while tracking objects, the data association includes a procedure to detect and solve intersection events.

At time t, the instantaneous observation of a moving object or *target* O_i, is a vector $\mathcal{M}_i^t = [TH_i^t, \mathbf{P}_i^t, S_i^t, \mathbf{V}_i^t, \mathbf{CH}_i^t, W_i^t, H_i^t]$ where

- $TH_i^t \in \mathbb{R}$ is the frame time-stamp at time t.
- $\mathbf{P}_i^t = [x_i^t, y_i^t] \in \mathbb{R}^2$ is the position of the target centroid on the image plane.
- $S_i^t \in \mathbb{R}$ is the spatial occupancy of the target, $S_i^t = \sum_{(i,j)\in CC_i^t} B^t(i,j)$, where CC_i^t is the connected component of the binary map at time t corresponding to O_i.
- $\mathbf{V}_i^t = [Vx_i^t, Vy_i^t] \in \mathbb{R}^2$ is the apparent velocity vector estimated by Kalman.
- $\mathbf{CH}_i^t$ is a weighted color histogram [4].
- $W_i^t \in \mathbb{R}$ and $H_i^t \in \mathbb{R}$ are, respectively, width and height of the bounding box of CC_i^t.

When target O_i exits from the scene, its trajectory is stored as a temporal series $\{\mathcal{M}_i^t\}_{t=1,\dots,k}$ of instantaneous observations.

3.2 The Choice of the Input Space

The temporal series of instantaneous observations measured according to the afore mentioned module includes features appropriate for the tracking phase. Instead, in the pipeline of behavior modeling, these observations should be appropriately translated into descriptions able to capture the degree of similarity among different behaviors. In general, the meaning of such observations may change according to the application domain under consideration.

In the literature, a common choice is to consider the target position into the image plane [20, 22]. However, modeling the observations as plain positions may be ambiguous in many situations (see, as an example, the scenario we adopted for our experimental evaluation—Fig. 4). A way to cope with these problems is to devise more complex input spaces where to measure, not only position, but also dynamic features (e.g., velocity or acceleration) or geometric features (as area, perimeter, or shape features—the latter useful if a variety of different objects may be moving in the scene). Notice that such features are often available, or easy to compute, from the object tracking phase. For this reason, it is advisable to map the trajectory description (as gathered by the tracking process) into an appropriate form, with the capability of capturing information semantically meaningful for the problem at hand. Our current description, that seems appropriate for the degree of complexity of the available test bed, takes inspiration from the seminal work by Stauffer and his coworkers [29] and may be summarized as a 5-dim vector

$$\mathbf{x}_i^t = \left[\mathbf{P}_i^t, S_i^t, M_i^t, D_i^t\right],$$

where, again, $\mathbf{P}_i^t$ represents the 2-dim object position on the image plane, S_i the object size at time t. M_i^t and D_i^t represent the object velocity at time t as module and direction and may be computed from the velocity vector $\mathbf{V}_i^t$. In the work reported here, appearance information is discarded, being not meaningful for the modeling experimental setting we consider (see Sect. 6 for details). The time-stamp is disregarded since temporal alignment among events is not relevant to our purpose. Finally, width and height of the bounding box, related to of the target dimension, are synthesized with the only size feature. The velocity is represented in the module-direction space as it is semantically more meaningful (e.g., the module of the velocity helps to discriminate among walking or running people, while the direction is an apparent and powerful feature that groups of coherent events probably have in common). As we will see in Sect. 6, these features appear to summarize a set of interesting properties among which building the behavioral models.

4 Temporal Series Representations

In this section, we provide a brief review of some popular representation schemes for temporal series. Indeed, the review is not exhaustive but refers to techniques which showed promising performances during our experimental analysis. The common underlying idea of the following approaches is that raw temporal data may be conveniently *translated* into a suitable low-rank parametric model. The subsequent behavioral analysis can be conveniently performed using effective methods for dealing with static data.

In the following, a data set of N temporal series is denoted by means of the set $\mathbf{X} = \{\mathbf{x}_i\}_{i=1}^N$, where each datum $\mathbf{x}_i$ is a sequence of k_i vectors in some Euclidean space $\mathbb{R}^d$, i.e. $\mathbf{x}_i = (x_i^1, x_i^2, \dots, x_i^{k_i})^T$ and $x_i^t \in \mathbb{R}^d$, $t = 1, \dots, k_i$. A natural way to interpret the index t is as the temporal index.

With a reference to the specific input space adopted in the experiments and anticipated in Sect. 3, in our case $d = 5$, unless otherwise stated.

4.1 Curve Fitting

In a parametric setting, one aims at representing an event by means of a fixed-length feature vector in order to overcome two crucial issues related to temporal series: (1) the fact that the events to be represented do not last the same amount of time, and (2) they do not have the same spatio-temporal evolution (sometimes even within events of the same class).

A popular and easy-to-implement parametric approach is based on *curve fitting*, in which a regression function is built for each sequence (see [23] and references therein). Here an important step is to choose a specific family of fitting functions. Such choice depends on the distribution of the points in the scene, which is strongly connected to the complexity of observed behaviors. Our early attempts to exploit curve fitting were based on second-degree polynomials fitted by means of a simple least squares-based method [20]. Common less naive alternatives are polynomials [15], which however have been shown successful for moderately complex series only. In a more general and articulated scenario, it is more appropriate to rely on a family of *B-spline* models, as proposed in [23].

In the following experimental analysis, we took inspiration from this last work to implement the representation schema based on curve fitting. More specifically, for each trajectory $\mathbf{x}_i$ of length k_i—represented as discussed in Sect. 3—we first normalize the components of the input vectors in the interval $[0, 1]$, and then we perform a maximum likelihood regression separately for each component to estimate the coefficients of a piecewise linear fixed-knot spline. The B-spline models have all the same fixed size according to the number m of knots. The normalization is required in order to make the coefficients of the B-spline models consistent among each other.

According to a common thread of previous works on the topic, the piecewise linear fixed knot splines are obtained using an intermediate feature space in which a point x is represented by a vector $\phi(x)$ defined as $\phi(x) = (1, x, |x - t_1|_+, \dots, |x - t_m|_+)$, where $|x - t_k|_+$ is equal to zero if $x < t_k$ and $(x - t_k)$ otherwise. In such space, it is possible to define a Mercer kernel

$$K(x_i, x_j) = 1 + x_i x_j + \sum_{k=1}^{m} |x_i - t_k||x_j - t_k|.$$

At the end of the above process, each temporal series $\mathbf{x}_i$ can be described by means of a single $d \times (m + 2)$ dimensional vector obtained by concatenating the regression coefficients for all the dimensions of the feature vector x_i^t.

4.2 Probabilistic Models

A possible alternative parametric approach to temporal series modeling may consist of building a probabilistic model of the present observation given all past observations: $p(x_{i,t}|x_{i,1}, x_{i,2}, \dots, x_{i,t-1})$. Because the history of observations can grow arbitrarily large, it is necessary to limit the complexity of such a model. A common approach to overcome this problem is to limit the window of past observations. The simplest model, $p(x_{i,t}|x_{i,t-1})$, is known as a first-order Markov model. One popular choice of probabilistic models that make this assumption, are the hidden

Markov models (HMMs), of which it is always possible to write down explicitly the likelihood function [24]:

$$p(\mathbf{x}|\theta) = \sum_{q_0,\dots,q_k} p(x_0|q_0)p(q_0)\prod_{t=1}^{k} p(x_t|q_t)p(q_t|q_{t-1}), \tag{1}$$

where the $\mathbf{q} = \{q_1, \dots, q_k\}$ are the hidden states corresponding to each element of $\mathbf{x}$ and can take one of the discrete values $\{1, \dots, M\}$. In order to specify a HMM, one has to specify the parameter vector $\theta = (\pi, \alpha, \mu, \Sigma)$, where $\pi \in \mathbb{R}^M$ represents the initial state probability distributions $\pi_i = p(q_0 = i)$ for all $i = 1, \dots, M$. $\alpha \in \mathbb{R}^{M\times M}$ is the state transition probability distribution, i.e., $\alpha_{ij} = p(q_t = j|q_{t-1} = i)$. $\mu = \{\mu_1, \dots, \mu_M\}$ and $\Sigma = \{\Sigma_1, \dots, \Sigma_M\}$ represents the mean and covariance of the Gaussian distributions for each state i, i.e., the emission density $p(x_t|q_t = i)$ is a normal distribution $\mathcal{N}(x_t|\mu_i, \Sigma_i)$.

In order to exploit such representation schema, given a data set of N trajectories the first step is to estimate the parameters of N distinct HMMs representing the probabilistic dynamical models that generated the observations. For example, such estimation may be achieved quite efficiently by choosing one off-the-shelf methods for maximizing the likelihood of the HMMs parameters given the observations. Once such estimation has been performed, each input temporal series $\mathbf{x}_i$ is translated into a single parameter vector $\theta_i = (\pi_i, \alpha_i, \mu_i, \Sigma_i)$ that identifies univocally its HMM. Consequently, the similarity among the trajectories can be computed using ad-hoc kernel functions defined for the probabilistic models, as discussed briefly in Sect. 5.2.

It is important to note that such approach is somehow robust with respect to the specific dimensionality of the instantaneous observations. That is, the algorithm used for generating the representative vectors θ_i does not depend on the number of components of the single vectors x_i^t. Although this is certainly true from an algorithmic viewpoint, in practice we noted that the higher is the number of dimensions the coarsest is estimation of the HMM parameters, and in many cases it may be troublesome to consider them as probabilistic generators of the observed trajectories.

4.3 String-Based Approach

This approach is based on the observation that a temporal sequence of discrete elements can be seen as a concatenation of symbols from a finite *alphabet* $\mathcal{A}$.

Intuitively, the alphabet could be associated with an appropriate partitioning of the input space. Since the choice of an appropriate alphabet is crucial and, as the input size d grows manual partitioning is not conceivable, we address the problem of partitioning the space guided by the available data. In the literature, methods for automatic space partitioning (usually based on vector quantization or clustering) have been proposed [9, 29].

In this chapter, we adopt an approach originally proposed in [19] and previously discussed in [20] which relies on clustering data in the d-dimensional input space

$\{\{x_i^t\}_{t=1,\dots,k_i}\}_{i=1,\dots,N}$. As for the specific clustering algorithm, we adopt *spectral clustering*, in accordance to the behavior modeling phase (more details are reported in Sect. 5). An appropriate kernel function allows us to build a similarity matrix that captures the internal structure of the data set. To combine different measurements, which could take values in different ranges, various choices are possible. A first way is to simply concatenate input vector, after an appropriate data normalization in the [0, 1] range, similar to the one adopted in the curve fitting approach. An alternative way (sometimes referred to as multi-cue integration in the supervised learning literature [30]), is based on a convex combination of similarity or kernel functions on sub-sets of coherent features. Given two observations in $\mathbb{R}^d$, x and y: $K(x, y) = \sum_{i=1}^{N_f} w_i K_i(x_i, y_i, \theta_i)$ where N_f is the number of sub-sets $N_f \leq d$, and w_i sum up to 1, θ_i are the parameters of kernel K_i.

In this work, where $d = 5$ while $N_f = 4$ (being the position a 2-dimensional measure), we adopt a linear combination of Gaussian kernels:

$$K(O_1, O_2) = W_P K_P(p_1, p_2) + W_S K_S(s_1, s_2) + W_M K_M(m_1, m_2) \\ + W_D K_D(d_1, d_2),$$

where O_1 and O_2 are two possible observations $O_1 = (p_1, s_1, m_1, d_1)$ and $O_2 = (p_2, s_2, m_2, d_2)$.

Once the alphabet is built, a temporal series $\mathbf{x}_i$ may be translated into a string s with an association of each element $x_i^t \in \mathbb{R}^d$ to the partition it belongs to. To obtain compressed descriptions that capture the peculiarities of each behavior, we consider only transitions between states, skipping the replicates of the same symbol, i.e., if a string contains replicates of the symbol $u \in \mathcal{A}$ it can be compressed as: $\alpha u^\lambda \beta \rightarrow \alpha u \beta$, for every arbitrary substrings α and β.

5 Learning Behaviors

Once data have been appropriately represented we rely on a learning from examples phase to learn how to discriminate among different behaviors. In this section, we consider both supervised and unsupervised settings.

5.1 The Learning Phase

In the supervised case, we assume we are given a set of labeled data to learn how to classify yet to be seen behaviors. In this case, the representation based on HMMs naturally leads to an approach well established in the literature, where a new trajectory is classified by computing the likelihood with all the learnt models (one for each known behavior) and considering it as realization of the one corresponding to the higher likelihood.

In all the other cases, the goal of classifying a trajectory depending on the behavior it belongs to is achieved by using a multi-class approach where the performance is evaluated on the classification error. We address it by means of classical binary learning tools and the use of a one-vs-all classification procedure. In particular, in our experiments we consider the regularized least-squares (RLS) algorithm, that can be formulated as follows: given a training set $\mathbf{z} = (\mathbf{x}, \mathbf{y}) = \{(x_1, y_1), \ldots, (x_n, y_n)\}$ of n-examples $(x_i, y_i) \in X \times \mathbb{R}$, regularized least squares amounts to

$$f_{\mathbf{z}}^{\lambda}(x) = \sum_{i=1}^{n} \alpha_i K(x, x_i) \quad \text{with } \alpha = (\mathbf{K} + n\lambda I)^{-1}\mathbf{y},$$

where $\lambda > 0$ is the regularization parameter, $\mathbf{K}$ is the $n \times n$-matrix $(\mathbf{K})_{ij} = K(x_i, x_j)$, with K is a Mercer's kernel. In the binary classification case $y \in \{-1, 1\}$. A more general class of algorithms which could be adopted in this framework, as well as references to the pertinent literature, can be found in [16].

In the unsupervised case, we assume the available data are not associated with a label, thus there is no explicit knowledge on common behavior. Unsupervised learning may be used to mine information from the data available. In this case, we apply clustering to group the available trajectories in coherent groups and associate with each group an estimated behavior. The obtained analysis allows us to both (i) estimate common behaviors, (ii) associate new data with the estimated behaviors.

The analysis we carry out is based on the use of the well established technique known as *spectral clustering*. Specifically, for the experiments, we implemented an efficient recursive algorithm for spectral clustering following [27].

The algorithms works as follow. Given a suitable similarity measure among data points,[1] the implemented algorithm clusters the data set by recursively computing the optimal bipartition of the weighted similarity graph G associated with a specific subset of data. It can be shown that an optimal bipartition of the graph into subgraphs A and B can be computed by minimizing the following measure called *normalized cut*:

$$\text{Ncut}(A, B) = \frac{\text{cut}(A, B)}{\text{vol}A} + \frac{\text{cut}(B, A)}{\text{vol}(B)},$$

where $\text{cut}(A, B)$ represents the sum of all edges linking points in A to points in B, while $\text{vol}(A)$ and $\text{vol}(B)$ represents the sum of the weights among points in A and B, respectively. The minimization of the Ncut can be achieved only approximately by analyzing the eigenstructure of the Laplacian matrix L associated with the graph. Such matrix is computed by a suitable normalization of the matrix W whose elements w_{ij} represent the similarity between the ith and jth node of the graph.

At each step of the recursive procedure, the algorithm computes the second smallest eigenvector $\mathbf{v} = (v_1, \ldots, v_N)$—i.e., the one corresponding to the first nonzero eigenvalue—of L. In [27], the authors showed that the closest approximation of the optimal graph partition is the one associated with the sign of the components v_i.

[1]Note that in the unsupervised setting, the similarity function plays the role of the kernel function of the RLS case—as they both model the similarity among available data.

Since our algorithm mimics a divisive hierarchical clustering, very few parameters are needed. The most important is the *stopping criterion*, which controls the level of granularity of the clustering. In our algorithm, this can be specified through either a simple threshold on the value of the normalized cut corresponding to the candidate partition or a minimum size allowed for a cluster. This means we do not have to decide the number of the clusters beforehand, but we have only to define the minimum size allowed for the small clusters.

In our approach, the choice of an appropriate similarity or kernel function is crucial, since it allows for the extraction of meaningful information from the data at hand [26]. We now review kernels available from the literature that may be profitably applied to time-series, with a reference to the representations. In the case of time-series, and depending on the chosen representation, it may be the case that kernels for variable-length descriptions are needed.

5.2 *Kernels for Time-Series*

It is well assessed that, in the learning from examples approaches, the obtained results strongly depend on the ability to capture the underlying notion of metric—or the similarity structure—over the input space. In turn, the similarity structure is related to the choice of a proper kernel function on the data. A function to be used as a kernel for RLS has to be a Mercer's kernel, that is *symmetric* and *positive definite*, while for spectral clustering should be *symmetric* and *positive*. Since in our experiments, we compare different representation schemes we have to choose appropriate kernels.

In the case of B-Splines, we refer to well established kernels such as the Gaussian kernels. The other two approaches require a more careful choice of the kernel able to deal with variable length data [26]. We considered the following:

5.2.1 Probability Product Kernel (PPK)

An efficient measure of similarity among temporal series represented by means of HMMs was introduced in [12], and used in [13] in tasks similar to ours.

Given two probability distributions representing two pairs of data sequences over the space of all potential observable sequences X, the generalized inner product can be computed by integrating the product of the distributions:

$$K\big(p(x|\theta), p(x|\theta)\big) = \int_X p^\beta(x|\theta)p^\beta(x|\theta)\,dx, \tag{2}$$

where β is a free parameter which allows for the specification of some properties of the kernel. For instance, if $\beta = 1/2$, the PPK becomes the classic Bhattacharyya affinity metric between two probability distributions, which is a Mercer kernel and can be conveniently used in our context. In addition, it is computable in closed form for a variety of distribution families. For the HMMs, we used the guidelines proposed in [13].

It is worth recalling here that different definitions of similarity among HMMs are present in the literature; therefore our comments below about the performance of the HMM-based approach in the unsupervised setting are relative to the specific case of the similarity measure (2).

5.2.2 Kernels for String-Based Representations

In this case the choice of an appropriate kernel is quite critical, considering the peculiarities of the available data. Indeed, even after the translation of each temporal series in a string, the event description is still variable length—depending on the number of transitions of a behavior instance from one state to another. We adopt a similarity measure from the family of *string kernels* the well known *P-spectrum* kernel, K_p [26]. Formally, the kernel may be defined as a feature map of strings followed by an appropriate dot product. The map makes explicit all sub-strings of length P of string s:

$$\phi_u^P(s) = \left|\{(v_1, v_2) : s = v_1 u v_2\}\right|, \quad u \in \mathcal{A}^P.$$

The associated kernel between two strings s and t is defined as:

$$K_P(s,t) = \langle \phi^P(s), \phi^P(t) \rangle = \sum_{u \in \mathcal{A}^P} \phi_u^P(s)\phi_u^P(t).$$

In other words, the *P-spectrum* relies on counting how many substrings of fixed length P the two strings have in common. If the focus is on simple transitions between atomic symbols of the alphabet, the natural choice is to consider basic substrings of length 2. P-spectrum kernel is a Mercer's kernel, therefore it is also symmetric and positive, and can be conveniently used in the spectral clustering domain.

5.3 Run Time Analysis

At run time, once a new dynamic event is observed, the objective of a behavior understanding module is to associate this new observation with one of the common behaviors learned by the system. In case none of the previously observed behaviors is appropriate for the new observation, we say that an *anomaly* is detected.

According to the adopted pipeline, the dynamic event is first represented with respect to the appropriate input space (for instance,, the one described in Sect. 3), then it is translated according to one of the parametric models discussed in Sect. 4 into the feature vector we use as a test datum.

At this point, in the supervised case, we estimate the most likely class—either with a multi-class RLS or by computing the highest likelihood of belonging to a given probabilistic model.

When dealing with unsupervised learning, an out-of-sample method on the clustering tree [6] may be applied. As an alternative, the membership of the new datum to a known behavior can be tested by comparing it and an appropriately chosen behavior candidate. The candidate may be computed via a voting strategy computed as follows: for each string s in cluster C we compute the similarity with all other elements of C. The most similar string to s is voted by s to be the candidate. By putting together the votes of all the strings in C, the string obtaining the higher number of votes is elected as a cluster candidate. In our experimental analysis, this technique has been preferred to out-of-sample, as it allows for a rapid visual inspection of the obtained results, since archetypical sequences can be easily visualized and compared with data waiting to be associated with clusters.

6 Experimental Analysis

The focus of this section is on the assessment of the discussed pipeline for behavioral modeling, comparing the various data abstraction modules proposed. After introducing data set and software tools, we discuss the experiments: first, we evaluate the different representation schemas in the context of supervised learning from examples, then we move to the unsupervised framework, where clustering is interpreted as a tool to associate real and estimated behavioral patterns.

As described in Sect. 3, the video analysis stage (low-level processing) is based on our previous work [21], as well as the computation of string-based representation and P-spectrum kernel [19].

All the experiments based on the HMMs models have been performed by learning from our data the parameters of the graphical model by means of well established *Bayes Net Toolbox* for Matlab[2] developed by Kevin Murphy and widespread used by machine learning practitioners.
In order to infer the parameters of the spline models we implemented our own module based on the Spline Matlab Toolbox.[3] The guidelines for designing such module are those described in [23], although there the authors rely on SVM for the spline-based regression, while we preferred to adopt a more standard least squares-based approach. For the latter, we exploited our own implementation.

6.1 Data Collection and Semi-automatic Labeling

Previous work [20] based experimental evaluations on synthetic data and benchmark data sets. In this work, we rely on a set of real data acquired by a video-

[2] Available for downloading at http://people.cs.ubc.ca/~murphyk/Software/BNT/bnt.html.

[3] Further details on: http://www.mathworks.com/products/splines/.

Fig. 2 Sample frames acquired by the surveillance camera, showing the potentially complex conditions of the observed environment: the crowd level differs depending on the temporal interval considered, while the illumination is strongly affected by physical properties of the environment, in particular the presence of windows along the right wall

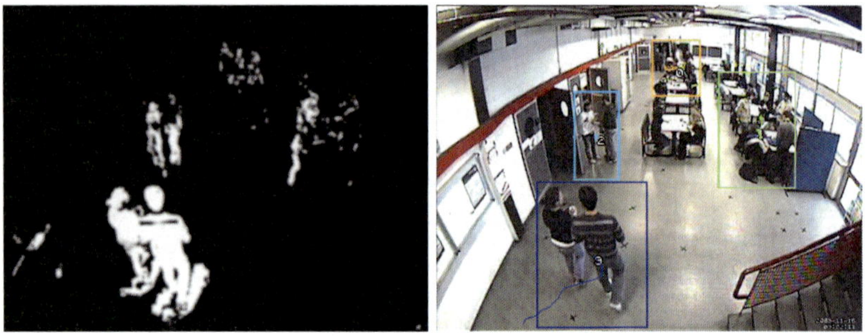

Fig. 3 An example of the resulting binary map after the low-level video analysis: the connected components (*left*) are first extracted and described with a features vector, then correlated over time (*right*) by means of tracking to model their dynamic evolution in the scene

surveillance system[4] and loosely annotated according to the procedure described in the following. The system monitors an indoor open space (one of the main halls of our Department), shown in Fig. 2, where a good amount of dynamic events occur during daytime. Only people are supposed to be moving in the scene: the monitored environment provides different complexity with respect to the crowd level, which in turns depend on several factors, such as day, temporal interval, period of the academic year (presence of lessons, examinations). The weather conditions strongly affect the scene appearance (see Fig. 2), being the hall illuminated by windows (on the right wall) as well as artificial lights. At each time instant, first the current frame has been segmented with respect to motion information (Fig. 3, left), the tracking procedure is applied (Fig. 3, right) then to gather the trajectories and finally map them into the dynamic events descriptors, as discussed in Sect. 3.2. Though the

[4]The Imanalysis suite, we obtained within a technology transfer program with the company Imavis srl, http://www.imavis.com/.

Table 1 Temporal ranges where video data have been considered for the experimental analysis

Starting time	Ending time
08:45	10:00
12:30	14:30
16:00	17:00

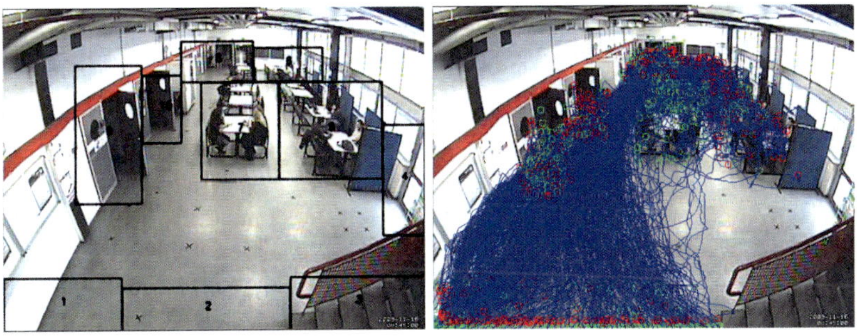

Fig. 4 *On the left*, the set of source-sink regions considered during the annotation process. The trajectories used to feed the high-level analysis module have been acquired during a week considering a set of interesting temporal ranges. After a cleaning procedure, the input data set results in the trajectories shown *on the right*. *Red* and *green circles* denote, respectively, *first* and *last points*

camera continuously acquires the video signal, we extracted trajectories from observations of a week, limited to the selection of temporal ranges listed in Table 1. The dynamic content of the included time intervals is highly representative of the totality of events typically occurring in the scene (people arriving and leaving, entering and exiting the doors, performing activities on the tables area, just to give some examples). After that short trajectories have been discarded, we annotated the resulting ones on the basis of the set of source-sink regions visualized on Fig. 4(left). The set of regions have been decided on the basis of physical properties of the environment, as presence of doors, tables, coffee, and drinks dispensers. We thus ended up with 8 general behaviors detailed in Table 2 and shown in Fig. 4(right), giving an impression of the richness of the observed events.

6.2 Model Selection

We now briefly review the selection of the parameters for all the methods included in the analysis.

6.2.1 Curve Fitting

The model selection with the representation schema based on spline curve fitting is straightforward. Indeed, the parametric model described in Sect. 4.1 requires, as

Table 2 The annotation process has been defined of the basis of a set of source-sink regions (see Fig. 4). We end up with a set of 8 behaviors, representing the most common and general dynamic events occurring in the monitored environment

Behavior	Source regions	Sink regions	# Trajectories
1	1	9, 10	281
2	1, 2	8	165
3	5, 6	7, 8	96
4	6	5	51
5	8	1, 2	210
6	8	7	104
7	9, 10	1, 2	215
8	9, 10	8	83

parameters, just the definition of the fixed knots m. Since the described approach relies on several mono-dimensional curve fitting problems—with the independent variable being the time—one has to set only the temporal spacing between two consecutive knots. In our experiments, we set such spacing as one third of the average length of the trajectories in our data set. In such way, on average, all the trajectories are represented by a fair number of nontrivial regression coefficients. We report the result obtained using both all the features ($d = 5$) and the position only ($d = 2$).

6.2.2 Probabilistic Model

In using the representation schema based on HMMs, we considered two different configurations. According to the first one, the parameters of the HMMs are learnt from the data using, at each time, input vectors comprising all the features (i.e., X and Y position on the image plane, area of the tracked object, and amplitude and modulus of the vector representing the approximate velocity). According to the second configuration, training data of each trajectory are based on only the two features corresponding to the position on the image plane. As already pointed out in [20], HMMs depend strongly on the initialization of their joint probability density function. Since in the unsupervised setting the HMM is used only as an intermediate step towards the computation of the similarity function between two trajectories, one is forced to train the HMMs using one trajectory only. Note that comments about this possibly troublesome choice are present in the original work [12] to which we refer the interested reader for further discussions. From a practical point of view, however, we noted that the bigger is the number of dimension of the feature vector the more difficult is for the training algorithm to reach the convergence. As a consequence, with the first configuration we have not been able to compute the similarity between several short trajectories. For this reason in the following, we report the results obtained with the second configuration only, with $d = 2$. The number of possible states of the HMM is three and the probability density function corresponding to each state is Gaussian.

6.2.3 String-Based Approach

When dealing with the string-based representation, we computed a family of possible alphabets $\mathcal{F} = \{\mathcal{A}_{\text{par}}\}_{\text{par} \in \mathbf{PAR}}$, where **PAR** is the set of allowed combination of parameters. To combine different features we considered different approaches: (1) a simple concatenation of the features in a vector, after that a normalization made the features comparable, and (2) the multi-cue integration, where each feature is processed singularly. Since in both cases spectral clustering is adopted to partition the input space (thus to build the alphabet), the parameters include a cut threshold, we called TH_{points}, chosen in the range [0.6, 1.0]. Concerning the kernel used, which is Gaussian in this case, we moreover had to set the σ (i.e., the standard deviations of the Gaussian kernel components, a single value for (1), a value for each feature in (2)). We derived the values from the definition of diameter of a points set S, which is defined as diam $= \max_{\mathbf{x},\mathbf{y} \in S} \text{dist}(\mathbf{x}, \mathbf{y})$, where dist is an appropriate measure of dissimilarity (the 2-norm in our case). The σ value(s) is then computed as a fraction of the diameter. Besides, the specific choice can be guided by the distribution of the input features. Focusing on the multi-cue integration approach, moreover, we selected the combination of features weights $\mathbf{W}$ such that $\mathbf{W} = \bigcup_k \{[W_P, W_S, W_M, W_D] | W_i \in \{0, k\}, \forall i \in \{P, S, M, D\}\}$ with $k \in \{0, \frac{1}{4}, \frac{1}{3}, \frac{1}{2}, 1\}$ and $\sum W_i = 1$ with $i \in \{P, S, M, D\}$. This selection constitutes a significant sampling of the totality of possible alphabets.

6.2.4 Kernel Choice

Let's now describe and motivate some of the choices relative to the parameters of the kernel functions. For what concerns the P spectrum, a natural choice is to fix $P = 2$ being the strings obtained by keeping the only transitions between states. The kernel used to compare two HMMs are the probability product kernels, described briefly in Sect. 5.2.1. According to the authors that first introduced such kernels [12], they are very robust with respect to the choice of the kernel parameters. In the specific case of HMMs, the kernel becomes the popular Bhattacharyya affinity matrix, and the only significative parameter T is called *mixing proportion*, and corresponds to the time interval considered for the evolution of the underlying dynamical systems. In our experiments, we used the same value ($T = 10$) proposed in [13], where the authors showed convincing evidence about the stability of such value. Our experiments confirmed such claim.

6.2.5 RLS Classification

Performing the multi-class classification based on regularized least squares and k-fold cross validation (k-CV), we needed to decide the number k of subsamplings (in our experiment $k = 5$) and the numerical range of the regularization parameter λ, which is chosen in the range $[10^{-3}, 1]$ and properly selected in correspondence of the minimum classification error.

6.2.6 HMMs Likelihood Estimation

The use of HMMs in the supervised setting refers to a widely accepted practice of learning different probabilistic models, each corresponding to the one that most likely generated one of different classes in which the data set can be decomposed. All the new trajectories are tested against all the generative models, and classified to belong to one specific class on the basis of the corresponding likelihood. In the experiments, we tried to be as close to the multi-class classification approach as possible. Therefore, we run $k = 5$ different times the learning algorithm for each class keeping separate a fraction of the all data set for the subsequent test.

6.2.7 Spectral Clustering

In the unsupervised analysis, spectral clustering is used to group coherent set of trajectories, it is thus necessary again to tune the cut threshold parameter, TH_{trj}: we chose such value in the range [0.5, 1.0].

6.3 Supervised Analysis

We now evaluate the results obtained for the supervised setting. In the string-based approach, we first need to estimate the most appropriate alphabet with respect to the different possible weight vectors $\mathbf{W}$ for the multiple cue integration of features. Table 3 reports the average hit rates obtained by 5-CV for a selection of weights. It is apparent how the presence of position leads, in this case, to the best results, with the highest performance obtained by for $\mathbf{W} = (1, 0, 0, 0)$. The association between input data and output labels seems to help coping with ambiguities in a data-set that, in principle, suffers from it, as shown in Fig. 3(right), where trajectories appear superimposed.

Table 4 compares the best result obtained with a string-based approach against the performances given by B-Splines. By using the representation schema based on B-spline, the performances are slightly higher.

It also reports a comparison between feature concatenation and multi-cue integration for the string-based case, highlighting a clear superiority of the latter.

In order to deal with HMM-based representation, we start from a 2-dim representation based on the positions only and use a standard approach of estimating one probabilistic model for each class, and then using the models as a composite classifier. Given a new trajectory, the output label can be assigned by computing the likelihood to belong to all the models and then choosing the one that maximizes such likelihood. Similarly with the previous 5-fold learning method, for the training of the HMMs models we divided the data set into 5 distinct subsets and run the inference algorithm for each of them, using the others as test set. The average performance is 64.98%, which is below the performances of the methods reported

Table 3 Hit rates in the string-based approach with a multi-label classification based on RLS

Alphabet	W_P	W_S	W_M	W_D	TH_{points}	%
1	0	0	0	1	0.90	48.38
2	0	0	0.5	0.5	0.95	52.70
3	0	0	1	0	0.95	38.68
4	0	0.5	0	0.5	0.95	50.29
5	0	0.5	0.5	0	0.95	41.18
6	0	1	0	0	0.90	24.57
7	0.5	0	0	0.5	0.95	63.9
8	0.5	0	0.5	0	1.00	63.91
9	0.5	0.5	0	0	0.95	63.64
10	1	0	0	0	0.95	**67.48**
11	0.25	0.25	0.25	0.25	1.00	62.58
12	0	0.33	0.33	0.33	0.95	52.54
13	0.33	0	0.33	0.33	1.00	64.33
14	0.33	0.33	0	0.33	1.00	62.5
15	0.33	0.33	0.33	0	1.00	62.16

Table 4 Supervised analysis based on RLS multi-label classification: comparison between methods

Representation schema	Success rate (%)
Strings (concatenation)	27.97
Strings (multi-cue integration)	67.48
B-Spline (5 dim)	71.93
B-Spline (2 dim)	**72.26**

in Table 4. However, by considering the confusion matrix (Fig. 5) of this classifier, one can easily see that the some of the classifiers—those corresponding to behaviors 1, 4, 5, and 8—achieve extremely good performances while most of the errors are done by the others; more specifically those corresponding to behaviors 2, 3, and 7 (a visual inspection to Fig. 6 suggests that these behaviors are not well discriminated to others).

From the obtained results, we may conclude that in the supervised case a B-spline abstraction module fitting the sole positions is the most appropriate data abstraction for the considered supervised scenario. However, as we will see below for the unsupervised scenario, such representation approach tends to be influenced by local properties of the data set, thus producing a large number of clusters consistent with the labels. More extensive experiments are needed in order to verify the robustness of the approach in the supervised case with respect to the increase of observed behaviors.

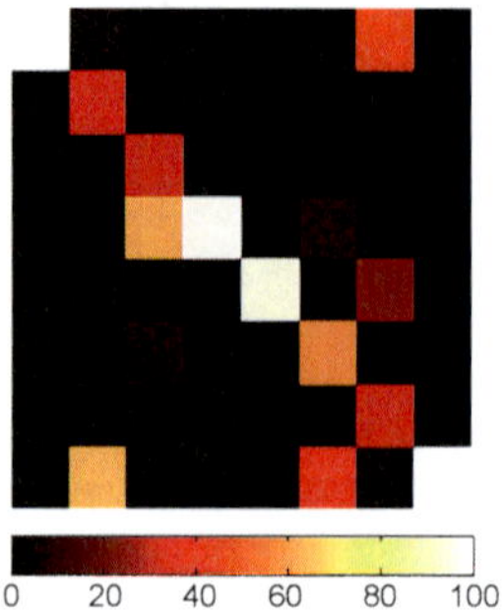

Fig. 5 Confusion matrix relative to the classifiers obtained by training on HMMs model for each behavioral pattern. The classes are ordered according to Table 2

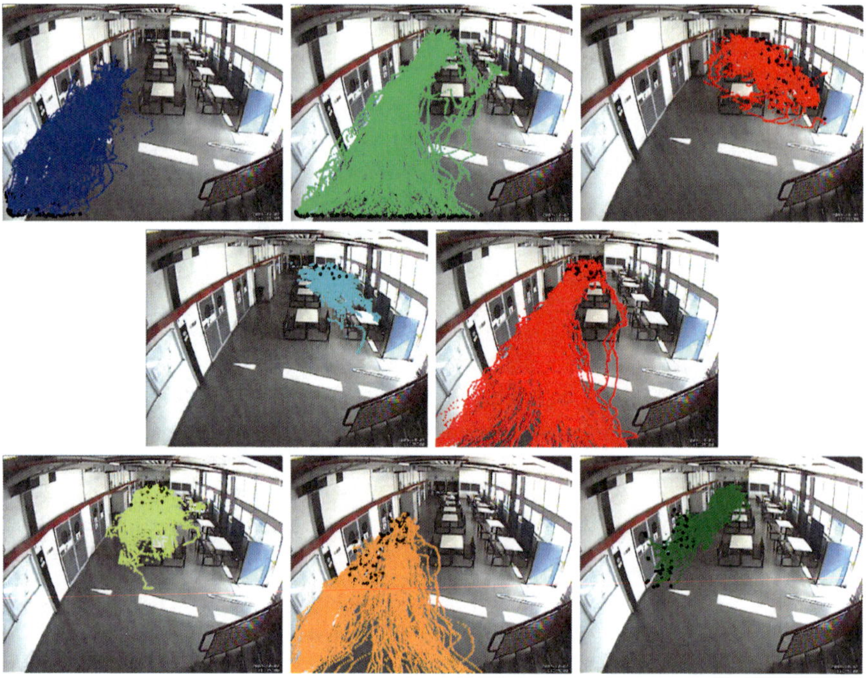

Fig. 6 The 8 behaviors resulting after the annotation process (see Table 2 for details on the annotation)

6.4 Unsupervised Analysis

In the unsupervised case, we evaluate the goodness of the results using the labels as a ground-truth to establish the correspondence between the true (annotated) behaviors and the estimated ones (i.e., the clusters). We say that an estimated cluster C corresponds to a true behavior B if the most of elements laying in C are labeled (in the ground-truth) as instances of B. By following this procedure, we then evaluate the results with respect to different granularity:

- With the *strict association to cluster* we assume that there exist a unique correspondence between true and estimated behaviors.
- Using *loose association to cluster* we admit that a true behavior could be split to more that one estimated clusters.

In the string-based approach, this procedure is also adopted to select the best alphabet from the family computed when adopting the multi-cue features integration: in particular, the choice of an appropriate selection of weights allows us to obtain a sub-set of alphabets which can be appropriate for a given environment and adaptable to its stable changes. This choice is guided by the set of available data, assuming to carry all meaningful information on common behaviors.

Our first experiment for the unsupervised case relies on performing such association assuming that each real behavior can correspond to just one estimated cluster (above, we called such approach *strict clusters association*). Since this could discourage solutions with higher numbers of clusters, we fix a constraint on the number of clusters by using the prior on the true number N_T ($N_T = 8$ in our experiments): we admit in the final evaluation only solutions whose estimated number of clusters N_C is in $[N_T - \delta, N_T + \delta]$ where $\delta \geq 0$ is an integer number.

In Table 5, we report the performances of the family of alphabets computed when adopting the multi-cue integration: the alphabet which results to perform the best is the number 13 (based on partitioning the input space by considering position and dynamic information of targets). The corresponding performance is then selected and compared against the other approaches in Table 6. Although the highest recognition rate corresponds to the B-Splines representation computed in the 2-dimensional

Table 5 Strict association rate for the alphabet computing with multi-cue integration, $\sigma = \frac{\text{diam}}{10}$ is computed independently for each feature

Alphabet	W_P	W_S	W_M	W_D	TH_{points}	TH_{trj}	# Clusters	%
1	0	0	0	1	1	0.85	6	50.53
2	0	0	0.5	0.5	1	0.75	7	52.61
3	0	0	1	0	1	0.9	7	34.85
4	0	0.5	0	0.5	0.95	0.9	8	48.96
5	0	0.5	0.5	0	1	0.7	6	21.90
6	0	1	0	0	–	–	–	–
7	0.5	0	0	0.5	0.95	0.5	7	55.51
8	0.5	0	0.5	0	–	–	–	–
9	0.5	0.5	0	0	–	–	–	–
10	1	0	0	0	1	0.5	7	59.00
11	0.25	0.25	0.25	0.25	0.95	0.85	8	49.37
12	0	0.33	0.33	0.33	0.9	0.75	7	49.29
13	0.33	0	0.33	0.33	0.95	0.6	7	**65.80**
14	0.33	0.33	0	0.33	0.9	0.7	7	55.60
15	0.33	0.33	0.33	0	–	–	–	–

Table 6 Summary and comparison among representation schema for the strict association of dynamic events

	# Cluster	%
String (concatenation)	2	34
String (multi-cue integration, $\sigma = \frac{\text{diam}}{10}$)	7	65.80
B-Spline (5 dim)	6	62.20
B-Spline (2 dim)	6	**67.94**
HMM (2 dim)	**8**	62.29

input space, one can notice that string-based and HMM-based representations better estimate the true number of clusters. The result obtained when the string-based representation is made upon an alphabet built on observations which are the plain concatenation of features is very poor: the spectral clustering fails in splitting the data in more than 2 subgroups, testifying the poor capability of the representation in characterizing the data-set. Again, this result speaks in favor of the advantages of multi-cue integration.

Looking at more practical scenarios, however, it is acceptable that, with respect to a manual annotation process which depends on the specific subject performing it, the structure learnt by the clustering process is slightly different from the annotated one. We thus consider the possibility to admit that a real behavior could correspond to different estimated clusters (the process we called *loose clusters association*). Moreover, trying to understand the reliability of the representation schemas, we ignore the constraint on the number of clusters, so that our best solution corresponds to the highest correct association rate. Since the method would tend to reward clustering instances with higher number of clusters, if coherent and compact in their contents, the results will allow to evaluate the capability of the representations of providing good associations rate while approaching the correct number of partitions. The obtained results are reported on Table 7.

Consistently with what previously observed, alphabet 13 performs better that the other. Notice that the very good performance provided by alphabet 7 (position and direction) is furthermore improved by adding the size (alphabet 14) or, more convincingly, the velocity module (alphabet 13). These considerations suggest that, on the specific input data-set that we considered, position, and direction are the most semantically meaningful feature, followed by velocity module and size.

We conclude our analysis with the comparisons among different methods against the loose association, reported in Table 8. When relying only on the percentage of correct associations, B-spline fitting results in an unreliable schema, highly overestimating the correct number of clusters. The performance of HMMs and string-based representations (the latter with multi-cue integration) are comparable.

Figure 7 (similarity matrices computed on the ordered data-set with respect to the 3 data representations) confirms this analysis: B-Splines tend to over-estimate similarities and then over-segment data; string-based approaches slightly underestimate similarities, but capture the expected diagonal block structure; HMMs appear to under-segment data and miss the expected block structure.

Table 7 Loose association rate for the alphabet computing with multi-cue integration and $\sigma = \frac{\text{diam}}{10}$ for each feature

Alphabet	W_P	W_S	W_M	W_D	TH_{points}	TH_{trj}	# Clusters	%
1	0	0	0	1	1	0.95	8	56.26
2	0	0	0.5	0.5	0.9	0.75	8	69.46
3	0	0	1	0	1	0.9	7	45.89
4	0	0.5	0	0.5	0.95	0.9	8	62.57
5	0	0.5	0.5	0	1	0.7	6	44.56
6	0	1	0	0	0.6	0.5	2	23.31
7	0.5	0	0	0.5	0.95	0.5	7	72.36
8	0.5	0	0.5	0	0.6	0.5	2	23.31
9	0.5	0.5	0	0	0.6	0.5	2	23.31
10	1	0	0	0	1	0.5	7	66.39
11	0.25	0.25	0.25	0.25	0.95	0.85	8	69.95
12	0	0.33	0.33	0.33	0.95	0.85	8	65.47
13	0.33	0	0.33	0.33	0.95	0.6	7	**76.18**
14	0.33	0.33	0	0.33	0.95	0.6	8	73.36
15	0.33	0.33	0.33	0	0.6	0.5	2	23.31

Table 8 Summary and comparison among representation schema for the loose association of dynamic events

	# Cluster	%
String (concatenation)	2	34
String (multi-cue integration, $\sigma = \frac{\text{diam}}{10}$)	7	76.18
B-Spline (5 dim)	30	83.38
B-Spline (2 dim)	31	91.94
HMM (2 dim)	8	74.30

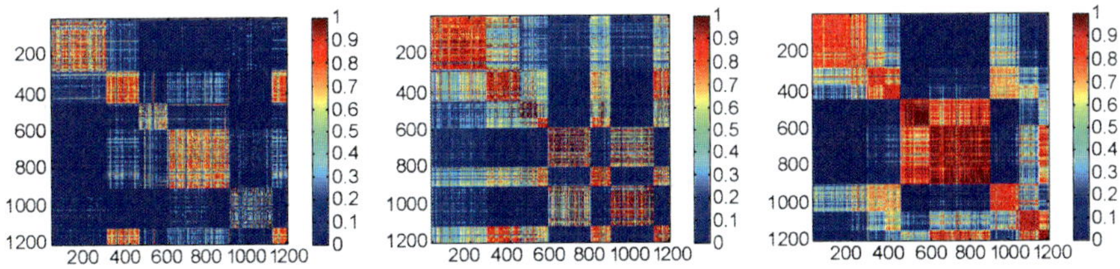

Fig. 7 Similarity matrices. *Form left to right*: best alphabet for string-based representation with multi-cue features integration ($W_P = 0.33$, $W_S = 0$, $W_M = 0.33$, $W_D = 0.33$, $TH_p = 0.95$, $TH_t = 0.6$; B-Splines based on 5 features; HMM-based representations on a 2-dimensional input space

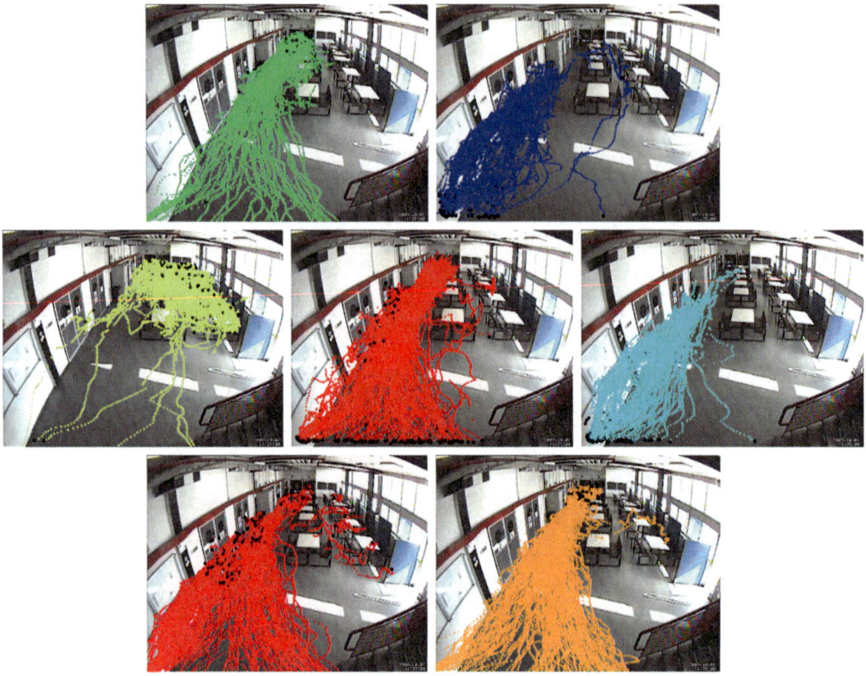

Fig. 8 Best clustering results obtained with the string-based representation schema with the loose evaluation. The resulting weights combination is $W_P = 0.33$, $W_S = 0.00$, $W_M = 0.33$, $W_D = 0.33$

Finally, Figs. 8 and 9 report the estimated behaviors for string-based and HMMs. It is clear in both cases they do not completely match the given annotations, but seem to reflect the fact the discriminative power of close range observations is higher than the one of far observations. This is not surprising, and simply suggests how, given the complexity of the scenario, a multi-camera video-surveillance system would be appropriate.

7 Discussion and Open Problems

This chapter dealt with the problem of learning common behavioral patterns from sets of dynamic events, within a video-surveillance framework. Assuming to dispose of a low-level video processing module that extracts dynamic events from a video stream, and to have selected an input vector of measurements appropriate for the specific domain and with a potential to discriminate among different behaviors, our analysis focused on finding a data abstraction and a learning procedure, as general as possible—ideally able to adapt to changes of the input measurements if required. The whole analysis considered both supervised and unsupervised settings, and was performed on a rather complex set of data acquired by a video-surveillance system. The analyzed methods presented different peculiarities that could make them appropriate in different circumstances.

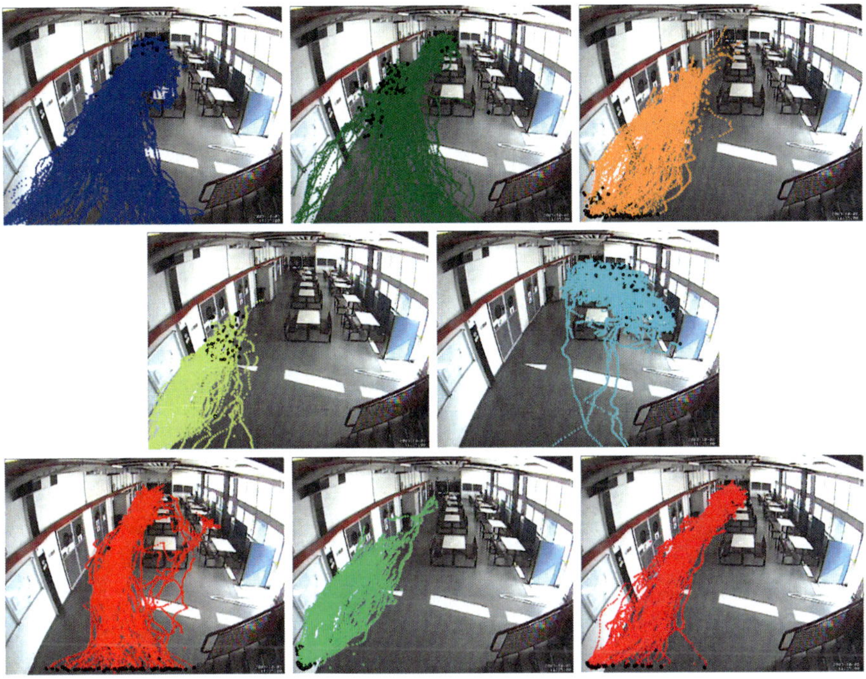

Fig. 9 Best clustering results obtained with the HMM-based representation schema with the loose evaluation

As for the choice of appropriate data abstractions, in the supervised case, the B-spline fitting approach lead to the best performance, regardless the choice of a specific input space. Similar conclusions were reached in the unsupervised case, if an estimate of the number of clusters is available. In this case, string-based approaches lead to comparable results. Finally, in the more realistic case the number of clusters is not available, B-Splines strongly over-segment the data, while string-based and HMMs (used in conjunction with the similarity measure defined in (2)) report more convincing results, with a better balance between the number of estimated clusters and the percentage of correct associations.

For what concerns the criticality of choosing a particular input representation, we observed how in a supervised setting, where labels are available, using the position as initial descriptions to represent the observations allows to achieve acceptable percentages of correct association to known behaviors. As opposite, with fully unsupervision, the position fails to disambiguate the data, while other information can be profitably inserted into the descriptions and significantly help to disambiguate the data.

Considering the ability of the data abstraction chosen to adapt to a change in the input representations (which could be useful if new input features were added to the initial representation), we observed how, in the case of curve fitting, the addition of a new measurement would require the estimation of a new fitting function;

string-based approaches would require the construction of a new set of alphabets, but the process is entirely data-driven and could be performed automatically. The HMMs method, in theory would scale nicely with the input representation change, but in practice, since we model one trajectory at a time, increasing the number of parameters, immediately degrades to performance (this was observed in the experimental analysis, where a full 5-*dimensional* input representation failed to produce convincing results).

All the reported experiments highlighted the fact that a specific labeling of dynamic events may be highly subjective and thus supervised approaches may not be in general appropriate for this application domain. Observing the manual annotation reported in Fig. 6, we see how behaviors 3, 4, and 6 appear to be very similar as they occur at a high distance from the viewing point. Not surprisingly, the estimated behaviors always fail to discriminate among these groups. This effect is magnified if we consider a more granular manual annotation.

To have a visual evidence of the capability of the representations to scale with the complexity, we specialize the data annotation as reported in Table 9.

Figure 10 reports the similarity matrices obtained by reordering the data with respect to the new 14 behaviors (we do not consider in this analysis the B-spline fitting for its poor performance when adopted into the clustering framework, see Sect. 6.4). It is apparent that the string-based representation (Fig. 10, left) provides a way to enhance this more detailed structures into the data, even if the compactness of classes might be improved. On the other hand, HMM-based approach (Fig. 10, right), although having the capability of enforcing the similarities among component of a same class, tends to see as similar events which are instances of different behaviors. We could thus conclude that the two schemas of representation provide

Table 9 Increasing the granularity of the interesting behaviors, the annotation enhances 14 different sets of coherent trajectories

Behavior	Source regions	Sink regions
1	9	8
2	2	8
3	6	5
4	1	8
5	8	7
6	8	1
7	9	1
8	6	8
9	1	9
10	1	10
11	8	2
12	10	8
13	5	6
14	10	2

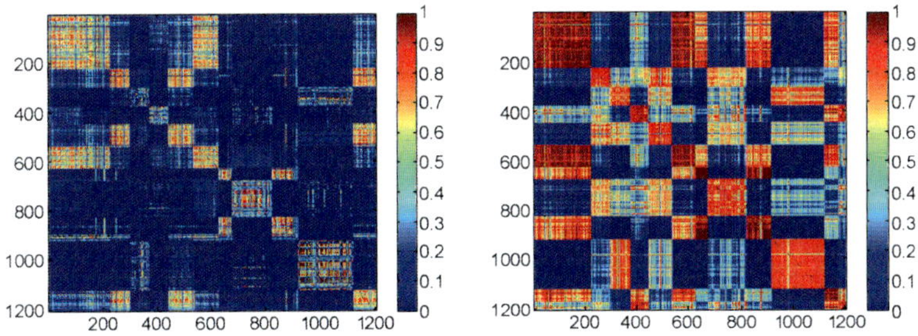

Fig. 10 Similarity matrices for the alphabet $W_P = 0.33$, $W_S = 0$, $W_M = 0.33$, $W_D = 0.33$ and the HMM—based representation considering 14 annotated behaviors

different views on the same data-set, with advantages and drawbacks in both cases. Future work will be devoted to investigate towards this direction.

Since the unsupervised case seem to be the most appropriate for the application under consideration, a critical open issue is how to perform model selection, and how to evaluate the quality of the obtained results in case labels are not available. As suggested in Sect. 2, the use of indices could be investigated, although it is not clear whether this approach could be satisfactory in such complex scenario. An interesting alternative worth investigating is the use of semi-supervised learning, where a limited number of labeled data could be extracted by means of a semi-automatic procedure as the one adopted in our experiments.

Another important point that will be included in future investigation is a more explicit use of the hierarchical nature of the adopted clustering methods, as it seems very appropriate to the data under analysis.

Acknowledgements The authors would like to thank Annalisa Barla and Luca Baldassarre for the code on multi-class RLS and the help given in running the experiments for the supervised case. The low-level processing modules are on going joint work with Augusto Destrero and Alberto Lovato.

Matteo Santoro is supported by Compagnia di San Paolo (Torino) through the Neuroscience Program for the project *Action representations and their impairment*.

References

1. Anjum, N., Cavallaro, A.: Multifeature object trajectory clustering for video analysis. IEEE Trans. Circuits Syst. Video Technol. **18**(11), 1555–1564 (2008)
2. Bashir, F., Khokhar, A., Schonfeld, D.: Object trajectory-based activity classification and recognition using hidden Markov model. IEEE Trans. Image Process. **16**, 1912–1919 (2007)
3. Bolshakova, N., Azuaje, F.: Cluster validation techniques for genome expression data. Signal Process. **83**(4), 825–833 (2003)
4. Comaniciu, D., Ramesh, V., Meer, P.: Kernel-based object tracking. IEEE Trans. Pattern Anal. Mach. Intell. **25**(5), 564–575 (2003)
5. Doretto, G., Chiuso, A., Wu, Y., Soatto, S.: Dynamic textures. Int. J. Comput. Vis. **51**(2), 91–109 (2003)
6. Fowlkes, C., Belongie, S., Chung, F.R.K., Malik, J.: Spectral grouping using the Nyström method. IEEE Trans. Pattern Anal. Mach. Intell. **26**(2), 214–225 (2004)

7. Günter, S., Bunke, H.: Validation indices for graph clustering. Pattern Recogn. Lett. **24**(8), 1107–1113 (2003)
8. Halkidi, M., Batistakis, Y., Vazirgiannis, M.: On clustering validation techniques. J. Intell. Inf. Syst. **17**, 107–145 (2001)
9. Hu, W., Xiao, X., Fu, Z., Xie, D., Tan, T., Maybank, S.: A system for learning statistical motion patterns. IEEE Trans. Pattern Anal. Mach. Intell. **28**(9), 1450–1464 (2006)
10. Ivanov, Y., Bobick, A.: Recognition of visual activities and interactions by stochastic parsing. IEEE Trans. Pattern Anal. Mach. Intell. **22**(8), 852–872 (2000)
11. Fei-Fei, L., Niebles, J.C., Wang, H.: Unsupervised learning of human action categories using spatial-temporal words. In: Proc. of BMVC (2006)
12. Jebara, T., Kondor, R.I., Howard, A.: Probability product kernels. J. Mach. Learn. Res. **5**, 819–844 (2004)
13. Jebara, T., Song, Y., Thadani, K.: Spectral clustering and embedding with hidden Markov models. In: Kok, J.N., Koronacki, J., López de Mántaras, R., Matwin, S., Mladenic, D., Skowron, A. (eds.) ECML. Lecture Notes in Computer Science, vol. 4701, pp. 164–175. Springer, Berlin (2007)
14. Liao, T.W.: Clustering of time series data: a survey. Pattern Recogn. **38**(11) (2005)
15. Lin, W., Orgun, M.A., Williams, G.J.: Temporal data mining using multilevel-local polynomial models. In: International Conference IDEAL. Lecture Notes in Computer Science, vol. 1983. Springer, Berlin (2000)
16. Lo Gerfo, L., Rosasco, L., Odone, F., De Vito, E., Verri, A.: Spectral algorithms for supervised learning. Neural Comput. **20**, 1873–1897 (2008)
17. Medioni, G., Cohen, I., Bremond, F., Hongeng, S., Nevatia, R.: Event detection and analysis from video streams. IEEE Trans. Pattern Anal. Mach. Intell. **23**, 873–889 (2001)
18. Morris, B., Trivedi, M.M.: Learning trajectory patterns by clustering: experimental studies and comparative evaluation. In: IEEE Conference on Computer Vision and Pattern Recognition, CVPR '09 (2009)
19. Noceti, N., Santoro, M., Odone, F.: String-based spectral clustering for understanding human behaviours. In: THEMIS-BMVC (2008)
20. Noceti, N., Santoro, M., Odone, F.: Unsupervised learning of behavioural patterns for video-surveillance. In: Workshop MLVMA-ECCV (2008)
21. Noceti, N., Destrero, A., Lovato, A., Odone, F.: Combined motion and appearance models for robust object tracking in real-time. In: AVSS (2009)
22. Piciarelli, C., Micheloni, C., Foresti, G.L.: Trajectory-based anomalous event detection. IEEE Trans. Circuits Syst. Video Technol. **18**(11), 1544–1554 (2008)
23. Pittore, M., Campani, M., Verri, A.: Learning to recognize visual dynamic events from examples. In: IJCV (2000)
24. Rabiner, L.R.: A tutorial on hidden Markov models and selected applications in speech recognition. Proc. IEEE **77**, 257–286 (1989)
25. Rieck, K., Laskov, P.: Linear-time computation of similarity measures for sequential data. J. Mach. Learn. Res. **9**, 23–48 (2008)
26. Shawe Taylor, J., Cristianini, N.: Kernel Methods for Pattern Analysis. Cambridge University Press, Cambridge (2004)
27. Shi, J., Malik, J.: Normalized cuts and image segmentation. IEEE Trans. Pattern Anal. Mach. Intell. **22**(8), 888–905 (2000)
28. Special issue on event analysis in videos. IEEE Trans. Circuits Syst. Video Technol. **18**(11) (2008)
29. Stauffer, C., Grimson, E.: Learning patterns of activity using real-time tracking. IEEE Trans. Pattern Recogn. Mach. Intell. **22**(8), 747–757 (2000)
30. Tommasi, T., Orabona, F., Caputo, B.: Discriminative cue integration for medical image annotation. Pattern Recognit. Lett. **29**(15), 1996–2002 (2008)
31. Vishwanathan, S.V., Smola, A.J., Vidal, R.: Binet–Cauchy kernels on dynamical systems and its application to the analysis of dynamic scenes. Int. J. Comput. Vis. **73**(1), 95–119 (2007)

Part IV
Gesture and Action Recognition

Recognition of Spatiotemporal Gestures in Sign Language Using Gesture Threshold HMMs

Daniel Kelly, John McDonald,
and Charles Markham

Abstract In this paper, we propose a framework for the automatic recognition of spatiotemporal gestures in Sign Language. We implement an extension to the standard HMM model to develop a gesture threshold HMM (GT-HMM) framework which is specifically designed to identify inter gesture transitions. We evaluate the performance of this system, and different CRF systems, when recognizing gestures and identifying inter gesture transitions. The evaluation of the system included testing the performance of conditional random fields (CRF), hidden CRF (HCRF) and latent-dynamic CRF (LDCRF) based systems and comparing these to our GT-HMM based system when recognizing motion gestures and identifying inter gesture transitions.

1 Introduction

Recognizing gestures which appear in sign language is a challenging problem. Gestures lack a clear categorical structure and similar gestures can happen at various timescales. Another difficulty with recognizing gestures are inter gesture transitions which occur between valid gestures. For example, when performing hand gestures, the hands must move from the end point of the previous gesture to the start point of the next gesture. These inter gesture transition periods are called movement epenthesis [20] and are not part of either of the gesture. As such, an accurate recognition system must be able to distinguish between valid sign segments and movement epenthesis.

As will be discussed in Sect. 1.1, the limitation of current methods of continuous gesture recognition is that explicit training of movement epenthesis models are required, explicit rules for gesture segmentation must be specified or unnatural constraints are put on the signer, such as unnatural pauses between words. The

D. Kelly (✉) · J. McDonald · C. Markham
Computer Science Dept., N.U.I. Maynooth, Maynooth, Co. Kildare, Ireland
e-mail: dankelly@cs.nuim.ie

L. Wang et al. (eds.), *Machine Learning for Vision-Based Motion Analysis*, Advances in Pattern Recognition,
DOI 10.1007/978-0-85729-057-1_12,

main contribution of the work detailed in this chapter is that we propose a gesture threshold hidden Markov model (GT-HMM), which is a spatiotemporal gesture recognition system which does not require explicit epenthesis training or specific rules to determine gesture boundaries.

1.1 Related Work

The research on spatiotemporal gesture and sign recognition has two main categories: isolated and continuous recognition. Isolated recognition focuses on the classification of a single hand gesture that is performed by the user. In continuous recognition, the user performs gestures one after the other and the aim is to spot and classify meaningful gesture segments from within the continuous stream of sign language.

1.1.1 Isolated Gesture Recognition

Agris et al. [32] propose an isolated sign recognition system based on a combination of Maximum Likelihood Linear Regression and Maximum A Posteriori estimation. Their method was developed to consider the specifics of sign languages, such as one-handed signs. They implement selected adaptation methods from speech recognition to improve the performance of their system when performing user independent recognition. A recognition rate of 78.6% was reported when recognizing 153 isolated signs.

Shanableh et al. [25] proposed an isolated temporal gesture technique for the recognition of Arabic Sign Language. They propose temporal features which are extracted through forward, backward, and bidirectional predictions. These prediction errors are thresholded and accumulated into one image that represents the motion of the sequence. Experiments, based on a database of isolated signs, show that their method achieved a classification performance ranging from 97% to 100% when classifying 23 different sign classes.

Cooper et al. [8] implement an isolated sign recognition system using 1st order Markov chains. In their model, signs are broken down in visemes (equivalent to phonemes in speech) and a bank of Markov chains are used to recognize the visemes as they are produced. Experiments, based on 5 unseen examples of each of the 164 signs in the vocabulary, shows a classification accuracy of 72.6%.

Kim et al. [18] evaluate a accelerometer and EMG based sign recognition system on 7 word level signs and results show an average accuracy of 99.8% when tested on a total of 560 isolated samples.

Ding et al. [10] develop a sign language recognition model which incorporates hand shape, motion and 3D position in single framework. Signs are recognized using a tree based classifier where for example, if two signs have a similar hand shape then the root of the tree would represent the hand shape and the branches would represent different motion based gestures. A recognition rate of 93.9% is reported for a vocabulary of 38 signs.

While these works propose promising gesture recognition techniques, the experiments are based on isolated gesture samples. Natural gestures which occur in sign language are continuous, therefore sign language recognition requires spotting of the gesture from continuous videos (i.e., determining the start and end points of a meaningful gesture pattern).

1.1.2 Continuous Gesture Recognition

Extending isolated recognition to continuous signing is a nontrivial task. It requires automatic detection of movement epenthesis segments so that the recognition algorithm can be applied to segmented signs.

One proposed solution to movement epenthesis detection is an explicit segmentation model were subsets, of features from gesture data, are used as cues for valid gesture start and end point detection. Oz et al. [23] propose a continuous recognition framework which detects "signing" and "not signing" segments using a velocity network. The velocity network classifies a "signing" segment from when the hand first shows a change in velocity until the time the velocity shows a series of low velocities. A Neural Network based classifier was trained to recognize 60 different 1 handed ASL signs. Experiments conducted on a total of 360 ASL words, using histograms of feature vectors, showed a recognition accuracy of 95%. The limitation of this explicit segmentation model arises from the difficulty in creating general rules for sign boundary detection that could be applied to all types of manual and nonmanual gestures [22]. For example, fluent signers perform sign language sentences in a very fluid and natural manner and sign boundaries often do not occur when there is a sharp change in hand velocity.

An approach to dealing with continuous recognition, without explicit segmentation, is to use hidden Markov models (HMM) for implicit sentence segmentation. Bauer et al. [2] and Starner et al. [26] model each word or subunit with a HMM and then train the HMMs with data collected from full sentences. In the latter, experiments were conducted on a vocabulary of 40 signs using a set of 478 sentences for training and testing. Results showed a word detection rate of 96.8%. Brashear et al. [5] propose an extension to the work of Starner et al. where a HMM based sign recognition system was implemented to recognize continuous sentences using camera and accelerometer inputs. Experiments conducted on a vocabulary of 5 signs were recognized with 90.5% accuracy. It was shown that combining accelerometer and vision data improved the performance when compared to vision only data (52.4%) and accelerometer only data (65.9%). A downside of these methods is that training on full sentence data may result in a loss in valid sign recognition accuracy due to the large variations in the appearance of all the possible movement epenthesis that could occur between two signs.

Other works deal with movement epenthesis by explicitly modeling the movements between signs. Gao et al. [11] propose a transition-movement model where transition HMMs are created to model the transitions between each unique pair of signs (TMMs). After dynamically clustering transition parts, to reduce the total number of TMMs, an iterative segmentation algorithm is applied for automatically segmenting the continuous sentences. Experiments, conducted on a set containing 3000 sentence samples with a vocabulary of 5113 signs from Chinese sign language (CSL), showed their method achieved an accuracy of 90.8%.

Vogler et al. [31] propose a system to incorporate hand motion and hand posture data into a single recognition framework. A set of parallel HMMs were implemented to recognize signs from a vocabulary of 22 signs. Separate HMMs were implemented to model movement epenthesis between each unique ending and starting location of signs. Experiments showed their framework achieved a sign detection rate of 87.88% when tested on 99 sentences containing a total of 312 signs.

While these works have had promising results in gesture recognition and movement epenthesis detection, the training of such systems involves a large amount of extra data collection, manual data labeling, and model training due to the extra number of HMMs required to detect movement epenthesis. Few researchers have addressed the problem of movement epenthesis without explicitly modeling these movements.

An interesting approach to gesture spotting was proposed by Junker et al. [15] where a combination of explicit motion segmentation and HMM gesture classification is carried out. A pre-selection stage is implemented in order to identify relevant motion events. These candidate motion segments are then classified in isolation using HMMs. Experiments conducted to evaluate the gesture spotting system showed that the method performed well when spotting gestures in 2 different activity scenarios. Results showed a total precision of 0.74 and a total recall of 0.93 for the first scenario and total precision of 0.73 and a total recall of 0.79 for the second scenario.

Another solution to segmenting signs from continuous streams of data, without modeling movement epenthesis, is to use grammar based information. Yang et al. [36, 38] proposed an ASL recognition method based on an enhanced Level Building algorithm and a Trigram grammar model. Their method was based on a dynamic programming approach to spot signs without explicit movement epenthesis models. The recognition rate was 83% with 39 signs, articulated in 150 different sentences. Their work is based on a two step approach for the recognition of continuous signs, where the first step recognizes the possible signs in the sentence and the second applies a grammar model to the possible signs. They report only the results obtained after the second step which applies a trigram grammar model to the signs. The reliance of the system to the grammar model was shown in the experiments where the recognition rate of the system decreased from 83% to 68% when a the trigram model was replaced by a bigram model. Holden et al. [14] develop an Australian sign language recognition system where each sign is modeled using a HMM model. The recognition model employs grammar rules, based on 21 distinct signs, to recognize continuous sentences. Experiments show their system achieved 97% recognition rate on 163 test sign phrases, from 14 different sentences. It was noted in the work that the sign vocabulary used in experiments consisted of signs which where mainly distinguishable from motion alone.

Yang et al. [37] propose a very promising technique without the need for explicit epenthesis training or grammar rules. They develop threshold models in a conditional random field model which performs an adaptive threshold for distinguishing between signs in a vocabulary and nonsign patterns. Experiments show their system can spot signs from continuous data with an 87.0% detection rate from a vocabulary of 48 signs where the system was trained on 10 isolated samples of each of the 48 signs. The system was then tested on continuous sentences which contained 480 samples of the signs in the sign vocabulary.

While these works have had promising results in gesture recognition and movement epenthesis detection, the training of explicit epenthesis models involve a large amount of extra data collection, manual data labeling, model training and recognition computation due to the extra number of HMMs required to detect movement epenthesis. Another approach employed is to utilize grammar rules which can be used to reduce the number of possible sign combinations which appear in signed sentences. Grammar rules will become an important aspect of sign language recognition when sign vocabularies grow to represent a large portion of the signs used in everyday sign language communication. State of the art sign recognition is at the stage where the main focus is on developing algorithms to model signs. It is difficult to evaluate sign recognition models, which employ grammar rules, on small sign vocabularies. For example, in a corpus of 30 signs, if a grammar rule is used to predict that the next sign is likely to be from the noun category, the number of possible signs the recognition model must choose from could be reduced to around 8 signs. With the overall goal of large corpus sign recognition in mind, experiments should be conducted in order to evaluate recognition models in their ability to distinguish one sign from as many other signs as possible. It is unclear how the works, which employ grammar rules, would perform if the grammar models were created from larger real world corpora.

Few researchers have addressed the problem of movement epenthesis without employing grammar or segmentation rules or explicitly modeling the epenthesis. We propose a solution to this by developing a spatiotemporal gesture model which addresses the problem of movement epenthesis detection without the need for explicit epenthesis training. We develop a HMM based gesture threshold training and recognition framework to classify spatiotemporal gesture and to identify movement epenthesis without explicitly training on movement epenthesis examples. In this work, we will discuss the implementation of our framework and will show that our proposed model can effectively recognize gestures from within sign sentences independent of any grammar rules.

1.2 Chapter Outline

Before discussing our GT-HMM framework, we give an overview of HMMs and Threshold HMMs in Sects. 2 and 3, respectively. In Sect. 4, we then detail the implementation of our proposed GT-HMM framework which is specifically designed to classify manual and nonmanual signals and to identify movement epenthesis. In Sect. 5, we carry out a comprehensive evaluation of the GT-HMM framework and compare with current state of the art temporal event modeling frameworks such as CRFs, HCRFs, LDCRFs and standard HMM systems. Since sign language involves not only hand gestures but also nonmanual signals, gesture recognition evaluations will be carried out on hand gestures and nonmanual signals conveyed through head and eye brow movements.

2 Hidden Markov Models

HMMs are a type of statistical model and can model spatiotemporal information in a natural way. HMMs have efficient algorithms for learning and recognition, such as the Baum–Welch algorithm and Viterbi search algorithm [24]. They have been utilized for the task of gesture recognition in a large number of works in the literature. HMMs where first used for task of gesture recognition by Yamato et al. [35] and for the task of sign language recognition by Starner et al. [26]. In these seminal works, the authors state that the key characteristics of HMMs, which make them suitable for gesture recognition, is their learning ability and time-scale invariability.

A HMM is a collection of states connected by transitions. Each transition (or time step) has a pair of probabilities: a transition probability (the probability of taking a particular transition to a particular state) and an output probability (the probability of emitting a particular output symbol from a given state).

To represent a gesture sequence, it is defined as a set of observations. An observation $\mathbf{f}_t$, is defined as an observation vector made at time t, where $\mathbf{f}_t = \{o_1, o_2, \ldots, o_M\}$ and M is the dimension of the observation vector. A particular gesture sequence is then defined as $G = \{\mathbf{f}_1, \mathbf{f}_2, \ldots, \mathbf{f}_T\}$. We will discuss the features used to represent different gesture types in Sect. 5.

An HMM is characterized in by the following:

1. N, the number of states in the model. We denote the individual states as $S = \{s_1, s_2, \ldots, _N\}$, and the state at time t as q_t.
2. M, the dimension of the observation vector.
3. $A = \{a_{ij}\}$, the state transition probability distribution. Where A is an $N \times N$ matrix and a_{ij} is the probability of making transition from state s_i to s_j.
4. $B = \{b_j(\mathbf{f})\}$, the observation symbol probability distribution. Where b_j is the probability distribution in state j and $1 \leq j \leq N$.
5. $\pi = \{\pi_i\}$, the initial state distribution.

The compact notation $\lambda = \{A, B, \pi\}$ is used to indicate the complete parameter set of the model where A is a matrix storing transitions probabilities a_{ij} between states s_i and s_j, B is a matrix storing output probabilities for each state and π is a vector storing initial state probabilities.

HMMs can use either a set of discrete observation symbols or they can be extended for continuous observations signals. In this work, we use continuous multidimensional observation probabilities calculated from a multivariate probability density function.

The observation probability is expressed in the form shown in (1) and (2), where $\mathbf{f}$ is the M-dimensional vector being modeled, b_j is the vector probability in state j and $\aleph$ is a probability density function (PDF) of an M-dimensional multivariate Gaussian, with mean vector μ_j and covariance Σ_j

$$b_j(\mathbf{f}) = \aleph(\mathbf{f}; \mu_j, \Sigma_j) \tag{1}$$

$$= (2\pi)^{-\frac{N}{2}} |\Sigma|^{-\frac{1}{2}} \exp\left(-\frac{1}{2}(\mathbf{f} - \mu_j)^T \Sigma_j^{-1} (\mathbf{f} - \mu_j)\right). \tag{2}$$

2.1 HMM Algorithms

Given the form of HMM, there are three algorithms that can be performed on the HMM that make HMMs useful in real-world applications:

- The forward backward algorithm [24] is used to calculate $P(G|\lambda)$, the probability of the observation sequence $G = \{\mathbf{f}_1, \mathbf{f}_2, \ldots, \mathbf{f}_T\}$ given the model $\lambda = \{A, B, \pi\}$.
- The Viterbi algorithm [24] is used to find the single best state sequence $Q = \{q_1, q_2, \ldots, q_T\}$, for the given observation sequence $G = \{\mathbf{f}_1, \mathbf{f}_2, \ldots, \mathbf{f}_T\}$.
- The Baum–Welch algorithm [24] is used to determine the model parameters (A, B, π) to maximize the probability of the observation sequence given the model.

2.2 Types of HMMs

There are two main types of HMMs:

1. Ergodic model: An HMM in which every state of the model could be reached from every other state of the model (see Fig. 1(a)).
2. Left–right model (Bakis model): An HMM in which the state sequence associated with the model has the property that as time increases the states index increases or stays the same (i.e., the states progress from left to right) (see Fig. 1(b)).

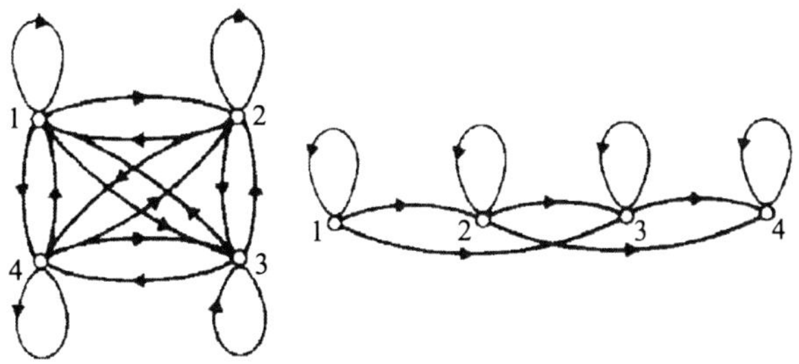

Fig. 1 HMM model types (a) Ergodic (b) Left-Right

The left–right model has the desirable property that it can readily model signals whose properties change over time and as such left–right models have been successfully applied to speech recognition tasks [24] and, more recently, to gesture recognition and sign recognition tasks [22].

3 Threshold HMM Model

Lee and Kim [19] proposed a threshold HMM to handle nongesture patterns. The threshold model was implemented to calculate the likelihood threshold of an input pattern and provide a confirmation mechanism for provisionally matched gesture patterns. We build on the work carried out by Lee and Kim to create a framework for calculating a probability distribution of a two hand input sign using continuous multidimensional observations. The computed probability distribution will include probability estimates for each pretrained sign as well as a probability estimate that the input sign is a movement epenthesis.

In general, a HMM recognition system will choose a model with the best likelihood as the recognized gesture if the likelihood is higher than a predefined threshold. However, this simple likelihood threshold often does not work, thus, Lee and Kim proposed a dynamic threshold model to define the threshold of a given gesture sequence.

If a set of left–right HMMs can be trained such that each state represents a particular gesture segment, then a self transition of a state represents a particular segment of a target gesture and the outgoing state transition represents a sequential progression of the segments within a gesture sequence. With this in mind, an ergodic model, with the states copied from all gesture models in the system, can be constructed as shown in Figs. 2(a) and 2(b). Figure 2(b) shows the threshold model as a simplified version of the ergodic model where dotted lines denote null transitions (i.e., no observations occur between transitions).

Threshold model states are created by copying all states from the left–right HMMs such that output observation probabilities and self transition probabilities are kept the same, but all outgoing transition probabilities are equally assigned as:

$$a_{ij} = \frac{1 - a_{ii}}{N - 1} \quad \forall j,\ i \neq j, \tag{3}$$

where a_{ij} is the transition probability from state s_i to s_j and N is the number of states excluding the start and end states. (The start and end states produce no observations.) If each left–right HMM can be trained such that each state represents a gesture subpattern, then, by maintaining the self-transition and output probabilities in the threshold states, a threshold model represents the set of all gesture subpatterns. Constructing the threshold model as an ergodic structure thus makes it match well with all patterns generated by combining any of the gesture subpatterns in any order. The likelihood of the threshold model, given a valid gesture pattern, would be smaller than that of the dedicated gesture model because of the reduced outgoing

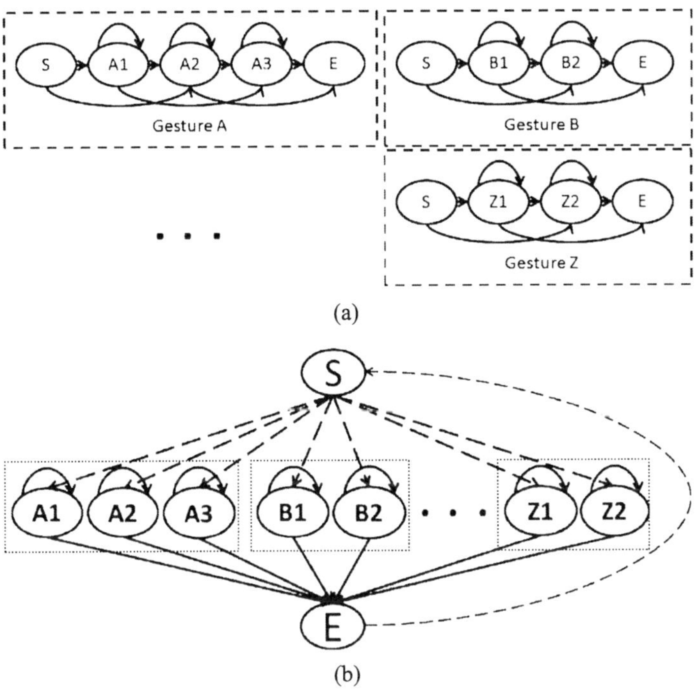

Fig. 2 (**a**) Dedicated gesture models; (**b**) threshold model

transition probabilities. However, the likelihood of the threshold model, given an arbitrary combination of gesture subpatterns, would be higher than that of any of the gesture models, thus the threshold model, denoted as $\bar{\lambda}$, can be used as a movement epenthesis likelihood measure.

4 GT-HMM Framework

In this work, we propose a GT-HMM as an improvement to the standard HMM, and Threshold HMM, to automatically train and model natural gestures and movement epenthesis. The GT-HMM comprises of two main improvements to the standard HMM framework. Firstly, we develop a gesture subunit initialization technique to create HMM states which model particular gesture subpatterns. Secondly, we implement a threshold HMM, which utilizes states which were initialized using the gesture subunits, to compute a dynamic epenthesis likelihood of input gestures. In this section, we will detail the implementation of the GT-HMM.

4.1 GT-HMM Training

In order to create a robust threshold HMM which models movement epenthesis, dedicated HMM models must be trained such that each state represents a particular gesture segment. We develop a technique to do this by expanding on the Threshold HMM framework to develop a GT-HMM framework which utilizes our proposed gesture subunit initialization and training techniques. Our GT-HMM framework models continuous multidimensional gesture observations within a HMM network to recognize motion based gestures and identify movement epenthesis. We now describe this framework.

Each dedicated gesture model is trained on isolated gestures performed by a fluent signer. Before training a HMM using the Baum–Welch algorithm, the model must first be initialized. Initialization includes the computation of an initial state transition matrix and calculation of each states' emission variables μ_j and Σ_j. In order to initialize these components of the HMM, an understanding of the gesture segmentation, or state transitions, must be built. One approach to achieving this would be to explicitly hand label different subunits or gesture phonemes [33]. Part of the goal of this work is to create a general data collection, training and recognition system. Data collection consists of a recording step and a labeling step. Labeling is an integral step in creating valid sign data, thus we envisage that all data will be labeled by fluent signers. Since movement and position of the hands are two of the four building blocks of sign language which Stokoe [27] identified, manually breaking these building blocks into smaller subunits would be an unintuitive and time consuming step for fluent signers to segment in a consistent manner. With this in mind, a training system was developed to initialize and train data with minimum human intervention where signs are labeled at a sign level and not at a phoneme level.

Kim et al. [17] implement an iterative HMM training procedure in order to estimate more accurate initial HMM parameters for automatic speech segmentation. In their method, hand labeled phone labels are use to initialize the HMMs. Following this, the Baum–Welch algorithm is run on the HMMs to tune the HMM parameters. The Viterbi algorithm is then run and the initialization data and phone labels are realigned to correspond with the Viterbi best paths. The HMM is then reinitialized using the realigned phone labels and this iterative procedure is repeated until no improvement is observed. We extend this iterative HMM training model proposed by Kim et al. [17] to develop an iterative gesture subunit initialization and training model. Our training model also includes an extra parameter selection layer which finds the best combination of (S, R), where S is the total number of states in the HMM and R is the reach of a state (i.e., in a left–right model, the reach is the number of states that it is possible to transition to from the current state). For a target sign, we extract data from a number of manually labeled isolated video sequences of a fluent signer performing that sign. Figure 3 shows a visualization of isolated examples of the "Alot" sign extracted from video sequences.

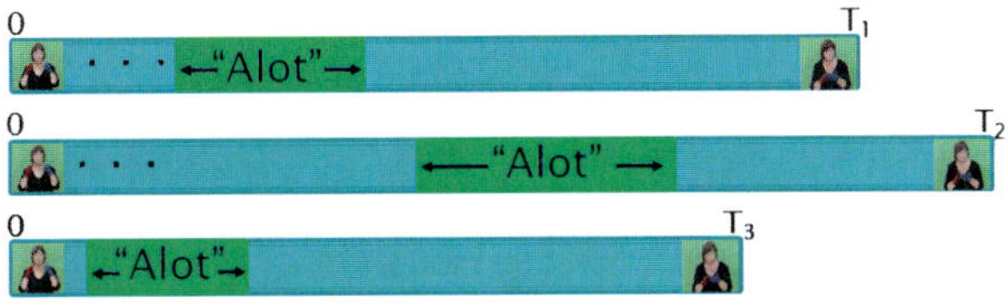

Fig. 3 Visualization of isolated examples of "Alot" sign extracted from video sequences

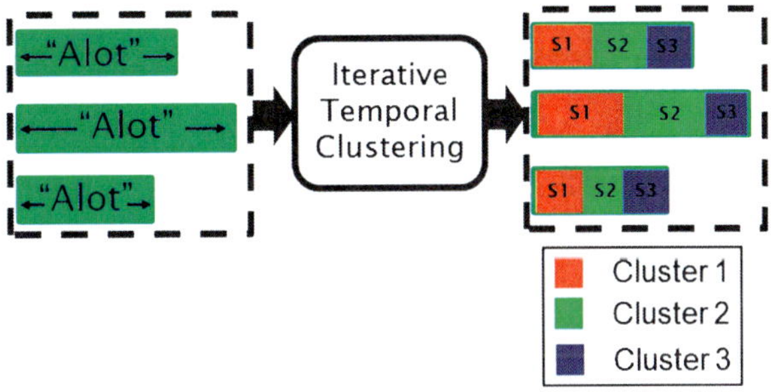

Fig. 4 Initial segmentation: visualization of isolated examples of "Alot" being segmented into $S = 3$ subgestures

4.1.1 Gesture Subunit Initialization

Extracting isolated examples of a sign produces a set of observation sequences $\Delta_y = \{G_y^1, G_y^2, \ldots, G_y^K\}$ where K is the total number of isolated examples. In HMM training, more accurate initial estimates of the HMM parameters produce more accurate classification results, as shown for example in [17]. The goal of our proposed initialization technique is to automatically find gesture subunits and the most accurate initial emission variables which describe the gesture subunits. Our technique determines a labeling of which state each observation vector most probably matches. These state labels are then used to determine the subset of observation vectors which are associated with a specific HMM state. The observation vectors within the subset can then be used to improve the estimate of the mean and covariance parameters associated with a state.

To initialize λ_y, the HMM which will model the sign indexed by y, we first calculate $S - 1$ indices of each G_y^i which best segment the gesture into S subgestures. We propose an iterative temporal clustering algorithm to calculate the S subgestures. Figure 4 shows a visualization of isolated examples of the "Alot" sign being segmented into $S = 3$ subgestures.

The objective of our temporal clustering algorithm is to cluster observations in a temporal structure such that, for each observation sequence, each cluster index, assigned to each observation, is either greater than or equal to the cluster index of the observation which occurred previously in the sequence (i.e., for $S = 3$, a cluster index sequence of '00110122' is considered an invalid temporal clustering whereas a cluster index sequence of '00011122' is considered a valid temporal clustering). Our iterative clustering algorithm attempts to cluster all observations, in the set of observation sequences Δ_y, such that all cluster index sequences have a valid temporal clustering.

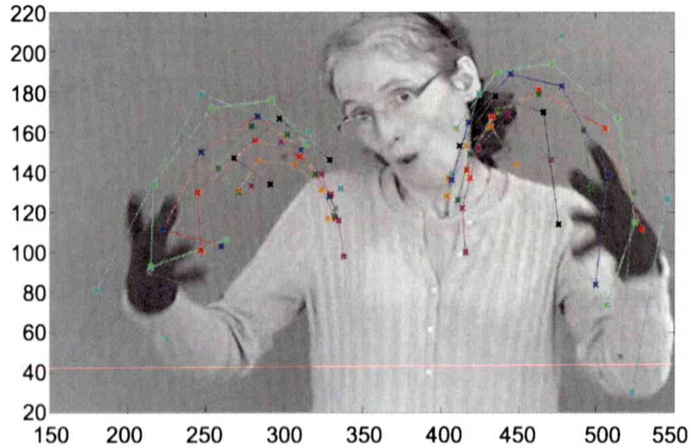

Fig. 5 Visualization of data extracted from 9 isolated examples of "Alot" sign

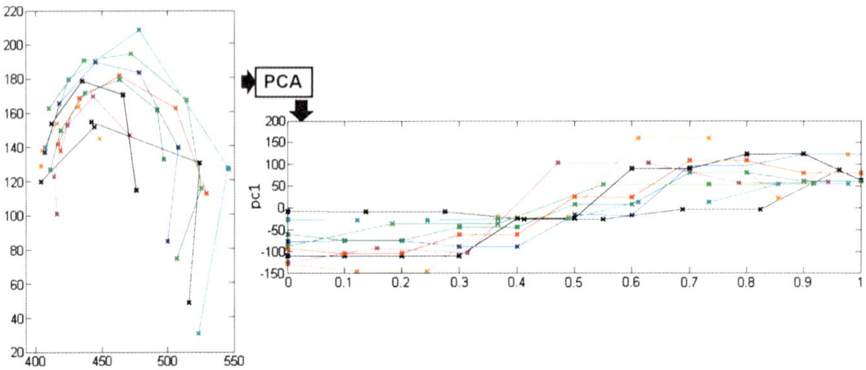

Fig. 6 Visualization of principal components of sign data for left hand

Our temporal clustering algorithm is based on a iterative time scaling procedure which incorporates a time variable into the observation vectors in order to temporally segment each observation sequence. We define a time augmented observation vector $\mathbf{f}_t(\eta) = \{o_1, o_2, \ldots, o_M, \eta\}$ and a time augmented observation sequence $G^{\neg} = \{f_1(\frac{1}{T}\Gamma^T), f_2(\frac{2}{T}\Gamma^T), \ldots, f_T(\Gamma^T)\}$, where Γ^T is a time scaling factor. Each iteration of the algorithm increases the time scaling factor and calculates cluster indices for each observation vector, using k-means clustering, until all time augmented observation sequences have a corresponding valid temporal clustering.

We use a sample set of nine sequences of the "Alot" sign in order to illustrate how our temporal clustering algorithm is used to initialize the HMMs. Figure 5 shows a visualization of the data extracted from the seven signs.

As will be discussed later, in Sect. 5.2.1, the observation vectors used to describe a sign in this work are 5-dimensional vectors. In order to illustrate the temporal clustering of these observation vectors we perform PCA on the data in order to reduce the dimensionality of the data to 1-dimension (see Fig. 6). Although we illustrate the clustering results using the principal component, it is important to note that all clustering, including the results shown in this section, were calculated using the 5-dimensional observation vector.

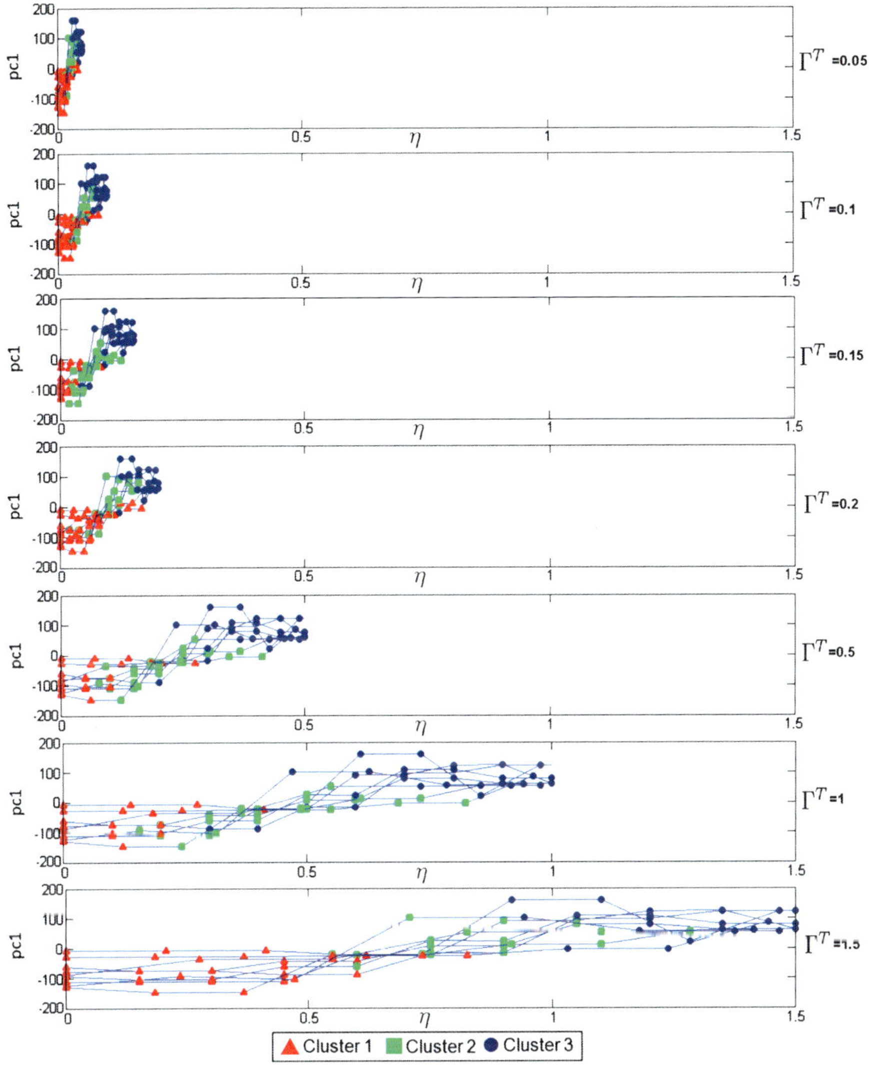

Fig. 7 Iterative clustering steps. *Each plot* represents a clustering step with a different time scaling factor Γ^T

We now illustrate an example of our temporal clustering algorithm applied to the left-hand observation sequences for the sign "Alot" when clustering the signs into $S = 3$ subgestures. For each iteration of the algorithm, all time augmented observation vectors are clustered and then each cluster index sequence is analyzed in order to determine if all clusters have a valid temporal clustering. This process is repeated until all observations sequences have a valid temporal clustering. Figure 7 shows each step of the clustering algorithm when applied to the 7 samples of the left-hand

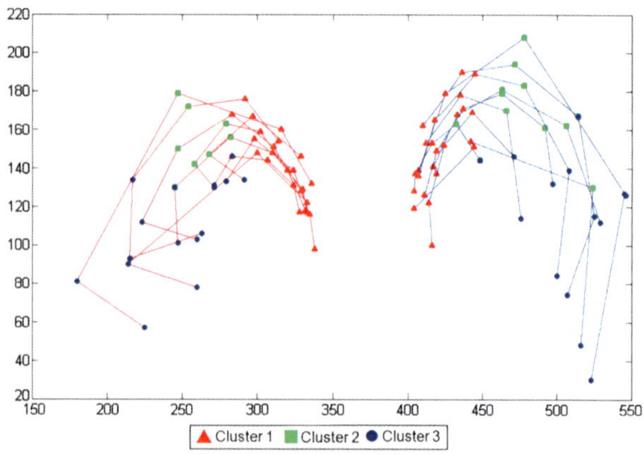

Fig. 8 Temporal clustering results after final iteration

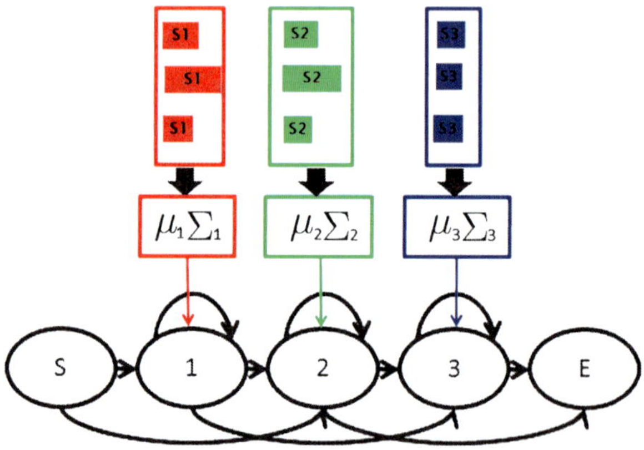

Fig. 9 Initialization: mean vector μ and the covariance matrix Σ calculated for each HMM state

data for the sign "Alot", while Fig. 8 shows the results of the clustering for both the left- and right-hands.

The final temporal clusters are then used to divide the observation sequences into the S subgestures and the mean vector μ_j and the covariance matrix Σ_j is calculated for each state (see Fig. 9).

The Baum–Welch algorithm [24] is then applied to λ_y using all training data Δ_y. After training, the Viterbi algorithm [24] is run on each of the training sequences in Δ_y to produce most probable state sequences. The initial S subgestures are then realigned to match the Viterbi paths. This re-estimation and realignment process is continued until the likelihood, produced by the Baum–Welch algorithm, converges. The overall process is repeated for different combinations of (S, R) to find the combination which produces the highest likelihood from the Baum–Welch re-estimation. Figure 10 gives an overview of the iterative training and parameter selection procedure.

After training, and finding the optimal parameters for each HMM λ_y, a threshold model $\overline{\lambda}$ is created using the method discussed in Sect. 3. Using each λ_y, and its corresponding set of states $S_y = \{s_{y1}, \ldots, s_{yN^y}\}$, the thresh-

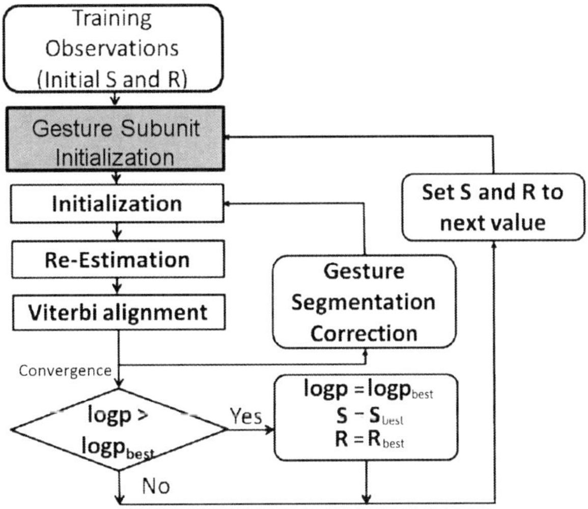

Fig. 10 HMM initialization and training procedure

old model states $\overline{S}$ are initialized by copying all the HMM states such that $\overline{S} = \{s_{11}, \ldots, s_{1N^1}, \ldots, s_{y1}, \ldots, s_{yN^y}, \ldots, s_{Y1}, \ldots, s_{YN^Y}\}$, where N^y defines the number of states in λ_y. The set of HMMs, to recognize the Y pretrained gestures, is then denoted as $\Lambda = \{\lambda_1, \lambda_2, \ldots, \lambda_Y, \overline{\lambda}\}$.

4.2 *GT-HMM for Gesture Recognition*

4.2.1 Gesture Classification

Given a sequence of gesture observations G, representing an unknown gesture, the goal is to accurately classify the gesture as an epenthesis or as one of the Y trained gestures. To classify the observations, the Viterbi algorithm is run on each model given the unknown observation sequences G, calculating the most likely state paths through each model y. The likelihoods of each state path, which we denote as $P(G|\lambda_y)$, are also calculated. The sequence of observations can then be classified as y if the maximum likelihood $P_{\text{ML}}(G|\lambda_y) \geq \tau_y$, where the maximum likelihood is defined in (4) and the movement epenthesis likelihood, τ_y, is defined in(9)

$$P_{\text{ML}}(G|\lambda_y) = \max_y P(G|\lambda_y), \tag{4}$$

$$\tau_y = P(G|\overline{\lambda})\Gamma_y, \tag{5}$$

where Γ_y is a constant scalar value used to tune the sensitivity of the system to movement epenthesis gestures.

4.2.2 Parallel Training

Vogler et al. [29, 30] show that parallel HMMs can improve recognition rates of two handed gestures when compared to standard HMMs. In order to recognize two hand gestures in sign language, we implement a parallel GT-HMM system. The parallel GT-HMM initializes and trains a dedicated parallel HMM denoted as $\lambda'_y = \{\lambda_{Ly}, \lambda_{Ry}\}$ where λ_{Ly} and λ_{Ry} are HMMs which model the left- and right-hand gestures, respectively. Each parallel GT-HMM is trained using the same gesture subunit initialization and training technique discussed in Sect. 4.1.

A weighting of ω_{Ly} and ω_{Ry} is applied to the left-hand HMM and right-hand HMM, respectively, to account for variations in information held in each of the hands for a particular sign. The weights are implemented to give more emphasis to the hand which conveys most information. For example, if a signer's dominant hand performs a waving gesture while the nondominant hand doesn't move, then more emphasis should be put on the dominant hand during the classification process. The weighting applied in our system is based on a hand variation ratio which calculates the relative variance between the left- and right-hand observation sequences. The weighs are calculated using training data from all observation sequences G^k_{Ly} and G^k_{Ry}, where $1 \leq k \leq K$, K is the total number of training examples and G_{Ly} and G_{Ry} are the left- and right-hand observations, respectively. The variance of the left- and right-hand observations are calculated by calculating the variance of each observation dimension $\sigma^2_{Ly}[i]$ and $\sigma^2_{Ry}[i]$, where $0 \leq i \leq M$ and M is the dimension of the observation vectors. The left HMM weight, ω_{Ly}, and right HMM weight, ω_{Ry}, are then calculated as using (6) and (7), where $\omega_{Ry} + \omega_{Ly} = 1$.

$$\omega_{Ly} = \sum_{i=0}^{M} \frac{\sigma^2_{Ly}[i]}{(\sigma^2_{Ly}[i] + \sigma^2_{Ry}[i]) \times M}, \tag{6}$$

$$\omega_{Ry} = \sum_{i=0}^{M} \frac{\sigma^2_{Ry}[i]}{(\sigma^2_{Ly}[i] + \sigma^2_{Ry}[i]) \times M}. \tag{7}$$

A parallel GT-HMM framework, $\overline{\lambda}' = \{\overline{\lambda}_L, \overline{\lambda}_R\}$ is then created using the network of trained parallel HMMs λ_y $(y \in Y)$.

4.2.3 Parallel Gesture Classification

To classify the parallel observations $G' = \{G_L, G_R\}$, the Viterbi algorithm is run on each model given the unknown observation sequences G_L and G_R, calculating the most likely state paths through each model y. The likelihoods of each state path, which we denote as $P(G_L|\lambda_{Ly})$ and $P(G_R|\lambda_{Ry})$, are also calculated. We calculate the overall likelihoods of a two handed gesture by computing the weighted sum of the left and right HMM likelihoods as defined in (8)

$$P(G'|\lambda'_y) = P(G_L|\lambda_{Ly})\omega_{Ly} + P(G_R|\lambda_{Ry})\omega_{Ry}. \tag{8}$$

The movement epenthesis likelihood is similarly calculated from a weighted sum of left and right threshold model likelihoods as defined in (9)

$$\tau'_y = \frac{P(G_L|\overline{\lambda_L})\Gamma_{Ly} + P(G_R|\overline{\lambda_R})\Gamma_{Ry}}{2}, \tag{9}$$

where Γ_{Ly} and Γ_{Ry} are constant scalar values used to tune the sensitivity of the system to movement epenthesis. Experiments, which will be discussed in Sect. 5, different scalar values were evaluated and results show that scaler values between 1.05 and 1.1 perform best when identifying epenthesis.

The observation sequence can then be classified as y if $P_{\mathrm{ML}}(G'|\lambda'_y) \geq \tau'_y$, where $P_{\mathrm{ML}}(G'|\lambda'_y)$ is the maximum likelihood defined as $\max_y P(G'|\lambda'_y)$.

4.2.4 Continuous Recognition

Thus far, we have described methods for classifying a given observation sequence as one of a number of pretrained gestures or as a movement epenthesis. We will now describe our system for spotting and classifying spatiotemporal gestures within continuous sequences of natural sign language.

The first step in our GT-HMM gesture spotting algorithm is gesture end point detection. To detect a gesture end point in a continuous stream of gesture observations $G = \{\mathbf{f}_1, \mathbf{f}_2, \ldots, \mathbf{f}_T\}$, we calculate the model likelihoods of observation sequence $G^* = \{\mathbf{f}_{T-L}, \mathbf{f}_{T-L-1}, \ldots, \mathbf{f}_T\}$ where G^* is a subset of G and L defines the length of the observation subset used. In the work, we report we set L to the average length of the observation sequences used to train the system.

A candidate gesture, κ, with end point, $\kappa_e = T$, is flagged when $\exists y : P(G^*|\lambda_y) \geq \tau_y$. Figure 11 illustrates the likelihood time evolution of the hand gesture model "Lost" when given an observation sequence where the signer performs the "Lost" sign. It can be seen from Fig. 11 that a number of candidate end points occur between $T = 16$ and $T = 21$.

For each candidate end point, we calculate a corresponding start point κ_s. Different candidate start points are evaluated using the measurement shown in (10) where

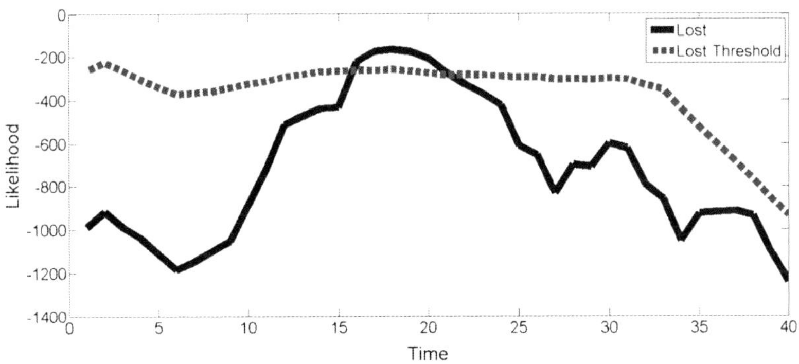

Fig. 11 Likelihood evolution of "Lost" gesture model and associated threshold model

$\beta_y(G)$ is a normalized metric ($0 \leq \beta_y(G) \leq 1$) which measures the likelihood of gesture y relative to the epenthesis likelihood given observations G

$$\beta_y(G) = \frac{P(G|\lambda_y)}{P(G|\lambda_y) + \tau_y}. \tag{10}$$

To find a candidate start point, the metric $\beta_y(G_{s\kappa_e})$ is calculated over different values of s, where $G_{s\kappa_e} = \{\mathbf{f}_s, \mathbf{f}_{s+1}, \ldots, \mathbf{f}_{\kappa_e}\}$ and $(\kappa_e - (L \times 2)) \leq s < \kappa_e$. The candidate gesture start point κ_s, is then found using (11)

$$\kappa_s = \arg\max_s \beta_y(G_{s\kappa_e}). \tag{11}$$

4.2.5 Candidate Selection

The start and end point detection algorithms may flag candidate gestures which overlap, and for this reason we expand on the continuous sign recognition algorithms with a candidate selection algorithm. The purpose of the candidate selection algorithm is to remove overlapping candidate gestures such that the single most likely gesture is the only remaining gesture for a particular time frame.

We will use a sample sign language sentence "I Lost Alot of Books" to illustrate our candidate selection algorithm in the context of our gesture and threshold likelihood evaluation, where the system was trained on the following 8 signs; "Paper", "Alot", "Bike", "Clean", "Paint", "Plate", "Lost" and "Gone". Figure 12 illustrates the difference between the HMM gesture model likelihood $P(G|\lambda_y)$ and its corresponding threshold τ_y, where positive values indicates $P(G|\lambda_y) \geq \tau_y$. We illustrate only 4 gesture model likelihoods as all other gesture model likelihoods never exceed their corresponding threshold.

The first step in the candidate selection algorithm is to cluster overlapping gestures, with the same gesture classification, together. Each of these candidate gestures, within the cluster, have an associated metric which we denote as $\kappa_p = \beta_y(G_{\kappa_s\kappa_e})$. We remove all but one candidate gesture from this cluster leaving only the candidate gesture, κ^B, with the highest κ_p value. We repeat this step for each

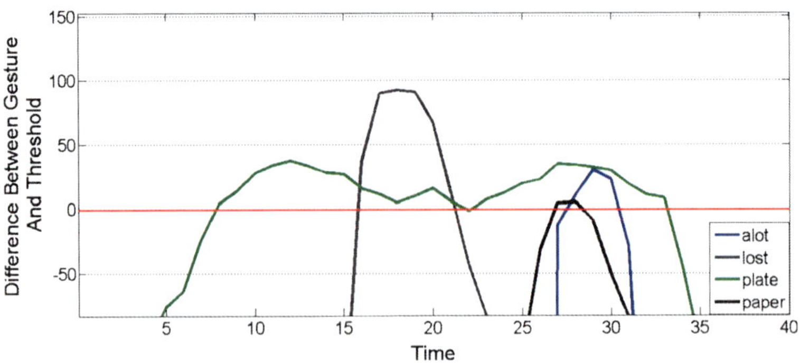

Fig. 12 HMM gesture models and corresponding HMM threshold model likelihood difference

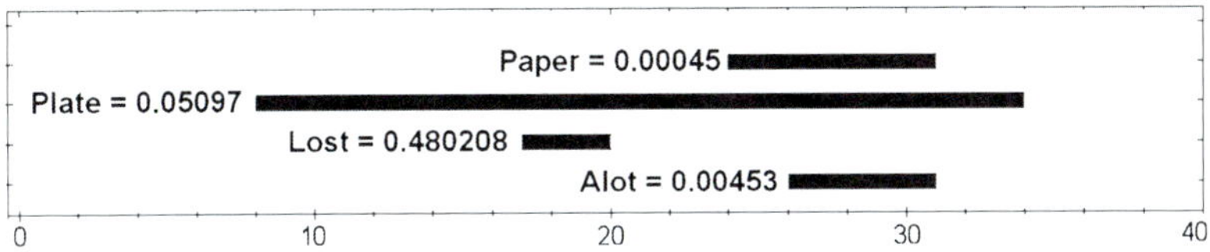

Fig. 13 Candidate gestures, Υ, after first candidate selection step

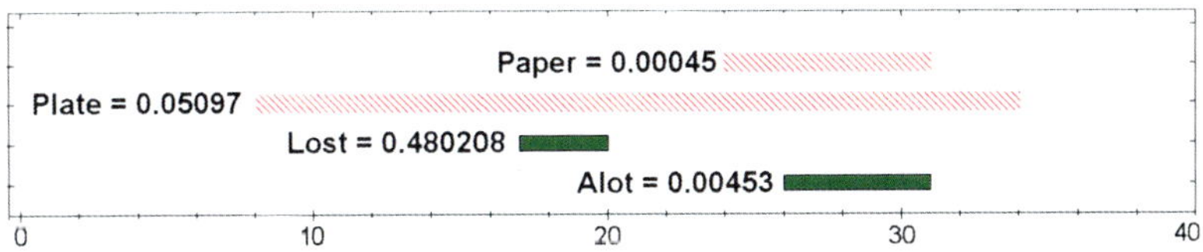

Fig. 14 Candidate gestures, Υ. Candidates marked in *red* (*dashed*) denote gestures which are removed by the second candidate selection step. Candidates in *green* (*solid*) denote the final recognized gestures

cluster to produce a set of candidate gestures $\Upsilon = \{\kappa^{B1}, \kappa^{B2}, \ldots, \kappa^{BK}\}$, where K is the total number of clusters created from grouping overlapping gestures, with the same gesture classification, together. Figure 13 shows the time segments and κ_p metrics of each candidate gesture after the first candidate selection step.

The second candidate selection step finds sets of overlapping candidates and removes the least probable candidates such that a maximum of only one candidate is detected for any given time frame. Figure 14 shows the time segments and gesture probabilities of the recognized gestures after the first and second candidate selection step where the signs "Lost" and "Alot" are correctly recognized from a sample sign language sentence "I Lost Alot of Books".

5 Experiments

In this work, we will evaluate our GT-HMM framework on a number of different gesture data sets and compare with different temporal event modeling frameworks including CRFs.

5.1 Feature Extraction

In this section, we conduct evaluations on different gesture recognition systems using data extracted from video sequences of sign language sentences being performed by a fluent Irish sign language (ISL) signer.

For completeness, prior to discussing the experiments, we briefly describe the feature extraction techniques implemented.

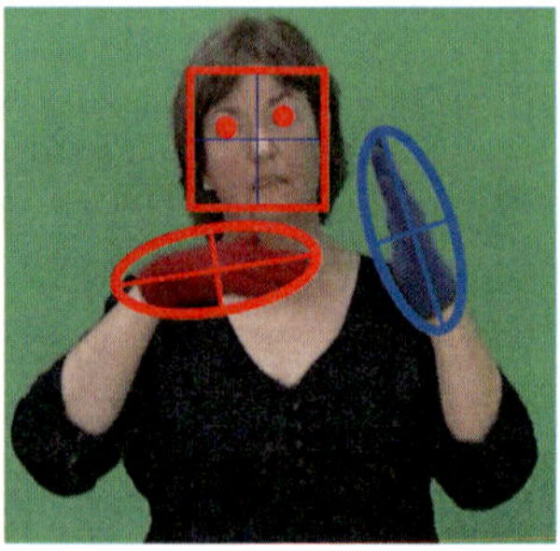

Fig. 15 Extracted features from image

Tracking of the hands is performed by tracking colored gloves using the Mean Shift algorithm [7]. Face and eye positions are used as features for head movement recognition and also used as hand gesture cues. Face and eye detection is carried out using a cascade of boosted classifiers working with haar-like features proposed by Viola and Jones [28]. A set of public domain classifiers [6], for the face, left eye and right eye, are used in conjunction with the OpenCV implementation of the haar cascade object detection algorithm. Figure 15 shows a visual example of the features of an ISL signer being tracked.

We define the raw features extracted from each image as follows; right-hand position (RH_x, RH_y), left-hand position (LH_x, LH_y), face position (FC_x, FC_y), face width (FW) (face region is square), left-eye position (LE_x, LE_y) and right-eye position (RE_x, RE_y).

In order to recognize nonmanual signals conveyed through facial expressions, we locate the facial features of interest using Cootes' implementation of Active Shape Model (ASM) [9]. In the context of facial feature localization, ASMs can be viewed as statistical models of the shapes of the face which deform iteratively to fit to new images. Since the ASM is constrained by a statistical shape model, the range of possible deformations is constrained by the variance which exists in the training set. As a consequence, the accuracy of the ASM depends on the range of facial movements included in the training set. For the experiments included in this paper, our data set consisted of 3,500 images in total. From which 300 key frames representing the variance in the data set were manually labeled with 46 points. Figure 16(b) shows the ASM which was trained on these image-points pairs.

During sign language communication, the face is frequently occluded by the hands. Our approach to overcoming this particular problem was to fit the ASM to the parts of the face that were visible, and use the previous points for occluded parts of the face. This can be seen in Fig. 16 where the position of the mouth from Fig. 16(c) is used in Fig. 16(d) when the mouth is occluded. This is a valid approach as the hands move rapidly and rarely cover the same portion of the face for multiple frames.

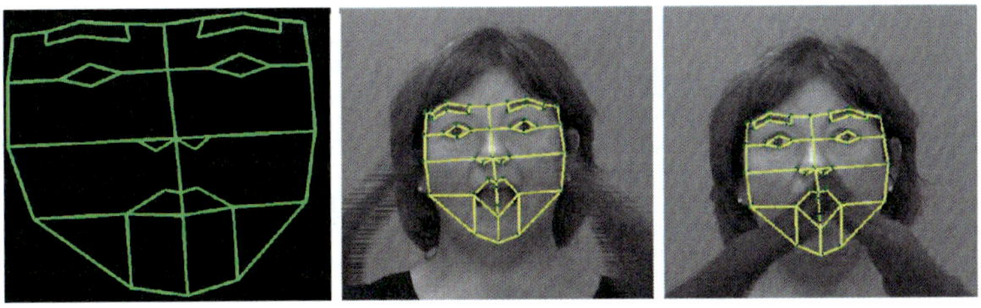

Fig. 16 (**a**) Sample ASM which was fitted to each image; (**b**) sample of an unoccluded image; (**c**) example of an occluded image

5.2 *Evaluation of Techniques on Isolated Gestures*

Wang et al. [34] perform experiments to show that the HCRF model performs better at classifying head and arm gestures than CRFs and HMMs. In their experiments, the models were evaluated on their ability to classify a given segmented gesture sequence as one of a number of pre trained gestures but the models were not tested on non-gesture sequences. In order to evaluate and assess the ability of a HCRF model to recognize gestures in sign language, the performance of the model must be evaluated when identifying non-gestures/epenthesis as well as being evaluated on the performance of classifying gestures.

Morency et al. [21] perform experiments to evaluate the performance of the LD-CRF model on three different data sets. The first data set was a head nod data set where the system was trained and tested on frames labeled as a head nod or labeled as not a head nod. The second data set, similar to the first data set, was trained and tested on positive and negative examples of heads nods. The final data set was an eye gaze data set, and the system was trained and tested on frames labeled as either an eye gaze-aversion gesture or a non gaze-aversion gesture. The LDCRF model was shown to out perform CRF, HCRF and HMM based classifiers (as well as a support vector machine based classifier). From these experiments, it is difficult to access whether or not the LDCRF model could be implemented to recognize a larger vocabulary of gestures or whether or not the LDCRF model could be used in a sign language based system. In the experiment Morency et al. carry out, each of the gesture data set experiments were trained to recognize a single gesture with positive and negative examples of the gesture. In order to evaluate the LDCRF model for a sign language recognition system, the model should be tested on a larger vocabulary of gestures. In their experiments, the gesture model was trained on positive and negative examples of the gesture. Training a model to recognize to recognize movement epenthesis in sign language is unfeasible due to the large number of possible epenthesis that can occur between signs.

The goal of the experiments conducted in this section is to evaluate the performance of the HMM, T-HMM, GT-HMM and the different CRF models when recognizing motion based gestures and identifying epenthesis which occur in sign language. The T-HMM model we evaluate in this work is a parallel threshold HMM

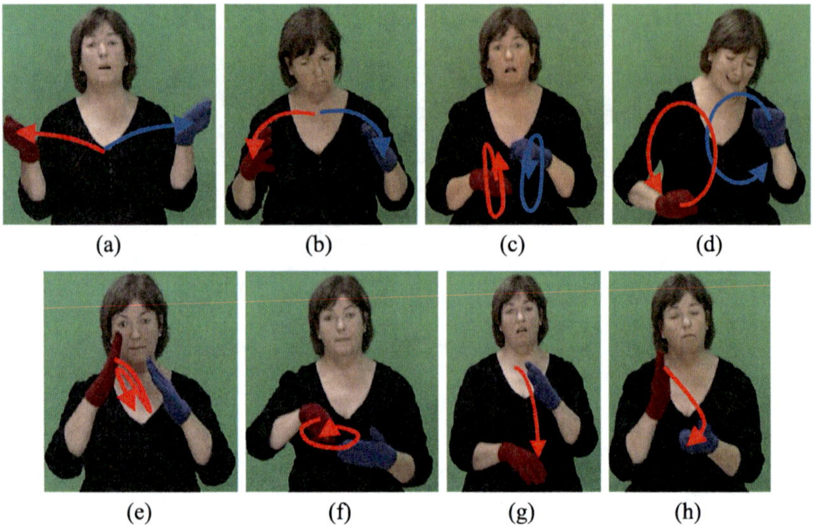

Fig. 17 Example of the eight different signs the system was tested on (**a**) newspaper, (**b**) a lot, (**c**) bike, (**d**) clean, (**e**) paint, (**f**) plate, (**g**) lost, (**h**) gone

framework where training and classification are carried out in the same manner as the GT-HMM model. The key difference between the GT-HMM and the T-HMM is that the T-HMM is not trained using the gesture subunit initialization technique described in Sect. 4.1.1. Instead, the T-HMM model is initialized using a standard segmentation method, utilized by Holden et al. [13], where the observation sequences are linearly segmented into S equal subsequences.

Since sign language communication is multimodal it involves not only hand gestures (i.e., manual signing) but also non-manual signals conveyed through facial expressions, head movements, body postures and torso movements [22]. In order to evaluate the use of HMMs and CRFs in recognizing motion based gestures in sign language, we evaluate the models on three data sets; a manual signing data set (i.e., two handed motion based gestures) and two nonmanual signal data set based on head motion gestures and eye brow gestures.

5.2.1 Manual Sign Experiments

The first data set we use to evaluate the models on is a set of two handed spatiotemporal hand gestures used in sign language. This data set consists of eight different manual signs extracted from videos of a fluent signer performing natural sign language sentences. Figure 17 illustrates an example of a signer performing each of the eight manual signs.

In order recognize manual signs, we must extract two observation channels from the video streams. The two observation channels correspond to the left-hand observations G_L and the right-hand observations G_R. The observations G_L and G_R are combined into a parallel observation sequence G' which is processed by the parallel models. We extract a set of observation sequences Δ'_y from the video sequences, where $y \in Y$, Y is the set of sign labels, $\Delta'_y = \{G'_{1y}, \ldots, G'_{Ty}\}$ and T is the number of sample observation sequences recorded for each gesture label y.

This set is then divided into a training set, Δ''^{t}_y, and a test set, Δ'^{ζ}_y. A set of 10 training signs and a set of 10 test signs were recorded for each sign (a total of 160 gesture samples). The HMM, T-HMM, GT-HMM, CRF, HCRF and LDCRF models were then trained on Δ''^{t}_y.

An additional set of observations Δ'_E, which represents a collection of movement epenthesis, were also extracted from the video sequences to test the performance of the threshold model. For each valid sign, 10 movement epenthesis, that occurred before and after the valid sign in different sign language sentences, were recorded. An additional set of 20 random movement epenthesis were also recorded, resulting is a test set of 100 samples to evaluate the models on.

To evaluate the performance of the models, we perform a ROC analysis on the different models and calculate the AUC for each model. The classification of a gesture is based on a comparison of a model probability and a threshold value. In our ROC analysis of each model, we vary the threshold and create a confusion matrix for each of the thresholds. In the case of the T-HMM and GT-HMM models, we vary the weighting of the threshold and in the case of the CRF models we vary the static threshold value Ω.

When implementing the HCRF model and LDCRF model we vary the number of hidden states and also vary the window parameter ω. The window parameter defines the amount of past and future history to be used when predicting the state at time t such that long range dependencies can be incorporated. In our experiments we test each model on a two different groups of data. The first data group, which we denote as data set 1, is a set which includes all test sequences Δ'^{ζ}_y and epenthesis sequences Δ'_E. The second data group, which we denote as data set 2, is a set which includes just the test sequences Δ'^{ζ}_y.

While an extensive evaluation of the models using different feature vectors was conducted, we report only the results of models using the best performing feature vectors. The best performing feature vector for the HMM models was the feature, $\mathbf{f} = \{RP_x, RP_y, V_x, V_y, D_H\}$, which describes the position of the hands relative to the eyes, the direction of the movement of the hand and the distance between the two hands. The best performing feature vector for the three different CRF models was the feature vector $\mathbf{f} = \{V_x, V_y\}$, which describes the direction of the movement of the hand. These are the feature vectors used for the evaluation of the HMM and CRF models.

Table 1 shows the AUC measurements of the HMM, T-HMM, GT-HMM and different variations of the CRF models when classifying gestures using their corresponding best performing feature vector.

The results show that the overall best performing model, with an AUC of 0.985, was the LDCRF model with 8 hidden states per label when tested on the data set 2. Since a sign language recognition system must be able to identify move-

Table 1 Manual signs: AUC measurements for different models

Model	Data set 1[†]	Data set 2[‡]
HMM	0.902	0.943
GT-HMM	0.976	0.977
T-HMM	0.941	0.944
CRF $\omega = 0$	0.833	0.876
CRF $\omega = 1$	0.794	0.828
HCRF $\omega = 0$, $S = 6$	0.909	0.944
HCRF $\omega = 1$, $S = 6$	0.957	0.983
HCRF $\omega = 2$, $S = 6$	0.944	0.971
HCRF $\omega = 0$, $S = 8$	0.947	0.965
HCRF $\omega = 1$, $S = 8$	0.934	0.968
LDCRF $\omega = 0$, $S^* = 1$	0.847	0.881
LDCRF $\omega = 0$, $S^* = 2$	0.806	0.842
LDCRF $\omega = 0$, $S^* = 3$	0.808	0.836
LDCRF $\omega = 0$, $S^* = 4$	0.863	0.901
LDCRF $\omega = 0$, $S^* = 8$	0.942	0.985
LDCRF $\omega = 1$, $S^* = 8$	0.899	0.928

[†]Data set which includes 100 epenthesis samples

[‡]Data set which does not include epenthesis samples

[*] $S*$ refers to number of hidden states per label for LDCRF

ment epenthesis as well as recognize gestures, the results of the tests performed on the data set 1 are more relevant to the evaluation of a sign language recognition model. The model which scores best when recognizing data set 1 is the GT-HMM which has an AUC of 0.976. Although the HCRF and LDCRF perform better than the GT-HMM when classifying gestures, the performance of both drop significantly when the epenthesis data is introduced. The performance of the GT-HMM drops a small amount compared to the relatively large drops of all the CRF models. This indicates that the GT-HMM is more robust when classifying gestures and identifying epenthesis.

The results of the experiments also reveal the influence our gesture subunit initialization has on the recognition performance. The GT-HMM framework initializes the HMM using our initialization method described in Sect. 4.1, while the T-HMM utilizes a linear segmentation method. Results showed that the GT-HMM had a 3.5% better AUC measurement when compared to the T-HMM model.

5.2.2 Head Gesture Experiments

The second data set we evaluate the HMM and CRF models on is a set of head movement gestures used to convey nonmanual information in sign language. The head gesture set consists of three different head movement gestures extracted from videos of a fluent signer performing natural sign language sentences.

A visual example of a signer performing each of the three different head movement gesture is in shown in Fig. 18.

Similar to the manual sign experiments described in Sect. 5.2.1, observation sequences $\Delta_y = \{G_{1y}, \ldots, G_{Ty}\}$ were extracted from the videos and divided into a training set, Δ_y^t, and a test set, Δ_y^ζ. For the nonmanual signal experiments, a set of 6 training signs and a set of 6 test signs were recorded for each sign (a total of 36 gesture samples). The HMM models and all CRF models were then trained on Δ_y^t

An additional set of 25 other head gesture sequences Δ_E, outside of the training set, were also extracted from the video sequences to test the performance of the system when identifying movement epenthesis. Similar to the hand gesture experiments, we test the head gesture models on two data groups; data group 1 includes the gesture test sequences and the nongesture sequences, while data set 2 includes only the gesture test sequences.

While a ROC analysis of the nonmanual models was conducted using the same procedure described in Sect. 5.2.1, we report only the results of models using the best performing feature vectors. The best performing feature vector for the HMM models, when classifying head gestures, was a 2-dimensional vector $\mathbf{f} = \{V_x, V_y\}$ describing the velocity of the head movement in the x and y directions. To calculate

(a)

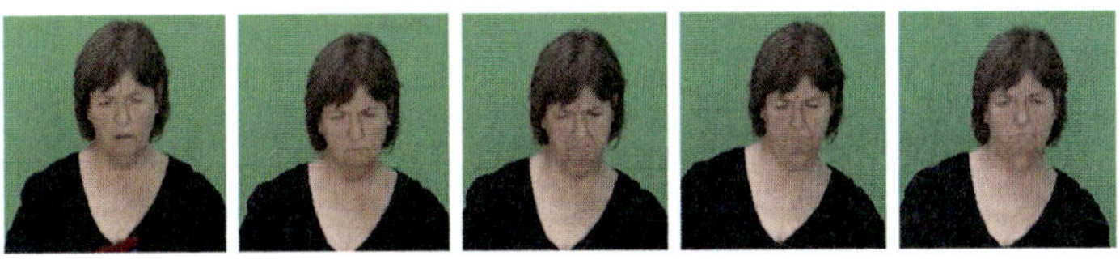

(b)

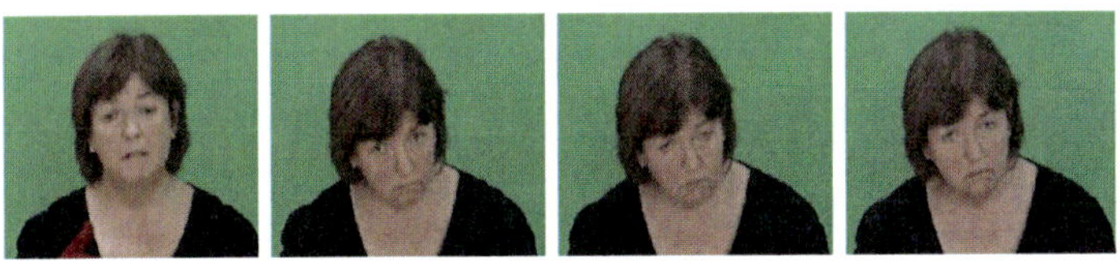

(c)

Fig. 18 Example of the three different head movement gestures the system was tested on (**a**) right movement, (**b**) left movement, (**c**) left forward movement

Table 2 Nonmanual signals: AUC measurements for different models

Model	Data set 1[†]	Data set 2[‡]
HMM	0.848	0.891
GT-HMM	0.936	0.947
T-HMM	0.873	0.882
CRF $\omega = 0$	0.736	0.768
CRF $\omega = 1$	0.527	0.545
HCRF $\omega = 0$, $S = 2$	0.698	0.801
HCRF $\omega = 1$, $S = 2$	0.786	0.911
HCRF $\omega = 2$, $S = 2$	0.702	0.816
HCRF $\omega = 0$, $S = 4$	0.784	0.927
HCRF $\omega = 1$, $S = 4$	0.719	0.811
HCRF $\omega = 0$, $S = 6$	0.743	0.850
HCRF $\omega = 1$, $S = 6$	0.736	0.893
HCRF $\omega = 0$, $S = 8$	0.715	0.838
HCRF $\omega = 1$, $S = 8$	0.708	0.788
LDCRF $\omega = 0$, $S^* = 3$	0.794	0.899
LDCRF $\omega = 1$, $S^* = 3$	0.763	0.880
LDCRF $\omega = 0$, $S^* = 6$	0.760	0.827
LDCRF $\omega = 1$, $S^* = 6$	0.717	0.791
LDCRF $\omega = 0$, $S^* = 9$	0.868	0.922
LDCRF $\omega = 1$, $S^* = 9$	0.837	0.901
LDCRF $\omega = 2$, $S^* = 9$	0.894	0.952
LDCRF $\omega = 3$, $S^* = 9$	0.795	0.861

[†]Data set which includes 25 nongesture samples

[‡]Data set which does not include nongesture samples

[*]S^* refers to number of hidden states per label for LDCRF

the velocity vector of the head, we use the mid point between the eyes and calculate the movement of the midpoint from frame to frame. As with the HMM models, the best performing feature vector for the CRF models was the 2-dimensional velocity vector $\mathbf{f} = \{V_x, V_y\}$. Table 2 shows the AUC measurements of the HMM, T-HMM, GT-HMM and different variations of the CRF models when classifying head movements using their corresponding best performing feature vector.

Table 2 shows the AUC measurements of models.

The results of this experiment repeat the same trend found in the results of the manual sign recognition experiment. The LDCRF model performs best when classifying gestures in data set 2. The recognition rate of the CRF models then decrease significantly when nongestures are introduced. The best performing model for data set 1 is again the GT-HMM model with an AUC of 0.936. There was a 4.2% decrease in performance between the data set 1 AUCs of the GT-HMM model and the 9 state LDCRF. This result further suggests that the GT-HMM model is a more robust model when recognizing gestures when epenthesis gestures are taken in to account.

There was a 5.8% decrease in performance between the LDCRF from data set 2 to data set 1, while there was only a 1.1% change in performance of the HMM threshold model. This result suggests that the performance of the LDCRF would decrease more than that of the HMM threshold model when the number of epenthesis gestures introduced into the system increased.

Similar to the results achieved on the manual gestures in Sect. 5.2.1, these results indicate that the gesture subunit initialization of the GT-HMM improved classification performance. Results showed that the GT-HMM had a 6.3% better AUC measurement than the T-HMM which was initialized using a standard HMM segmentation method.

5.2.3 Eye Brow Gesture Experiments

Research on American Sign Language conducted by Grossman et al. [12] has linked eyebrow gestures to certain affective states and questions. Anger, wh-questions (who, where, what, when, why, how) and quizzical questions exhibited lowered brows and squinted eyes, while surprise and yes/no questions showed raised brows and widened eyes. In this paper, we focus on identifying these lowered brow gestures and raised brow gestures.

The third data set we evaluate the HMM and CRF models on is a set of eye brow movement gestures used to convey non manual information in sign language. The eye brow gesture set consists of two different eye brow movement gestures extracted from videos of a fluent signer performing natural sign language sentences.

Similar to the hand gesture and head gesture experiments described in Sect. 5.2.1, observation sequences $\Delta_y = \{G_{1y}, \dots, G_{Ty}\}$ were extracted from the videos and divided into a training set, Δ^t_y, and a test set, Δ^ζ_y. For the eye brow experiments, a set of 5 training signs and a set of 5 test signs were recorded for each sign (a total of 20 gesture samples). The HMM models and all CRF models were then trained on Δ^t_y.

An additional set of 20 other eye brow gesture sequences Δ_E, outside of the training set, were also extracted from the video sequences to test the performance of the system when identifying movement epenthesis.

We carry out a ROC analysis of the nonmanual models using the same procedure described in Sects. 5.2.1 and 5.2.2. Table 3 shows the AUC measurements of models. The best performing feature vector for the HMM models and CRF models, when classifying eye brow movements, was a 2-dimensional vector $\mathbf{f} = \{\phi_{LR}, D^\Delta\}$. The value ϕ_{LR} is the average angle between ϕ_L and ϕ_R shown in Fig. 19, and D^Δ is the change in distance between the two eyes.

The results of this experiment repeat the same trend found in previous results of the hand gestures and head gestures. The LDCRF model performs best when

Table 3 Nonmanual eye brow signals: AUC measurements for different models

Model	Data set 1[†]	Data set 2[‡]
HMM	0.882	0.911
GT-HMM	0.948	0.951
T-HMM	0.905	0.912
CRF $\omega = 0$	0.859	0.899
CRF $\omega = 1$	0.878	0.921
CRF $\omega = 3$	0.889	0.933
CRF $\omega = 6$	0.866	0.901
HCRF $\omega = 0$, $S = 1$	0.525	0.533
HCRF $\omega = 0$, $S = 3$	0.858	0.922
HCRF $\omega = 2$, $S = 3$	0.809	0.899
HCRF $\omega = 5$, $S = 3$	0.781	0.912
HCRF $\omega = 0$, $S = 6$	0.825	0.922
HCRF $\omega = 2$, $S = 6$	0.802	0.922
LDCRF $\omega = 0$, $S^* = 1$	0.788	0.891
LDCRF $\omega = 1$, $S^* = 1$	0.892	0.911
LDCRF $\omega = 5$, $S^* = 1$	0.913	0.944
LDCRF $\omega = 10$, $S^* = 1$	0.893	0.922
LDCRF $\omega = 0$, $S^* = 3$	0.888	0.932
LDCRF $\omega = 3$, $S^* = 3$	0.912	0.931
LDCRF $\omega = 5$, $S^* = 3$	0.918	0.955
LDCRF $\omega = 10$, $S^* = 3$	0.905	0.933
LDCRF $\omega = 0$, $S^* = 6$	0.864	0.896
LDCRF $\omega = 5$, $S^* = 6$	0.912	0.944
LDCRF $\omega = 10$, $S^* = 6$	0.890	0.912

[†]Data set which includes 20 nongesture samples

[‡]Data set which does not include nongesture samples

[*] S^* refers to number of hidden states per label for LDCRF

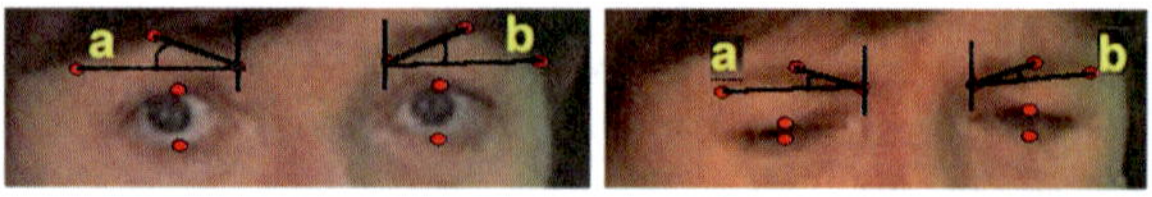

Fig. 19 Example of subject performing a raised brow gestures (*left*) and a lowered brow gesture (*right*). a and b represent the angles ϕ_L and ϕ_R, respectively

classifying gestures in data set 2. The recognition rate of the CRF models then decrease significantly when nongestures are introduced. The best performing model for data set 1 is again the HMM threshold model with an AUC of 0.948. There was a 3% decrease in performance between the data set 1 AUCs of the HMM threshold model and the 3 state LDCRF. This result further suggests that the GT-HMM model is a more robust model when recognizing the eye brow movement gestures when epenthesis gestures are taken in to account.

There was a 3.7% change in performance between the LDCRF from data set to data set 1, while there was only a 0.3% change in performance of the HMM threshold model. This result suggests that the performance of the LDCRF would decrease more than that of the GT-HMM model when the number of epenthesis gestures introduced into the system increased.

Experiments also show that the GT-HMM model, initialized using our automated initialization technique, had a 4.3% better AUC measurement than the T-HMM initialized using the standard HMM segmentation method.

5.2.4 Benchmark Data-Set: Marcel InteractPlay Database

In this section, we discuss user independent experiments which were conducted to evaluate the performance of the models when recognizing gesture performed by signers not represented in the training set. The user independent data set we utilize is a benchmark spatiotemporal hand gesture data set. While this data set does not include sign language gestures or movement epenthesis data, we utilize this data set in order to evaluate the performance of the GT-HMM framework when performing user independent gesture recognition and compare to with additional gesture recognition techniques.

The database contains 3D hand trajectories, in addition to 3D coordinates of the head and the torso, of isolated hand gestures. The database consists of 16 gestures carried out by 20 different persons. For each person, there exists 50 samples of each gesture resulting in a total of 1,000 gesture samples for each of the 16 gesture classes. 3D trajectories of the hands were obtained using stereo vision system which tracks colored gloves, a colored t-shirt and the face in real-time using the EM algorithm (see Fig. 20). A detailed description of the 3D blob tracking can be found in [4].

Just et al. [16] conduct a comparative study, using the InteractPlay data set, to compare two state-of-the-art techniques for temporal event modeling. In their work, HMMs and IOHMMs, an extension to HMMs first proposed by Bengio and Frasconi [3], are evaluated on the InteractPlay data-set. A 12-dimensional feature vector, comprising of left- and right-hand positions and velocities, is used to describe gesture sequences. In order to improve user-independent recognition, each user performs a calibration pose and all other feature vectors are then normalized using the calibration pose. Training was carried out using all 8,000 gesture sequences from 10 of the 20 people in the data set. Testing was then carried out on 8,000 gesture

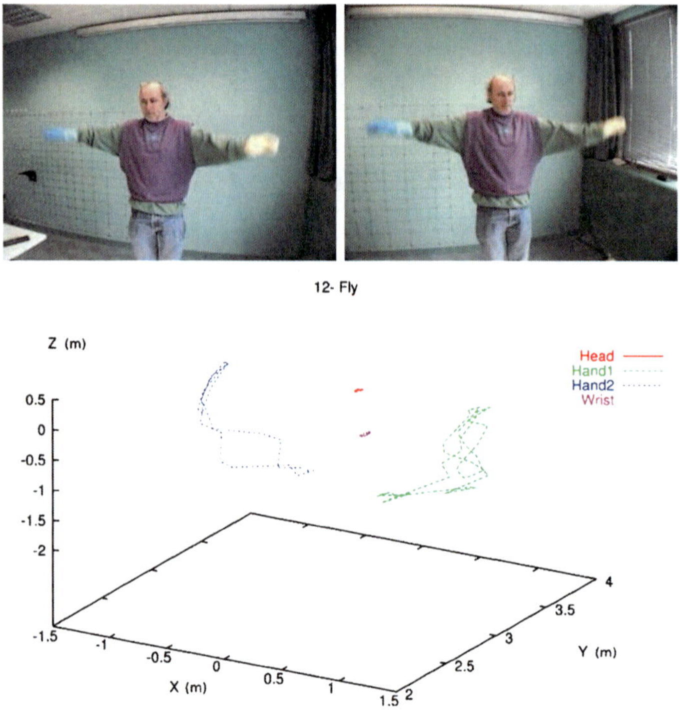

Fig. 20 *Top*: Example of images from a gesture sequence from the point of view of the left camera (*on the left*) and from the point of view of the right camera (*on the right*). *Bottom*: 3D coordinates of the center of each blob (head, torso, left-hand and right-hand) for a "fly" gesture. The z-axis is the vertical axis of the person

sequences from the remaining 10 people not used in training. Results of their experiments show that the HMM and IOHMM achieved an average recognition rate of 75% and 63%, respectively.

We carry out a number of evaluation protocols when testing the GT-HMM framework on the Interactplay data set:

- P1—To investigate the influence the number of training subjects used has on the overall recognition performance, we first train the gesture modeling frameworks using data from only 1 subject. This training set consisted of 50 samples of each gesture class. We then test the models on gesture sequences obtained from 10 subjects. Feature vectors are comprised of 3D hand positions and velocities.
- P2—In order to evaluate the different gesture modeling frameworks, and directly compare results with the work of Just et al., we use the same training and testing protocol as described in their work. This consists of training the framework using data from 10 subjects and testing the framework on the remaining 10 subjects.

One of the overall goals of this work is to create a gesture recognition framework with minimal supervision during training. Just et al. utilize calibration poses to normalize different user inputs and improve recognition of user independent ges-

Table 4 InteractPlay database performance results

	Protocol	#Training subject	#Test subject	Accuracy	ROC AUC
Just et al. IOHMM	P2	10	10	63%	–
Just et al. HMM	P2	10	10	75%	–
GT-HMM	P1	1	10	69.5%	0.745
T-HMM	P1	1	10	69.1%	0.721
LDCRF	P1	1	10	72.2%	0.785
HCRF	P1	1	10	71.8%	0.772
CRF	P1	1	10	68.9%	0.711
GT-HMM	P2	10	10	79.1%	0.831
T-HMM	P2	10	10	74.3%	0.772
LDCRF	P2	10	10	80.8%	0.857
HCRF	P2	10	10	80.1%	0.849
CRF	P2	10	10	75.3%	0.781

tures. The process of creating a set of calibration poses for each new user requires further supervision and thus all evaluations of our GT-HMM framework are carried out without using any calibration data.

Table 4 details the results achieved by the models implemented by Just et al. and the results achieved by the LDCRF, T-HMM and GT-HMM frameworks when tested on the different evaluation protocols. Isolated gesture recognition results reported in this Chapter thus far have been presented using the ROC AUC measurement. While we present both ROC AUC and accuracy measurements in Table 4, in this section we will discuss accuracy measurement results in order to directly compare performance with the results reported by Just et al.

Results of evaluation protocol P1 report classification accuracies of 69.5%, 69.1% and 72.2% for the GT-HMM, T-HMM and LDCRF models, respectively. While the performance of the models are lower than the performances reported by Just et al., the training set was comprised of gesture samples from only one signer.

Results of the evaluation protocol P2, which uses a training set of 10 subjects, show that our GT-HMM framework achieves a classification accuracy of 79.1%, a respective increase of 4.1% and 16.1% when compared to the HMMs and IOHMMs implemented by Just et al. The LDCRF and HCRF performed with classification accuracies of 80.7% and 80.1%, an improvement of 1.6% and 1%, respectively when compared to the GT-HMM model. In the isolated gesture recognition experiments discussed in the previous Sects. 5.2.1–5.2.3, the GT-HMM performs better than the LDCRF and HCRF when classifying isolated gestures and identifying movement epenthesis. However, the LDCRF and HCRF models performed better that the GT-HMM when classifying gestures from data sets which did not include movement epenthesis. Since the InteractPlay data set does not contain movement epenthesis data, the LDCRF and HCRF results reported on the InteractPlay data set are consistent with those in the previous sections.

Results of evaluations on the T-HMM framework, which does not use our gesture subunit initialization, show a decrease in classification accuracy of 4.8% when compared to the GT-HMM. This result is a further indication of the importance of the gesture subunit initialization technique in creating a robust HMM threshold framework which models natural human gestures.

A comparison of the results achieved on Protocols P1 and P2 show that all models, including our GT-HMM model, perform with an increased user independent classification accuracy when trained on gesture samples from a larger set of subjects.

5.3 Continuous Gesture Recognition Experiments

Thus far, we have discussed experiments conducted on isolated gestures. We now discuss experiments conducted on our GT-HMM framework and the best performing CRF model, the LDCRF model, to evaluate the performance of the respective models when spotting and classifying gestures within continuous sequences of natural gestures.

In order to find candidate start and end points using the LDCRF model, we flag segments of the observations sequences which have a corresponding probability greater than a predefined threshold Ω_y. The predefined thresholds used are the best performing thresholds calculated from the ROC analysis performed on the isolated gestures discussed in Sect. 5.2. Given a continuous sequence of observations, $G = \{\mathbf{f}_1, \mathbf{f}_2, \ldots, \mathbf{f}_T\}$, the conditional probability $P(y|\mathbf{f}_t, \theta)$ for all observations $\mathbf{f}_t$, where $0 < t < T$, is calculated for each gesture class y, where θ denotes the trained CRF parameters. A candidate gesture, κ, is flagged for each contiguous sequence where $P(y|\mathbf{f}_t, \theta) > \Omega_y$ $(0 \leq t < T)$. In order to remove overlapping candidate gestures flagged by the LDCRF model, we use the same candidate selection algorithm described in Sect. 4.2.5. The candidate selection algorithm utilizes a gesture likelihood metric to select the final set of recognized gesture. The metric used for the LDCRF system is defined as $\kappa_p = \frac{1}{\kappa_e - \kappa_s} \sum_{i=\kappa_s}^{\kappa_e} P(y|\mathbf{f}_i, \theta)$.

To evaluate the performance of the recognition models, we test the continuous recognition models on the same set of eight hand gestures, three head movements and two eyebrow gestures used in the isolated experiments in Sect. 5.2. We use the best performing models trained during the isolated recognition experiments, discussed in Sect. 5.2, to evaluate the continuous recognition performance of the GT-HMM framework and LDCRF model. Thus, we use an LDCRF model with 8 hidden states per label for hand gesture recognition. We use an LDCRF model with 9 hidden states per label for head gesture recognition and an LDCRF with 3 hidden states per label for eye brow gestures.

A total of 160 additional video clips of full unsegmented sign language sentences being performed by a fluent signer were recorded to test the performance of the continuous recognition systems. Each video clip contained at least one of the eight chosen manual signs. The three head movement gestures occurred a total of 30 times within the 160 videos while the two eye brow gestures occurred a total of 35 times. Videos were recorded at 25 frames per second with an average length of 5 seconds. In order to robustly evaluate the performance of our system, each of the 160 different sign language sentences, used to test the system, was performed in a mixture of different styles. The variations in the style of signs performed are similar to the types of variations that can occur in sign language in real world situations and thus testing the systems on these signs gives a good indication of how the systems will perform in real world scenarios.

Observation sequences G_L, G_R, G^H and G^E, for the left-hand, right-hand, head and eye brow respectively, were extracted from each video clip. Both the continuous LDCRF and GT-HMM recognition frameworks, described in Sect. 5.3, were used to process the observation sequences to spot and classify manual signs and head movement gestures from within the videos. The candidate selection algorithm, described in Sect. 4.2.5, was used to process candidate gestures for both the GT-HMM framework and LDCRF model.

In the gesture spotting and classification task, there are three types of errors: *an insertion error* occurs when the spotter reports a nonexistent gesture, *a deletion error* occurs when the spotter fails to detect a gesture, and *a substitution error* occurs when the spotter falsely classifies a gesture. From these error measures, we define two performance metrics shown in (12) and (13)

$$DetectionRatio = \frac{\#CorrectlyRecognizedGestures}{\#InputGestures}, \tag{12}$$

$$Reliability = \frac{\#CorrectlyRecognizedGestures}{\#InputGestures + \#InsertionErrors}. \tag{13}$$

5.3.1 Continuous Experiment Results

Tables 5 and 6 show the performance of our GT-HMM framework and the LDCRF model, respectively, when spotting and classifying gestures within continuous sequences of video. The experiment shows an overall detection rate of 95.1% and an overall reliability of 87% for our GT-HMM framework and an overall detection rate of 82.7% and an overall reliability of 74.2% for the LDCRF model when spotting and classifying gestures in continuous sign language sentences.

We also evaluate the performance of the start and end point detection relative to ground truth data labeled by a human sign language translator. Tables 5 and 6 show the start and end point performance results for our GT-HMM framework and the LDCRF model, respectively. Each table details the average absolute difference between the spotters start and end points and the human interpreters start and end points for signs that were correctly spotted and classified. The average start and end point error for the GT-HMM framework was 11.6 frames and 9.5 frames, respectively while the average start and end point error for the LDCRF model was 10.1 frames and 8.0 frames, respectively.

The results of the continuous experiments show that the GT-HMM threshold model performs better than the LDCRF model. The experiments showed that our

Table 5 Continuous GT-HMM performance results

Gesture	#Correct	#Ins‡	#Del†	#Sub††	Detection	Reliability	Start error	End error
Gone	20	2	0	0	1.0	0.9	±2.5	±8.4
Alot	20	1	0	0	1.0	0.95	±1.5	±1.6
Lost	20	2	0	0	1.0	0.9	±1.5	±3.5
Plate	19	3	1	0	0.95	0.83	±8.1	±12.2
Bike	20	1	0	0	1.0	0.95	±12.1	±12.0
Paint	20	0	0	0	1.0	1.0	±26.1	±20.7
Paper	16	1	1	3	0.8	0.76	±5.9	±1.6
Clean	18	1	1	1	0.9	0.85	±4.8	±5.2
Head left	11	1	1	0	0.91	0.84	±10.1	±7.7
Head right	10	2	0	0	1.0	0.83	±4.0	±4.3
Head left forward	8	2	0	1	0.88	0.72	±12.9	±6.5
EyeBrowDown	18	3	0	2	0.9	0.78	±19.2	±15.3
EyeBrowUp	15	2	0	0	1.0	0.88	±17.1	±24.9
Total	215	21	4	7	0.951	0.87	±11.6	±9.5

†Number of deletion errors

‡Number of insertion errors

††Number of substitution errors

GT-HMM framework achieved an 12.4% higher detection ratio and a 12.8% higher reliability measure than the LDCRF model.

The start and end point performance experiments reveal that both the GT-HMM framework and the LDCRF model perform with similar accuracy. Since these accuracies are calculated from correctly classified gestures, the results conform with the results on the classification of isolated gestures discussed in Sect. 5.2 where the LDCRF model was shown to perform well when positively classifying gestures in the absence of epenthesis.

5.3.2 Continuous User Independent Experiment Results

In this section, we conduct a second continuous sign language recognition experiment. We evaluate the same set of continuous recognition models, used in the experiments in Sect. 5.3.1, on a set of sign language sentences performed by a second signer not used in the training set (Fig. 21 shows an example of the eight signs being performed by the second signer).

A total of 160 additional video clips of full unsegmented sign language sentences being performed by the second signer were recorded. Each video clip contained at least one of the eight chosen manual signs.

Table 6 Continuous LDCRF performance results

Gesture	#Correct	#Ins‡	#Del†	#Sub††	Detection	Reliability	Start error	End error
Gone	16	2	0	4	0.8	0.72	±8.0	±6.1
Alot	20	2	0	0	1.0	0.9	±3.0	±1.7
Lost	9	1	2	9	0.45	0.42	±3.8	±1.5
Plate	20	1	0	0	1.0	0.95	±15.7	±8.4
Bike	16	2	3	1	0.8	0.72	±12.2	±9.8
Paint	17	3	1	2	0.85	0.74	±3.0	±6.7
Paper	18	2	1	1	0.9	0.81	±0.6	±1.6
Clean	17	2	1	2	0.85	0.77	±7.1	±5.4
Head left	11	3	0	1	0.91	0.73	±9.4	±9.2
Head right	8	1	0	2	0.8	0.72	±18.2	±10.2
Head left forward	9	1	0	0	1.0	0.9	±24.3	±5.1
EyeBrowDown	14	4	0	6	0.7	0.58	±13.3	±10.2
EyeBrowUp	12	2	0	3	0.8	0.70	±14.2	±28.2
Total	187	26	8	31	0.827	0.742	±10.1	±8.0

†Number of deletion errors

‡Number of insertion errors

††Number of substitution errors

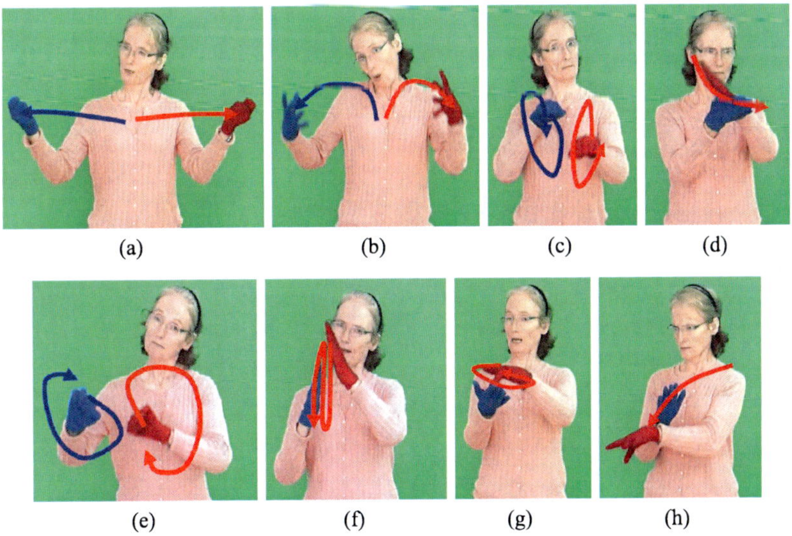

Fig. 21 Example of the eight different performed by the new signer (**a**) newspaper, (**b**) a lot, (**c**) bike, (**d**) gone, (**e**) clean, (**f**) paint, (**g**) plate, (**h**) lost

Table 7 User independent continuous GT-HMM performance results

Gesture	#Correct	#Ins‡	#Del†	#Sub††	Detection	Reliability	Start error	End error
Gone	10	3	0	7	0.58	0.5	6.3	2.8
Alot	18	4	0	8	0.69	0.6	2.2	1.2
Lost	16	3	0	2	0.88	0.61	1.5	3.5
Plate	12	4	2	2	0.75	0.6	9.6	13.1
Bike	28	2	1	2	0.90	0.84	3.0	8.3
Paint	20	2	0	0	1.0	0.9	6.1	4.0
Paper	15	3	1	5	0.71	0.62	5.1	3.6
Clean	13	3	0	2	0.86	0.72	7.9	8.3
User independent total	132	24	4	28	0.804	0.702	5.2	5.6
User dependent total	153	11	3	4	0.956	0.894	6.6	8.1

† Number of deletion errors

‡ Number of insertion errors

†† Number of substitution errors

Table 8 User independent continuous LDCRF performance results

Gesture	#Correct	#Ins‡	#Del†	#Sub††	Detection	Reliability	Start error	End error
Gone	9	4	0	8	0.52	0.36	4.2	2.5
Alot	22	2	0	4	0.84	0.78	5.0	2.3
Lost	9	2	0	9	0.5	0.45	3.4	5.5
Plate	11	5	1	4	0.68	0.52	8.9	14.1
Bike	27	5	0	4	0.87	0.75	10.8	2.6
Paint	13	3	0	7	0.65	0.56	6.6	4.5
Paper	8	3	2	11	0.38	0.33	5.9	1.6
Clean	12	4	1	2	0.8	0.63	4.8	5.2
User independent total	111	28	4	49	0.676	0.578	6.2	10.9
User dependent total	133	15	8	19	0.831	0.76	6.6	8.1

† Number of deletion errors

‡ Number of insertion errors

†† Number of substitution errors

Observation sequences, G_L and G_R, for the left-hand and right-hand, respectively, were extracted from each video clip. Both the continuous GT-HMM and LDCRF recognition frameworks were then used to process the observation sequences to spot and classify hand signs from within the videos of the second signer.

Tables 7 and 8 show the performance of our GT-HMM framework and the LDCRF model respectively when recognizing signs performed by the second signer. The experiment shows an overall detection rate of 80.4% and an overall reliability of 70.2% for our GT-HMM framework and an overall detection rate of 67.6% and an overall reliability of 57.8% for the LDCRF model.

When compared to the detection rates achieved in the experiments in Sect. 5.3.1, the overall detection rate of the GT-HMM framework dropped by 15.2% and, similarly, the overall detection rate of the LDCRF dropped by 15.5%. These results show that our GT-HMM framework consistently performs with a higher gesture detection rate than that of the LDCRF when recognizing signer independent signs.

In previous works which have used a small number of signers in the training set, results of user independent recognition evaluations have seen large decreases when compared to user dependent recognition results [22]. For example, in Assan et al. [1], accuracy for training on one signer and testing on another was 51.9% compared to 92% when the same signer supplied both training and test data. As shown in the experiments on the InteractPlay data-set in Sect. 5.2.4, user independent recognition performance using our GT-HMM threshold framework can be improved by utilizing a larger number of subjects. While the experiments conducted in this experiment show a decrease in the signer independent detection rate, the fact that only one signer was represented in the training set means that a decrease of only 15.2% can be interpreted as a promising result.were utilized to achieve recognition rates of 85% and 90.6% respectively.

5.4 Multimodal Recognition Examples

Incorporating nonmanual signals such as eyebrow gestures and head movement gestures into a sign language recognition framework provides the necessary foundations for differentiating between different types of questions, and also recognizing the start of sign language sentences. In this section, we will show recognition results from a number of manual and nonmanual signals and discuss how the combination of these signals can be utilized to create a more comprehensive understanding of sign language sentences.

Figure 22 shows the gestures spotted by our system in three different sentences. Figures 22(a) and 22(b) show gestures spotted from two sentences where the signer performs the words “CAR PETROL ALL GONE” in both sentences.

In the first sentence, the signer is asking a question “CAR PETROL ALL GONE HOW?”, but in the second sentence the signer is asking a yes/no question “CAR PETROL ALL GONE?”. The manual signs for both these sentences are the same but the difference can only be recognized from the head movement and eyebrow gestures. It can be seen that our system spots an eyebrow down gesture coinciding with a left-head movement followed by right-head movement. This indicates that the signer is asking a “wh” question. In the second sentence, our system spots a eyebrow down gesture at the beginning of the sentence followed by a head forward

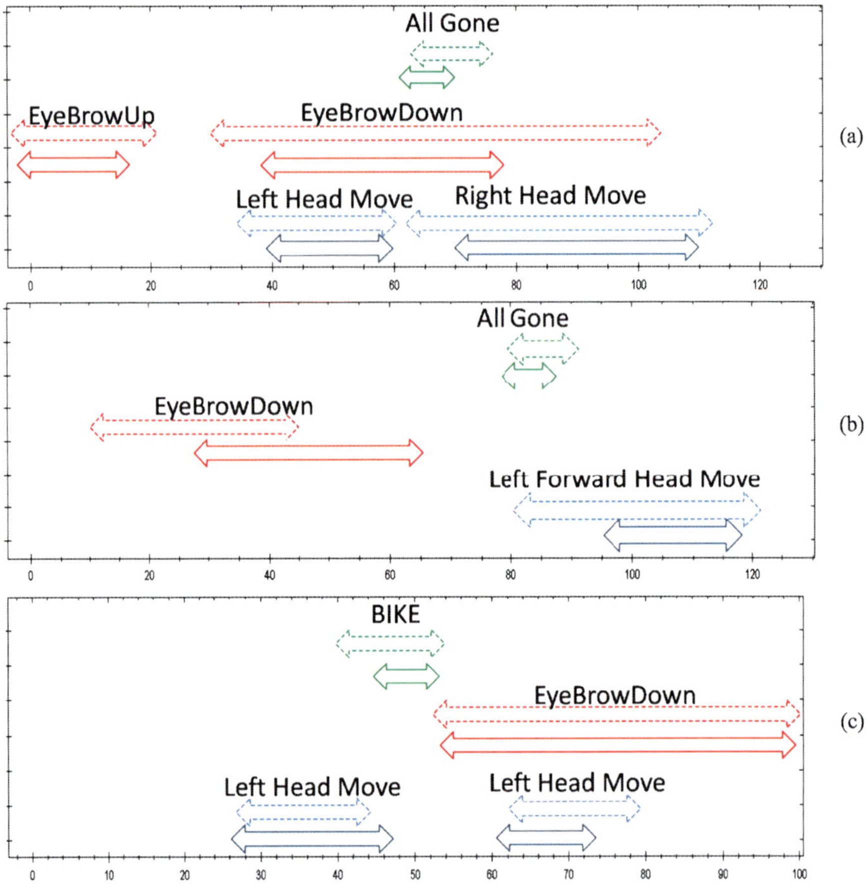

Fig. 22 Multimodal gesture labeling comparison of a human interpreter vs. our recognition system (*dotted arrows* represent hand labeled gestures, while *solid arrows* represent labels generated by our system)

movement, indicating the signer is asking a yes/no question. Figure 22(c) shows gestures from a sentence "WHO BIKE BROKE?", where our system spots an eyebrow down gesture coinciding with a left-head movement. Similar to the gestures in Fig. 22(a), the eyebrow down gesture coinciding with a head movement gesture indicates a "wh" question. Also from Fig. 22(a), it was observed that an eyebrow up gesture occurred at the start of the sequence. This is an interesting observation as the eyebrow up gesture is linked to the start of a new sentence or sequence.

6 Conclusion

In Sect. 1.1, we have discussed current methods of continuous gesture recognition. The downside of the majority of these methods is that explicit training of movement epenthesis models are required or unnatural constraints are put on the signer, such as unnatural pauses between words. The main contribution of the work proposed in this chapter is that we have developed a robust framework for the recognition of spatiotemporal based gestures which does not require explicit epenthesis training or require the need for specific rules to determine sign boundaries. Our framework requires only that the dedicated gesture models be trained, and as a result of this training a single epenthesis model can be implemented. We have expanded on the work of Lee and Kim [19] to develop a GT-HMM framework, utilizing our novel gesture subunit initialization technique, which models continuous multidimensional gesture observations within a parallel HMM network.

We evaluated our GT-HMM framework and compared it to current models for recognizing human motion. HMMs, CRFs, HCRFs and LDCRFs have recently been implemented in systems for automatically recognizing different human actions. We evaluate these techniques in the domain of sign language gesture recognition. In order to evaluate the performance of the models when recognizing sign language gestures, it was important to evaluate each model when identifying movement epenthesis as well as evaluating the performance of the models when classifying gestures. Experiments were conducted on an isolated data set of motion based manual signs and on an isolated data set of nonmanual head motion and eye brow motion gestures.

In experiments carried out on all three gesture types, the best performing model was the LDCRF when tested on a data set which did not include movement epenthesis. The results of this experiment were consistent with previous experiments on HCRFs and LDCRFs which Wang et al. [34] and Morency et al. [21] show that HCRFs and LDCRF perform better than the standard HMM model when classifying gestures. When a data set, which included movement epenthesis, was introduced to the experiments, the performance of the standard HMM model, and all CRF models, dropped significantly in relation to the performance of GT-HMM model. The GT-HMM model performed best in all three experiments, with movement epenthesis data, with an AUC of 0.976, 0.936 and 0.948 for the hand gesture, head gesture and eye brow gesture evaluations, respectively.

An important aspect of our GT-HMM framework is the gesture subunit initialization technique which is used to automatically initialize the parameters of GT-HMM framework. Experiments conducted on the manual and nonmanual data sets, showed that initializing the HMMs using our technique consistently improved hand gesture, head gesture and eye brow gesture classification performance by 3.5%, 6.3% and 4.3%, respectively when compared to a standard initialization technique. Moreover, experiments on the Interactplay data-set also indicate that our gesture subunit initialization improves classification performance with results showing a 4% increase in performance.

Experiments, to evaluate the performance of our GT-HMM framework and the LDCRF model when spotting and classifying gestures from continuous sequences, were also carried out. The continuous experiments showed that our GT-HMM framework achieved an 12.4% higher detection ratio and a 12.8% higher reliability measure than the LDCRF model when tested on 220 different hand, head and eye brow gestures. Additionally, a set of signer independent experiments where conducted to evaluate the performance of the models when recognizing signs performed by a signer not represented in the training set. Signer independent evaluations conducted on the InteractPlay, which contained no epenthesis data, showed that the GT-HMM and LDCRF models performed best when compared to HMMs and IOHMMs implemented by Just et al. [16]. Furthermore, signer independent evaluations on continuous sign language sentences showed that our GT-HMM framework consistently performs with a higher gesture detection rate than that of the LDCRF. The overall detection rate of the GT-HMM and LDCRF models decreased by only 15.2% and 15.5% respectively, when compared to signer dependent results.

References

1. Assan, M., Grobel, K.: Video-based sign language recognition using hidden Markov models. In: Proc. of the International Gesture Workshop on Gesture and Sign Language in Human-Computer Interaction, pp. 97–109. Springer, London (1998)
2. Bauer, B., Kraiss, K.F.: Towards an automatic sign language recognition system using subunits. In: GW'01: Revised Papers from the International Gesture Workshop on Gesture and Sign Languages in Human-Computer Interaction, pp. 64–75. Springer, London (2002)
3. Bengio, Y., Frasconi, P.: An input/output HMM architecture. Adv. Neural Inf. Process. Syst. **7**, 427–434 (1995)
4. Bernier, O., Collobert, D.: Head and hands 3d tracking in real time by the EM algorithm. In: RATFG-RTS'01: Proc. of the IEEE ICCV Workshop on Recognition, Analysis, and Tracking of Faces and Gestures in Real-Time Systems, p. 75. IEEE Comput. Soc., Washington (2001)
5. Brashear, H., Starner, T., Lukowicz, P., Junker, H.: Using multiple sensors for mobile sign language recognition. In: Proc. of the 7th IEEE International Symposium on Wearable Computers, pp. 45–52 (2003). doi:10.1109/ISWC.2003.1241392. http://ieeexplore.ieee.org/lpdocs/epic03/wrapper.htm?arnumber=1241392
6. Castrillon-Santana, M., Deniz-Suarez, O., Anton-Canalis, L., Lorenzo-Navarro, J.: Performance evaluation of public domain haar detectors for face and facial feature detection. In: VISAPP (2008)
7. Comaniciu, D., Ramesh, V., Meer, P.: Real-time tracking of non-rigid objects using mean shift. In: Proc. of the 2nd Conference on Computer Vision and Pattern Recognition, vol. 2, pp. 142–149 (2000). doi:10.1109/CVPR.2000.854761
8. Cooper, H., Bowden, R.: Large lexicon detection of sign language. In: CVHCI07, pp. 88–97 (2007)
9. Cootes, T., Taylor, C.: Statistical models of appearance for computer vision. Technical Report, Wolfson Image Analysis Unit, Imaging Science and Biomedical Engineering, University of Manchester, Manchester M13 9PT (2001)
10. Ding, L., Martinez, A.: Modelling and recognition of the linguistic components in American sign language. J. Image Vis. Comput. (2), 257–286 (2009, in press). doi:10.1109/5.18626
11. Gao, W., Fang, G., Zhao, D., Chen, Y.: Transition movement models for large vocabulary continuous sign language recognition. In: IEEE FG 2004, pp. 553–558 (2004). doi:10.1109/AFGR.2004.1301591
12. Grossman, R.B., Kegl, J.: To capture a face: a novel technique for the analysis and quantification of facial expressions in American sign language (2006)
13. Holden, E.J., Robyn, O.: Visual sign language recognition. Mutli-Image Anal. **2032**(6), 270–287 (2001). doi:10.1109/TPAMI.2005.112
14. Holden, E.J., Lee, G., Owens, R.: Australian sign language recognition. Mach. Vis. Appl. **16**(5), 312–320 (2005). doi:10.1007/s00138-005-0003-1
15. Junker, H., Amft, O., Lukowicz, P., Tröster, G.: Gesture spotting with body-worn inertial sensors to detect user activities. Pattern Recognit. **41**(6), 2010–2024 (2008). doi:10.1016/j.patcog.2007.11.016

16. Just, A., Marcel, S.: A comparative study of two state-of-the-art sequence processing techniques for hand gesture recognition. Comput. Vis. Image Underst. **113**(4), 532–543 (2009). doi:10.1016/j.cviu.2008.12.001. http://www.sciencedirect.com/science/article/B6WCX-4V70NKB-2/2/c6918c4235d7d44044cf6f612bfd6ea0
17. Kim, Y.J., Conkie, A.: Automatic segmentation combining an HMM-based approach and spectral boundary correction. In: ICSLP-2002, pp. 145–148 (2002)
18. Kim, J., Wagner, J., Rehm, M., Andre, E.: Bi-channel sensor fusion for automatic sign language recognition. In: 8th IEEE International Conference on Automatic Face & Gesture Recognition, 2008, pp. 1–6 (2008). doi:10.1109/AFGR.2008.4813341. http://ieeexplore.ieee.org/lpdocs/epic03/wrapper.htm?arnumber=4813341
19. Lee, H.K., Kim, J.H.: An HMM-based threshold model approach for gesture recognition. IEEE Trans. Pattern Anal. Mach. Intell. **21**(10), 961–973 (1999). doi:10.1109/34.799904
20. Liddell, S.K., Johnson, R.: American sign language: the phonological base. Sign Lang. Stud. **64**(6), 195–278 (1989). doi:10.1109/TPAMI.2005.112
21. Morency, L.P., Quattoni, A., Darrell, T.: Latent-dynamic discriminative models for continuous gesture recognition. In: CVPR, pp. 1–8 (2007). doi:10.1109/CVPR.2007.383299
22. Ong, S.C.W., Ranganath, S.: Automatic sign language analysis: a survey and the future beyond lexical meaning. IEEE Trans. Pattern Anal. Mach. Intell. **27**(6), 873–891 (2005). doi:10.1109/TPAMI.2005.112
23. Oz, C., Leu, M.: Linguistic properties based on American sign language isolated word recognition with artificial neural networks using a sensory glove and motion tracker. Neurocomputing **70**(16–18), 2891–2901 (2007). doi:10.1016/j.neucom.2006.04.016. http://linkinghub.elsevier.com/retrieve/pii/S0925231207001646
24. Rabiner, L.: A tutorial on hidden Markov models and selected applications in speech recognition. Proc. IEEE **77**(2), 257–286 (1989). doi:10.1109/5.18626
25. Shanableh, T., Assaleh, K., Al-Rousan, M.: Spatio-temporal feature-extraction techniques for isolated gesture recognition in Arabic sign language. IEEE Trans. Syst. Man Cybern. Part B, Cybern. **37**(3), 641–650 (2007). doi:10.1109/TSMCB.2006.889630
26. Starner, T., Pentland, A., Weaver, J.: Real-time American sign language recognition using desk and wearable computer based video. IEEE Trans. Pattern Anal. Mach. Intell. **20**(12), 1371–1375 (1998). doi:10.1109/34.735811
27. Stokoe, W.C.: Sign language structure: an outline of the visual communication systems of the American deaf. J. Deaf Stud. Deaf Educ. **10**(1), 3–37 (2005)
28. Viola, P., Jones, M.: Rapid object detection using a boosted cascade of simple features. In: CVPR, IEEE, vol. 1, p. 511 (2001). http://doi.ieeecomputersociety.org/10.1109/CVPR.2001.990517
29. Vogler, C., Metaxas, D.: Parallel hidden Markov models for American sign language recognition. In: ICCV, pp. 116–122 (1999)
30. Vogler, C., Metaxas, D.: A framework for recognizing the simultaneous aspects of American sign language. Comput. Vis. Image Underst. **81**, 358–384 (2001)
31. Vogler, C., Metaxas, D.: Handshapes and movements: multiple-channel ASL recognition. In: Gesture-Based Communication in Human-Computer Interaction. Lecture Notes in Computer Science, pp. 1–13. Springer, Berlin (2004)
32. von Agris, U., Schneider, D., Zieren, J., Kraiss, K.F.: Rapid signer adaptation for isolated sign language recognition. In: CVPRW'06: Proc. of the 2006 Conference on Computer Vision and Pattern Recognition Workshop, p. 159. IEEE Comput. Soc., Washington (2006). doi:10.1109/CVPRW.2006.165
33. Wang, C., Shan, S., Gao, W.: An approach based on phonemes to large vocabulary Chinese sign language recognition. In: IEEE FG 2002, p. 411. IEEE Comput. Soc., Washington (2002)
34. Wang, S.B., Quattoni, A., Morency, L.P., Demirdjian, D., Darrell, T.: Hidden conditional random fields for gesture recognition. In: CVPR, pp. 1521–1527 (2006). doi:10.1109/CVPR.2006.132
35. Yamato, J., Ohya, J., Ishii, K.: Recognizing human action in time-sequential images using hidden Markov model. In: CVPR, pp. 379–385 (1992). doi:10.1109/CVPR.1992.223161

36. Yang, R., Sarkar, S., Loeding, B.: Enhanced level building algorithm for the movement epenthesis problem in sign language recognition. In: CVPR07, pp. 1–8 (2007)
37. Yang, H.D., Sclaroff, S., Lee, S.W.: Sign language spotting with a threshold model based on conditional random fields. IEEE Trans. Pattern Anal. Mach. Intell. **99**(1) (2009)
38. Yang, R., Sarkar, S., Loeding, B.: Handling movement epenthesis and hand segmentation ambiguities in continuous sign language recognition using nested dynamic programming. IEEE Trans. Pattern Anal. Mach. Intell. **32**(3), 462–477 (2010). doi:10.1109/TPAMI.2009.26

Learning Transferable Distance Functions for Human Action Recognition

Weilong Yang, Yang Wang, and Greg Mori

Abstract Learning-based approaches for human action recognition often rely on large training sets. Most of these approaches do not perform well when only a few training samples are available. In this chapter, we consider the problem of human action recognition from a single clip per action. Each clip contains at most 25 frames. Using a patch based motion descriptor and matching scheme, we can achieve promising results on three different action datasets with a single clip as the template. Our results are comparable to previously published results using much larger training sets. We also present a method for learning a transferable distance function for these patches. The transferable distance function learning extracts generic knowledge of patch weighting from previous training sets, and can be applied to videos of new actions without further learning. Our experimental results show that the transferable distance function learning not only improves the recognition accuracy of the single clip action recognition, but also significantly enhances the efficiency of the matching scheme.

1 Introduction

The ability to generalize from a small training set is an important feature of any recognition system. While this statement can be made of recognition problems in general, in this chapter we focus on human action recognition from video data. There has been much recent progress in this field, with high performance on standard benchmark datasets. However, this level of performance has been achieved

W. Yang (✉) · Y. Wang · G. Mori
School of Computing Science, Simon Fraser University, Burnaby, BC, Canada
e-mail: wya16@sfu.ca

Y. Wang
e-mail: ywang12@cs.sfu.ca

G. Mori
e-mail: mori@cs.sfu.ca

L. Wang et al. (eds.), *Machine Learning for Vision-Based Motion Analysis*, Advances in Pattern Recognition,
DOI 10.1007/978-0-85729-057-1_13, © Springer-Verlag London Limited 2011

with techniques that utilize a large amount of training data. We argue that while this may be appropriate for certain tasks (e.g., action fingerprinting, or distinguishing subtle differences in action), it is problematic to assume that such large training sets exist for the task of discriminating between broadly different categories of actions. In this work, we focus on learning to recognize actions from a single clip, leveraging knowledge acquired from previous action categories. We focus on experiments on standard action recognition datasets, though the same principles could be applied to video retrieval and surveillance tasks.

As has been noted in the object recognition literature [17], humans are adept at learning new categories given small numbers of examples. Attempting to endow computer vision algorithms with similar abilities is appealing. This line of research is also of practical importance—gathering volumes of training data for unusual actions, for example, for surveillance or video retrieval tasks, is an expensive, labour-intensive process.

In this chapter, we attempt to push the boundaries of action recognition performance with a single, short video clip as training data (called the *target dataset*) for a particular action. Of course, it is impossible to learn a recognition system with a single video clip per action, so we assume that we also have access to a much larger dataset called the *source dataset*. However, the action categories of the source dataset are different from those of the target dataset. For example, the source dataset might contain a large number of video clips of "regular" actions (e.g., running, walking, hand-waving, etc.) readily available from many benchmark datasets (e.g., KTH dataset [38], Weizmann dataset [6], etc.). The target dataset might contain video clips of more "interesting" actions (e.g., spinning). We assume there is only one single clip for each action class on the target dataset. Our goal is to classify a new test video clip into one of the action labels of the target dataset. Since the source and target datasets have different action categories, we cannot naively combine them together and use a standard machine learning technique to learn our classification model. Instead, we will learn certain "generic knowledge" from the source dataset that can be "tranferred" to the target dataset. Though the target dataset only contains very small amount of training data, we might still be able to learn a good model by exploiting the "transferable knowledge" learned from the source dataset.

We work with a *figure-centric* representation in which a human detection and tracking algorithm has been run as a preprocessing step. The main contributions of this paper involve the development of a parametrized distance function for comparing video clips. The distance function is defined as a weighted sum of distances between a densely sampled set of motion patches on frames of the video clips. This distance function is effective for action recognition in the impoverished training data setting. We further develop an algorithm for learning these weights, that is, the distance function parameters. We develop a novel margin-based *transfer learning* algorithm, inspired by the work of Frome et al. [20] and Ferencz et al. [19]. The learned weights are a function of patch features and can be generically transferred to a new action category without further learning. This learning greatly improves the efficiency of our algorithm, and can improve recognition accuracy.

An earlier version of this work appeared in [43]. An extension of this work for action detection appeared in [42]. The rest of the chapter is organized as follows.

Section 2 reviews those work related transfer learning and human action recognition. Section 3 describes the motion feature and the basic matching scheme. Section 4.1 gives the details of our approach for learning transferable distance functions. We present experimental results in Sect. 5.2 and conclude in Sect. 6.

2 Previous Work

In this section, we review related work in both machine learning and computer vision literature. Section 2.1 gives an overview of different learning scenarios related to transfer learning. It also gives a brief review of commonly used approaches in transfer learning. Section 2.2 reviews related work from the computer vision literature, in particular those work related to human action recognition.

2.1 Related Work in Learning

Here we give a brief summary of transfer learning and other related learning problems that have been studied under various names (e.g., semi-supervised learning, self-taught learning, multi-task learning, domain adaptation, etc.). Interested readers are referred to [34] for a more detailed survey.

The standard scenario of machine learning problems is *supervised learning*. We are given a set of labeled training data $\mathcal{D} = \{(x_n, y_n) \in \mathcal{X} \times \mathcal{Y} : 1 \leq n \leq N\}$, where $\mathcal{X}$ is the input space and $\mathcal{Y}$ a finite label set. It is assumed that each (x_n, y_n) is drawn independently from a fixed, but unknown distribution p, that is, $(x_n, y_n) \sim p(x, y)$. Our goal is to learn a function $f : \mathcal{X} \to \mathcal{Y}$ that can be used to predict the class label y for a new instance x. For example, if we want to learn to recognize images of "dogs" versus "cats", an instance x_n will be an image, and the label y_n is "dog" or "cat". Unfortunately, it is very expensive and time-consuming to collect a large number of labeled training instances. One possible approach to deal with this issue is to use *semi-supervised learning* [46], which aims to learn a classification model from both labeled data $\{(x_i, y_i)\}_{i=1}^{N}$ and unlabeled data $\{x_j\}_{j=N+1}^{N+M}$, where $N \ll M$. In semi-supervised learning, the class labels and generative distributions of the unlabeled data are assumed to be the identical to those on the labeled data. In the "dog vs. cat" example, an unlabeled data instance x_j corresponds to an image of either *dog* or *cat*, but we simply do not know which label it is. A more challenging learning problem is *self-taught learning* [36], where the unlabeled data can have different class labels or generative distributions from the labeled data. Using previous example, an unlabeled data instance in self-taught learning can be an image of anything, that is, not limited to dogs or cats.

In our work, we are interested in *transfer learning* [34]. Unlike previously mentioned learning scenarios, all the data in transfer learning are labeled. The goal of transfer learning to learn a predictive model by applying knowledge learned previously from other different but related task, that is, it transfer knowledge from one supervised learning task to another. For example, if we want to learn to recognize "dogs" versus "cats" from two different kinds of labeled training datasets, called the source dataset $\mathcal{D}^{(s)} = \{(x_n^{(s)}, y_n^{(s)}) \in \mathcal{X} \times \mathcal{Y}^{(s)}\}$ and the target data $\mathcal{D}^{(t)} = \{(x_m^{(t)}, y_m^{(t)}) \in \mathcal{X} \times \mathcal{Y}^{(t)}\}$. The target dataset $\mathcal{D}_t$ contains instances labeled with either dogs or cats, that is, $\mathcal{Y}^{(t)} = \{\text{dog}, \text{cat}\}$. The source dataset $\mathcal{D}^{(s)}$ contains instances with labeled with other categories (e.g., $\mathcal{Y}^{(s)} = \{\text{horse}, \text{tiger}, \ldots\}$). The goal of transfer learning is to exploit $\mathcal{D}^{(s)} \cup \mathcal{D}^{(t)}$ to build a model that recognize a new instance with an unknown label in $\mathcal{Y}^{(t)}$. Transfer learning is related to another learning problem called *domain adaptation* [8, 10, 29]. In domain adaptation, the source dataset and the target dataset have the same label set, i.e. $\mathcal{D}^{(s)} = \mathcal{D}^{(t)}$. But the source dataset and target dataset are drawn from two different distributions, that is, $(x_n^{(s)}, y_n^{(s)}) \sim p^{(s)}(x, y)$, $(x_m^{(t)}, y_m^{(t)}) \sim p^{(t)}(x, y)$, and $p^{(s)} \neq p^{(t)}$.

Another closely related learning problem is *multi-task learning* [7]. The goal of multi-task learning is to learn different tasks together and assume there is some "relatedness" between different tasks. Ben-David and Schuller [5] provided a theoretical justification for multi-task learning. The problem setting of multi-task learning is almost identical to that of transfer learning. The only difference is that in multi-task learning, the goal is to learn a model that simultaneously do well on all the tasks. Using the previous example, the multi-task learning might aim to learn to classify all the possible animals, not just "dogs" and "cats".

Now we review the approaches commonly used in transfer learning. Since transfer learning and multi-task learning are very closely related, most techniques developed for one of those two problems can be easily adapted to the other one. For ease of presentation, we will loosely call both problems "transfer learning" in the rest of this section.

An important assumption of transfer learning is that various tasks involved in the learning are somehow related. Otherwise, it will be impossible to transfer the knowledge learned from one task to another one. Depending on the assumption of "relatedness", various techniques proposed in the literature roughly fall into two categories.

2.1.1 "Relatedness" via Features

A lot of work in transfer learning assumes that different tasks are related by some intermediate feature representations. Argyrious et al. [2–4] learn a low-dimensional representation which is shared across multiple related tasks. If the intermediate representation shared across tasks are semantically informative, we can even perform zero-data [27] or zero-shot [33] learning, where the target task does not have any training data. This idea of transferring intermediate representations via related tasks can also be applied to solve regular classification problem (i.e., single task). For example, Ahmed et al. [1] demonstrate that classification can be improved by learning a intermediate feature representation from so-called "pseudo-tasks". Those "pseudo-tasks" are auxiliary tasks constructed to help with learning a good feature representation for the target task.

2.1.2 "Relatedness" via Model Parameters

Another popular approach in transfer learning is to assume that the model parameters associated in different tasks are related in some way. Let us denote the model parameters of the tth task as $\mathbf{w}_t$ $(1 \leq t \leq T)$, where T is the number of tasks involved. Evgeniou et al. [13] assume $\mathbf{w}_t = \mathbf{w}_0 + \mathbf{v}_t$, where $\mathbf{w}_0$ are shared among all tasks and $\mathbf{v}_t$ are the parameters specific to the tth task. In [44, 45], $\mathbf{w}_t$ $(1 \leq t \leq T)$ are related by assuming they are draw from the same prior distribution $p(\mathbf{w})$, for example, $p(\mathbf{w})$ can be a Gaussian process [44], or a t-process [45].

A limitation of previous work in transfer learning is that the underlying classification model for each task is always assumed to be in a parametric form, for example, a linear classification with parameters $\mathbf{w}$. It is not clear how to generalize those transfer learning techniques to other non-parametric classifiers, e.g. K-nearest neighbor (KNN). In this work, we develop a technique for transferring "distance functions", which can be used in KNN classifiers. To the best of our knowledge, there has not been any previous work on transferring learning in this setting.

2.2 *Related Work in Vision*

A variety of action recognition algorithms have obtained high recognition accuracy on the standard KTH [38] and Weizmann [6] benchmark datasets. The literature in this area is immense; we provide a brief set of closely related work here. The vast majority of these methods use large amounts of training data, with either a leave-one out (LOO) strategy or other splits of the data involving large amounts of training data for each action.

Efros et al. [12] recognize the actions of small scale figures using features derived from blurred optical flow estimates. Fathi and Mori [16] learn an efficient classifier on top of these features using AdaBoost. Our method uses the same figure-centric representation, and defines patch distances using blurred optical flow. We learn a generic transferable distance function rather than individual classifiers, on smaller training sets.

A number of methods run interest point detectors over video sequences, and describe this sparse set of points using spatial and/or temporal gradient features [11, 30, 38]. In contrast with these methods, we use a densely sampled set of patches in our distance function. Our transfer learning algorithm places weights on these patches, which could be interpreted as a type of interest point operator, specifically tuned for recognition.

Shechtman and Irani [39] define a motion consistency designed to alleviate problems due to aperture effects. Distances between pairs of video clips are computed by exhaustively comparing patches centered around every space-time point. In our work we learn which patches are important for recognition, leading to a more efficient algorithm—though one could use motion consistency in place of blurred optical flow in a distance function.

Ke et al. [24, 25] define a shape and flow correlation based on matching of segmentations. Classification is done using a parts-based model [24] and an SVM trained on template distances in a LOO setting [25].

Jhuang et al. [22] describe a biologically plausible model containing alternating stages of spatio-temporal filter template matching and pooling operations. Schindler and Van Gool [37] examine the issue of the length of video sequences needed to recognize actions. They build a model similar to Jhuang et al. [22] and show that *short* snippets can be effective for action recognition. Both of these methods use large splits for training data. Our work focuses instead on the amount of data needed, rather than the temporal length of the clips. Weinland and Ronfard [41] classify actions based on distances to a small set of discriminative prototypes selected in a LOO experiment.

Tran and Sorokin [40] propose a metric learning method for action recognition from small datasets. Our experiments use fewer frames (25 per training clip), and compare favorably in terms of accuracy.

Our approach of learning distance functions is inspired by the work of Frome et al. [20, 21] and Ferencz et al. [19]. The work by Ferencz et al. [19] introduces the notion of "hyper-features" for object identification. In a nutshell, hyper-features are properties of image patches that can be used to estimate the salience of those patches. These salines measurements can be used later in matching based object identification. In our work, we use a similar idea to estimate the relative weights (i.e., salient) of motion patches extracted from video frames. We define the hyper-feature of a motion patch using a codebook representation. The main difference of our hyper-feature model with Ferencz et al. [19] is that our model is directly tied to the distance function used for the matching.

We use Frome et al.'s *focal learning* framework which considers similar-dissimilar triplets of training data points. Rather than learning distance functions specific to each image, we learn them generically based on patch *hyper-features*. This allows us to transfer them to new videos without retraining, and to use them even in cases where we have only one training example for a class (focal learning requires at least 2).

In computer vision, the subarea of face identification has a long tradition of using ideas similar to transfer learning. In face identification, a system is trained from faces of thousands of people (source dataset). During the testing, a well-trained system can easily identify the face which does not exist in the source dataset. Other applications of transfer learning in computer vision include the one-shot learning of Fei-Fei et al. [17], in which object recognition models are built using priors learned from previously seen object classes. Farhadi et al. [14, 15] use comparative features for transferring distances between templates for sign language and multi-view action recognition. Quattoni et al. [35] perform transfer learning using kernel distances to unlabeled prototypes. Lampert et al. [26] learn to detect unseen object classes by considering object attributes as intermediate feature representation that can be transferred.

3 Motion Descriptors and Matching Scheme

We will classify the test video (we will call it *query video*) using the nearest neighbor (NN) classifier after computing the distances between the query video and each clip in the template set. The query video will be assigned to the action label according to the best matched template clip. The reason to use the NN classifier is that most other learning based approaches rely on complicated models with a large number of parameters, and thus cannot deal with the situation of very small training sets. In the following, we first introduce the motion descriptors used for representing a video clip (Sect. 3.1). Then we describe our patch-based matching scheme for comparing two video clips Sect. 3.2).

3.1 Motion Descriptors

In this work, we use a figure-centric representation of motion in which a standard human detector and tracking algorithm has been applied. The motion descriptors in Efros et al. [12] are used to represent the video frames. We first compute the optical flow at each frame. The optical flow vector field F is then split into two scalar fields, F_x and F_y corresponding to the x and y components of the flow vector. F_x and F_y are further half-wave rectified into four nonnegative channels F_x^+, F_x^-, F_y^+, F_y^-, so that $F_x = F_x^+ - F_x^-$ and $F_y = F_y^+ - F_y^-$. Then, those four channels are blurred using a Gaussian kernel to obtain the final four channels Fb_x^+, Fb_x^-, Fb_y^+, Fb_y^- (see Fig. 1).

3.2 Patch Based Action Comparison

We compute the distance between two video clips by comparing the patches from both clips. Patch based methods are very popular in object recognition, due to the fact that local patches are more robust to pose variation than the whole object. We represent each patch using the four channel motion descriptor. Suppose the four channels for patch i are $\mathbf{a}_1$, $\mathbf{a}_2$, $\mathbf{a}_3$, $\mathbf{a}_4$, and each channel has been concatenated to a vector. Similarly, the four channels for patch j are $\mathbf{b}_1$, $\mathbf{b}_2$, $\mathbf{b}_3$, $\mathbf{b}_4$. We denote $\hat{\mathbf{a}}_k = [a_k^1 - \bar{a}_k, a_k^2 - \bar{a}_k, \ldots, a_k^n - \bar{a}_k]$, and $\hat{\mathbf{b}}_k = [b_k^1 - \bar{b}_k, b_k^2 - \bar{b}_k, \ldots, b_k^n - \bar{b}_k]$, where $\bar{a}_k$ and $\bar{b}_k$ are the mean values of channel $\mathbf{a}_k$ and $\mathbf{b}_k$ respectively, a_k^i denotes the ith element in channel vector $\mathbf{a}_k$. The similarity between patch i and j is computed using normalized correlation, and the distance is given by

$$d(i,j) = C - \sum_{k=1}^{4} \frac{\hat{\mathbf{a}}_k^T \hat{\mathbf{b}}_k + \varepsilon}{\sqrt{(\hat{\mathbf{a}}_k^T \hat{\mathbf{a}}_k + \varepsilon)(\hat{\mathbf{b}}_k^T \hat{\mathbf{b}}_k + \varepsilon)}}, \tag{1}$$

where C is a positive constant to make the distance nonnegative, and ε is a small constant.

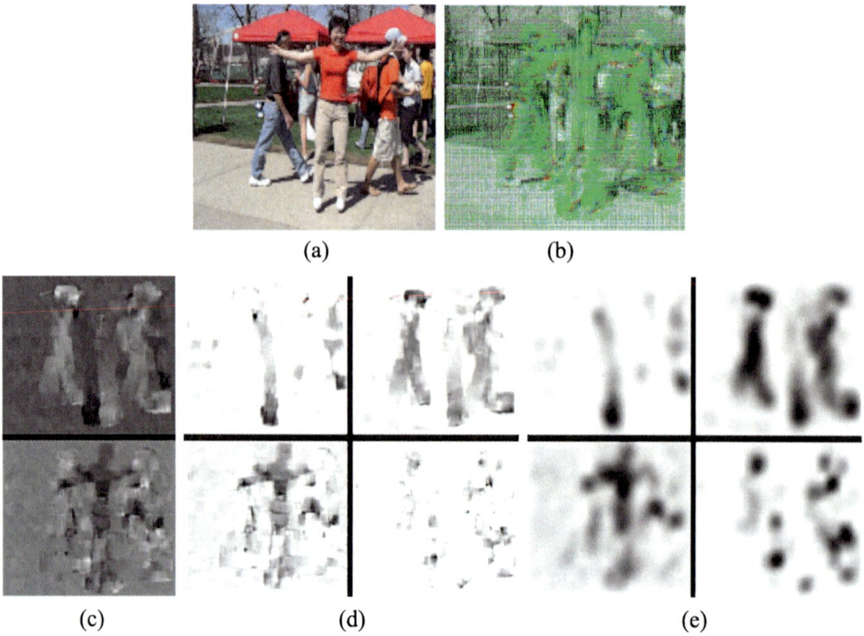

Fig. 1 Construction of the motion descriptor. (**a**) Original image; (**b**) optical flow; (**c**) x and y components of optical flow vectors F_x, F_y; (**d**) half-wave rectification of x and y components to obtain 4 separate channels F_x^+, F_x^-, F_y^+, F_y^-; (**e**) final blurry motion descriptors Fb_x^+, Fb_x^-, Fb_y^+, Fb_y^-

Different people may perform the same action differently. Take the walking action as an example, different people may have different strides, so the legs may appear in different positions of cropped frames. In order to alleviate the effect of such variations, we choose a local area search scheme. It is illustrated in Fig. 2. The distance between query and template clips is:

$$D_{\mathrm{qt}} = \sum_{i=1}^{M} \min_{j\in[1,N]} \left\{ \sum_{s=1}^{S} \min_{r\in R_s} d(q_{is}, t_{jr}) \right\} \tag{2}$$

where q_{is} denotes the sth patch on the query frame i, and t_{jr} denotes the rth patch on the template frame j. R_s is the corresponding search region of s-patch (the blue rectangle in Fig. 2). M and N are the frame numbers of query clip and template clip, respectively. S is the total number of patches on the query frame.

In order to compute the clip-to-clip distance D_{qt} from query to template, we need to know the frame correspondence first. By considering temporal constraints, one can apply dynamic time warping or other dynamic programming methods. However, in this work, for simplicity, we correspond each query frame to its closest neighbor among the template frames. This can result in several query frames corresponding to the same template frame. But it is reasonable since the query clip may contain repetitive actions and have variations in speed.

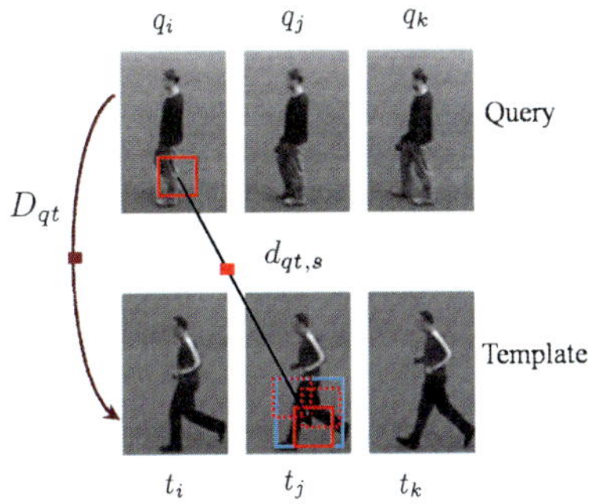

Fig. 2 The comparison process between the query and template clips. $d_{\mathrm{qt},s}$ denotes the distance between the sth patch on the query clip to its corresponding patch on the template clip. D_{qt} denotes the distance between query and template clips. The distance between clips is the sum of the distance from query frames to their best matched template frames. The frame-to-frame distance is the sum of the distance between best matching patches

After obtaining the frame correspondence and local patch correspondence, D_{qt} is the sum of the elementary patch-to-patch distance as $D_{\mathrm{qt}} = \sum_{s=1}^{M \times S} d_{\mathrm{qt},s}$, where $M \times S$ is the total number of patches on the query clip over space and time, $d_{\mathrm{qt},s}$ denotes the distance from the sth patch on the query clip to its corresponding patch on the template clip.

In Sect. 5.2, we will show that even with such a simple motion descriptor and matching scheme, we can achieve very good results on three different datasets by only using one clip as template per action. The results are comparable to previously published results using large training sets.

4 Learning a Transferable Distance Function

As mentioned earlier, our primary goal is to deal with the scenario where only one clip is available for each action in the template. This scenario is of practical interests because for some specific human action like sports or dancing actions, it is costly to collect a very training set. In particular, for the task of action video retrieval, normally we only have one clip video provided by the user. On the other hand, for some simple actions such as walking and hand-waving, a large number of clips can be easily obtained from standard benchmark datasets, that is, KTH and Weizmann datasets. Although the direct comparison of the query and template clips is able to achieve relatively good results, we would still like to exploit the available labeled datasets to assist in action recognition from a single template clip, even when the action of the template clip is totally different from the actions in the labeled datasets. In machine learning, this is known as transfer learning. The goal is to leverage the knowledge from related tasks to aid in learning on a future task.

Following the terminology of transfer learning, we denote the fully labeled action data we already have at hand as the *source training set*. Note that the class labels of actions in the source training set and the template set are different.

4.1 Transferable Distance Function

The human visual system is amazingly good at learning transferable knowledge. For example, humans are adept at recognizing a person's face after seeing it only once. One explanation for this amazing ability is that people have learned to focus on discriminative features (e.g., eyes, nose, mouse) of a face, while not being distracted by other irrelevant features [19]. This idea of knowledge transfer has been exploited in the context of object recognition and identification [17, 19, 31]. In particular, Ferencz et al. [19] propose to predict the patch saliency for object identification by its visual feature called *hyper-feature*. The relationship between the hyper-feature and the patch saliency is modeled using a generalized linear model.

Similarly, in human action recognition, we believe there exists a certain relationship between the saliency and the appearance of a patch. For example, for a boxing action, the region around the punching-out arm is much more salient than the still leg. In a hand-waving action, the arm parts are salient too. Given a source training set, our goal is to learn the knowledge, such as "stretched-arm-like" or "bent-leg-like" patches are more likely to be salient for action recognition. This knowledge will be "transferable" to unknown actions in the template and query datasets, since the algorithm will look for these patches and assign them high weights for the matching based recognition.

Inspired by work on learning distance function [20], we formulate our problem of learning the relationship into the framework of max-margin learning of distance functions. But the goal of our learning problem is different from that of Frome et al. [20]. The output of Frome et al. [20] is the weight associated with each image patch in the training data. In our problem, although we do get the weight as a by-product, we are more interested in learning the relationship between the patch appearance and its saliency.

We define the hyper-feature of the ith patch as $\mathbf{f}_i$, the weight assigned to this patch as w_i. The construction of the hyper-feature will be discussed in Sect. 4.4. We assume that $\mathbf{f}_i$ and w_i have the following relationship via the parameter $\mathbf{P}$:

$$w_i = \langle \mathbf{P} \cdot \mathbf{f}_i \rangle. \tag{3}$$

Then we will have $\mathbf{w} = \mathbf{P}^T \mathbf{F}$, where each column of $\mathbf{F}$ refers to the hyper-feature vector of a patch, $\mathbf{w}$ denotes the vector which is the concatenation of the weights w_i. Our goal is to learn $\mathbf{P}$ from the source training set. Then given any new action video, even if its action does not exist in the source training set, we will be able to compute the weight (i.e., saliency) of each patch in the new video by (3). In our work, we would like to estimate the saliencies of patches on the query video.

Combined with the learned distance function, the final clip-to-clip distance D_{qt} is defined as a weighted sum of all the elementary distances

$$D_{\text{qt}} = \sum_{s=1}^{S} w_{q,s} d_{\text{qt},s} = \langle \mathbf{w}_q \cdot \mathbf{d}_{\text{qt}} \rangle, \tag{4}$$

where $\mathbf{d}_{\text{qt}}$ is the distance vector, and each element denotes the elementary patch-to-patch distance $d_{\text{qt},s}$. Note $w_{q,s}$ is the weight of the sth patch on the query clip.

4.2 Max-Margin Formulation

The learning of **P** follows the focal learning framework in [20]. The distance function obtained by $\mathbf{w} = \mathbf{P}^T\mathbf{F}$ will satisfy the constraint that the distance between similar actions is smaller than dissimilar actions by the margin 1, that is

$$\begin{aligned} &\big\langle \mathbf{w}_i \cdot (\mathbf{d}_{ij} - \mathbf{d}_{ik})\big\rangle > 1, \\ &\big\langle \mathbf{P}^T\mathbf{F}_i \cdot (\mathbf{d}_{ij} - \mathbf{d}_{ik})\big\rangle > 1, \end{aligned} \tag{5}$$

where $\mathbf{d}_{ik}$ is the distance vector between the similar action i and k, and $\mathbf{d}_{ij}$ is the distance vector between the dissimilar action i and j. To avoid the problem of large patch-to-patch distances implying a high similarity, we enforce the nonnegativity of the weights, $\langle \mathbf{P} \cdot \mathbf{f}_m \rangle \geq 0$. For simplicity, we replace $\mathbf{d}_{ij} - \mathbf{d}_{ik}$ as $\mathbf{x}_{ijk}$.

The max-margin optimization problem can be formulated as

$$\begin{aligned} \min_{\mathbf{P},\xi} \quad & \frac{1}{2}\|\mathbf{P}\|^2 + C\sum_{ijk} \xi_{ijk} \\ \text{s.t.} \quad & \big\langle \mathbf{P}^T\mathbf{F}_i \cdot \mathbf{x}_{ijk}\big\rangle \geq 1 - \xi_{ijk}, \quad \forall i, j, k, \\ & \langle \mathbf{P} \cdot \mathbf{f}_m \rangle \geq 0, \quad \forall m, \\ & \xi_{ijk} \geq 0, \quad \forall i, j, k, \end{aligned} \tag{6}$$

where ξ_{ijk} is the slack variable and C is the trade-off parameter, similar to those in SVM. The hyper-feature $\mathbf{F}_i$ is known so we can write $\mathbf{Y}_{ijk} = \mathbf{F}_i \cdot \mathbf{x}_{ijk}$. The first constraint can be re-written as $\langle \mathbf{P} \cdot \mathbf{Y}_{ijk} \rangle \geq 1 - \xi_{ijk}$.

If we remove the second constraint, the optimization problem in (6) will be similar to the primal problem of the standard SVM. The optimization problem is very similar to the one in Frome's work [20], but differs in the second constraint. Instead of the simple nonnegative constraint $\mathbf{P} \geq \mathbf{0}$, like the one in [20], our constraints involve linear functions of the hyper-feature vectors.

The Lagrangian formulation is:

$$\begin{aligned} \mathcal{L} = \frac{1}{2}\|\mathbf{P}\|^2 + C\sum_{ijk} \xi_{ijk} - \sum_{ijk} \alpha_{ijk}\big[\langle \mathbf{P} \cdot \mathbf{Y}_{ijk} \rangle - 1 + \xi_{ijk}\big] \\ - \sum_{ijk} \lambda_{ijk}\xi_{ijk} - \sum_{m} \mu_m \langle \mathbf{P} \cdot \mathbf{f}_m \rangle. \end{aligned}$$

We can gather all the dual variables

$$\begin{aligned} \mathcal{L} = \frac{1}{2}\|\mathbf{P}\|^2 + \sum_{ijk} \alpha_{ijk}\big[\langle \mathbf{P} \cdot \mathbf{Y}_{ijk} \rangle - 1\big] + \sum_{ijk} \xi_{ijk}[C - \alpha_{ijk} - \lambda_{ijk}] \\ - \sum_{m} \mu_m \langle \mathbf{P} \cdot \mathbf{f}_m \rangle. \end{aligned}$$

Since Lagrangian is linear with ξ_{ijk}, either ξ_{ijk} or $C - \alpha_{ijk} - \lambda_{ijk}$ must be zeros. So, we can remove ξ_{ijk} from Lagrangian and obtain one constraint

$$0 \leq \alpha_{ijk} \leq C. \tag{7}$$

By taking the derivative of the remaining part of Lagrangian with respect to **P** and setting to zeros, we can get

$$\begin{aligned} \frac{\partial \mathcal{L}}{\partial \mathbf{P}} &= \mathbf{P} - \sum_{ijk} \alpha_{ijk} \mathbf{Y}_{ijk} - \sum_{m} \mu_m \mathbf{f}_m = 0 \\ \Longrightarrow \quad \mathbf{P} &= \sum_{ijk} \alpha_{ijk} \mathbf{Y}_{ijk} + \sum_{m} \mu_m \mathbf{f}_m. \end{aligned} \tag{8}$$

Substituting (8) back to the Lagrangian we can get:

$$\begin{aligned} \Theta(\alpha, \mu) &= \frac{1}{2} \left\| \sum_{ijk} \alpha_{ijk} \mathbf{Y}_{ijk} + \sum_{m} \mu_m \mathbf{f}_m \right\|^2 \\ &\quad - \sum_{ijk} \alpha_{ijk} \left[\left\langle \left(\sum_{ijk} \alpha_{ijk} \mathbf{Y}_{ijk} + \sum_{m} \mu_m \mathbf{f}_m \right) \cdot \mathbf{Y}_{ijk} \right\rangle - 1 \right] \\ &\quad - \sum_{m} \mu_m \left[\left\langle \left(\sum_{ijk} \alpha_{ijk} \mathbf{Y}_{ijk} + \sum_{m} \mu_m \mathbf{f}_m \right) \cdot \mathbf{f}_m \right\rangle \right] \\ &= \frac{1}{2} \left\| \sum_{ijk} \alpha_{ijk} \mathbf{Y}_{ijk} + \sum_{m} \mu_m \mathbf{f}_m \right\|^2 - \left\| \sum_{ijk} \alpha_{ijk} \mathbf{Y}_{ijk} \right\|^2 - \left\| \sum_{m} \mu_m \mathbf{f}_m \right\|^2 \\ &\quad - 2 \left\langle \sum_{ijk} \alpha_{ijk} \mathbf{Y}_{ijk} \cdot \sum_{m} \mu_m \mathbf{f}_m \right\rangle + \sum_{ijk} \alpha_{ijk} \\ &= -\frac{1}{2} \left\| \sum_{ijk} \alpha_{ijk} \mathbf{Y}_{ijk} + \sum_{m} \mu_m \mathbf{f}_m \right\|^2 + \sum_{ijk} \alpha_{ijk}. \end{aligned}$$

Then, the dual problem of (6) can be written as follows

$$\max_{\alpha,\mu} \quad -\frac{1}{2} \left\| \sum_{ijk} \alpha_{ijk} \mathbf{Y}_{ijk} + \sum_{m} \mu_m \mathbf{f}_m \right\|^2 + \sum_{ijk} \alpha_{ijk} \tag{9}$$

$$\text{s.t.} \quad 0 \le \alpha_{ijk} \le C, \quad \forall i, j, k, \tag{10}$$

$$\mu_m \ge 0, \quad \forall m, \tag{11}$$

where the α_{ijk} and μ_m are the dual variables corresponding to the first and second constraints in (6), respectively. The primal variable **P** can be obtained from the dual variables by (8).

$$\mathbf{P} = \sum_{ijk} \alpha_{ijk} \mathbf{Y}_{ijk} + \sum_{m} \mu_m \mathbf{f}_m. \tag{12}$$

4.3 Solving the Dual

Similar to [20], we solve the dual problem by iteratively performing updating on two dual variables. By taking the derivative of the dual with respect to one of the dual variables α_{abc} and then setting it to zero,

$$\frac{\partial \Theta}{\partial \alpha_{abc}} = \left\langle \left(m_{ijk}\alpha_{ijk}\mathbf{Y}_{ijk} - \sum_m \mu_m \mathbf{f}_m \right) \cdot \mathbf{Y}_{abc} + 1 \right\rangle \tag{13}$$

$$= -\sum_{ijk} \alpha_{ijk} \langle \mathbf{Y}_{ijk} \cdot \mathbf{Y}_{abc} \rangle - \left\langle \sum_m \mu_m \mathbf{f}_m \cdot \mathbf{Y}_{abc} \right\rangle + 1 \tag{14}$$

$$= -\sum_{ijk \neq abc} \alpha_{ijk} \langle \mathbf{Y}_{ijk} \cdot \mathbf{Y}_{abc} \rangle - \alpha_{abc} \|\mathbf{Y}_{abc}\|^2 - \left\langle \sum_m \mu_m \mathbf{f}_m \cdot \mathbf{Y}_{abc} \right\rangle + 1. \tag{15}$$

After setting (15) to zero, we can obtain the updating rule for the dual variable α_{abc}. Similarly, we can get the updating rule for the dual variable μ_a. The two updating rules are as follows:

$$\alpha_{abc} \leftarrow \frac{1 - \sum_{ijk \neq abc} \alpha_{ijk} \langle \mathbf{Y}_{ijk} \cdot \mathbf{Y}_{abc} \rangle - \sum_m \mu_m \langle \mathbf{f}_m \cdot \mathbf{Y}_{abc} \rangle}{\|\mathbf{Y}_{abc}\|^2}, \tag{16}$$

$$\mu_a \leftarrow \frac{-\sum_{ijk} \alpha_{ijk} \langle \mathbf{Y}_{ijk} \cdot \mathbf{f}_a \rangle - \sum_{m \neq a} \mu_m \langle \mathbf{f}_m \cdot \mathbf{f}_a \rangle}{\|\mathbf{f}_a\|^2}. \tag{17}$$

After each round of update, we can simply clip the dual variables to their feasible regions. α_{abc} will be clipped to 0 if negative and to C if larger than C. μ_m will be clipped to zero if negative. See [21] for more details. After solving this dual problem, we can obtain $\mathbf{P}$ through (12).

4.4 Hyper-Features

Inspired by codebook approaches in object and scene categorization, we represent the hyper-feature of each patch as a $|V|$-dimensional vector $\mathbf{f}$, where $|V|$ is the codebook size. The ith element of $\mathbf{f}$ is set according to the distance between the feature vector of this patch and the ith visual word. The feature vector of each patch consists of histogram of oriented gradient (HOG) [9] and patch positions in the form of $h = \{g, x, y\}$, where g denotes the HOG descriptor of the patch. x and y are the coordinates of the patch in the frame. To construct the codebook vocabulary, we randomly select a large number of patches from the source training set, then run k-means clustering. The center of each cluster is defined as a codeword. The hyper-feature $\mathbf{f}_m$ for the mth patch is constructed as follows

$$\mathbf{f}_m(v_i) = \frac{K_\sigma(D(v_i, h_m))}{\sum_{j=1}^{|V|} K_\sigma(D(v_j, h_m))}, \tag{18}$$

where $\mathbf{f}_m(v_i)$ denotes the ith element in the hyper-feature vector $\mathbf{f}_m$. $D(v_i, h_m)$ denotes the Euclidean distance between the ith codeword and the patch m. K_σ is the Gaussian-shape kernel as $K_\sigma(x) = \frac{1}{\sqrt{2\pi}\sigma} \exp(-\frac{x^2}{2\sigma^2})$. Note that (18) leads to a generalized linear patch weighting model using Gaussian radial basis functions.

5 Experiments

We test our algorithms on three different datasets: KTH human action dataset [38], Weizmann human action dataset [6], and the cluttered action dataset [24]. We first give a brief overview of these three datasets, then present the experimental results.

5.1 Datasets

5.1.1 KTH Dataset

The KTH human action dataset contains six types of human actions (boxing, hand-waving, hand-clapping, jogging, running and walking) performed several times by 25 subjects in four different scenarios: outdoors, outdoors with scale variation, outdoors with different clothes and indoors. In total, there are 599 videos. Following the original setup, each video is divided into four sequences. After computing the motion descriptor, we run the human detection and tracking using the code provided by Felzenszwalb et al. [18]. All the frames and motion descriptors have been cropped to 90×60 and the human figure is put in the center of the frame.

On this dataset, the performance is saturating, with results from 90–94% [22, 37]. However, most of those methods choose either a half-half or leave-one-out cross validation scheme to split the training and testing sets. For example, in each round of the leave-one-out testing, 575 videos are used for training, and the remaining 24 videos are used for testing. Besides, for each video, there are 300–500 frames in which the actor repeatedly performs one single action. If we assume one complete action lasts 30 frames, the actual training set for the above leave-one-out scheme contains at least 5750 samples, and for each action category, there are 960 samples. In many real-world applications, it is impossible to collect equivalently large training sets for any given action.

5.1.2 Weizmann Dataset

The Weizmann human action dataset contains 93 sequences of nine actors performing ten different actions. Each sequence contains about 40–120 frames. In the figure-centric representation, all frames have been normalized to 90×60. The best performance published is 100% by using the large training set [16].

Fig. 3 Sample frames of cluttered human action dataset [24]

5.1.3 Cluttered Human Action Dataset

The cluttered human action dataset is a variant of the dataset collected by Ke et al. [24], which was initially designed for action detection in the crowed environment. It contains not only cluttered static backgrounds, but also cluttered dynamic backgrounds, such as moving cars or walking people. In order to test the robustness of our action recognition methods, we use it for recognition. From each raw video sequence in the original dataset, we manually crop out the actions of interest. This dataset contains 95 sequences with five actions, jumping jack, pushing elevator button, picking-up, one-hand waving, and two-hand waving. Each sequence contains about 30–50 frames. Representative frames are shown in Fig. 3.

5.2 *Experimental Results*

We perform the following experiments to evaluate our patch based comparison method and the transferable distance function learning:

1. evaluate the patch based comparison method on all three datasets
2. train the transferable distance function on Weizmann, and test on KTH
3. train the transferable distance function on KTH, and test on the cluttered action dataset

5.2.1 Direct Comparison on KTH

In this experiment, we evaluate the patch based direct comparison method on the KTH dataset. We first randomly select one actor, then randomly choose one clip per

Table 1 The accuracy of five rounds of experiments on KTH. The top row denotes the round index. The row of **Dc** refers to the results of direct comparison, and the row of **Tr** refers to the results of training on Weizmann and testing on KTH. Std. denotes the standard deviation

	1	2	3	4	5	Avg.	Std.
Dc	0.776	0.709	0.829	0.564	0.746	0.725	0.100
Tr	0.784	0.767	0.829	0.617	0.789	0.757	0.073

Table 2 Comparison of different reported results on KTH. We remark the setup of the training set. LOO refers to the "Leave-one-out" cross validation. Split refers to other split strategies of training and testing sets. Note that these numbers are not directly comparable due to variations in training/testing setup

Methods	Accuracy	Remark
Liu and Shah [28]	0.9416	LOO
Schindler and Van Gool [37]	0.9270	LOO
Jhuang et al. [22]	0.9170	Split
Nowozin et al. [32]	0.8704	Split
Neibles et al. [30]	0.8150	LOO
Dollar et al. [11]	0.8117	LOO
Ours (**Tr**)	**0.7571**	**One clip**
Ours (**Dc**)	**0.7248**	**One clip**
Schuldt et al. [38]	0.7172	Split
Ke et al. [23]	0.6296	Split

action from this actor as the template set. The clip contains at most 25 frames,that is, 1–1.5 complete action cycles. The sequences of the remaining actors are used as the query set. We decompose each frame into 40 patches. The patch size is 20×20 and the length of strides is 10.

We run the experiment five times and for each round we select a different actor as the template. The results are shown in the row of **Dc** of Table 1. The average result over the five rounds is 72.48%, which is comparable to the previously published results using large training set, as shown in Table 2. Note that due to the action variation in person, the performance depends on how distinguishable the templates are.

5.2.2 Training on Weizmann and Testing on KTH

In this experiment, we train a transferable distance function from Weizmann and test it on KTH. In order to meet the requirement of the transfer learning scenario, that is, the source training set does not contain the actions of the template set, we remove walking, running, and two-hand waving from the Weizmann dataset. We build the codebook vocabulary on the remaining sequences of Weizmann as described in Sect. 4.4. The number of codewords is set to 100. We used other codebook sizes and found they do not affect the performance substantially. In training, the parameters

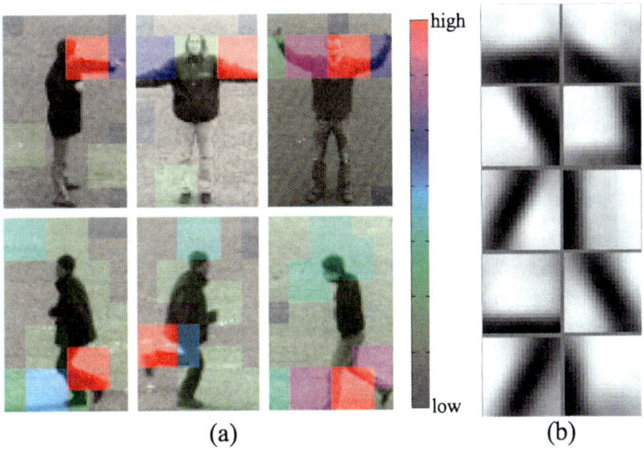

Fig. 4 (**a**) Illustration of the learned weights on the six actions of KTH. (**b**) The learned **P** allows us to rank the visual words in the vocabulary. The top ten words are visualized. Note that our visual words consist of appearance and spatial location features. Only appearance is illustrated. Please refer to text for more details

are set as, $\sigma = 0.5$ and $C = 0.0001$. Through training on Weizmann, we can obtain the relation **P**, which parameterizes the transferable distance function.

After the training, we first compute the hyper-features of the query videos in KTH using the codebook constructed from Weizmann. Then, we can obtain the distance function through (3). For the purpose of illustration, we visualize the learned weights in Fig. 4(a). The red patches refer to high weights. Note that patches on the frames are overlapping, we only show the highest weight for an overlapping region. For the six actions in KTH, we can see most of the patches with high weights lie on the most salient human parts, such as out stretched arms or legs. Unlike other motion based interest point detection methods [30], the learned weight for the moving body is lower than moving legs. This is more intuitive since the moving body does not help to distinguish running, jogging and walking. Moreover, the learned **P** allows us to rank the visual words in the codebook vocabulary. We visualize the appearance feature of top ten words in Fig. 4(b). We can observe that these words are all "out-stretched-limb like".

The recognition accuracies of five rounds of experiments are given in the row of **Tr** of Table 1. Note that for each round, we use the same templates as the direct comparison experiments. The largest improvement made by the transferable distance function is almost 6%. We can observe that in experiment round 1 and 3, the improvements made by the transferable distance function are minor. This is reasonable since the direct comparison has already achieved very good results. We also show the confusion matrices of experiment round 2 in Fig. 5. We can see that the transferable distance function significantly mitigates the confusion of the most difficult actions, such as hand-clapping versus hand-waving, and jogging versus running. In particular, we see an improvement of almost 30% for the hand-waving. The comparison with previously published results are given in Table 2.

Another benefit of learning transferable distance function is that it can be used to speedup the comparison. In the patch based direct comparison method, for each patch on the query frame, we need to search its corresponding area on the template frame and find the best matched one. This process is time-consuming since there

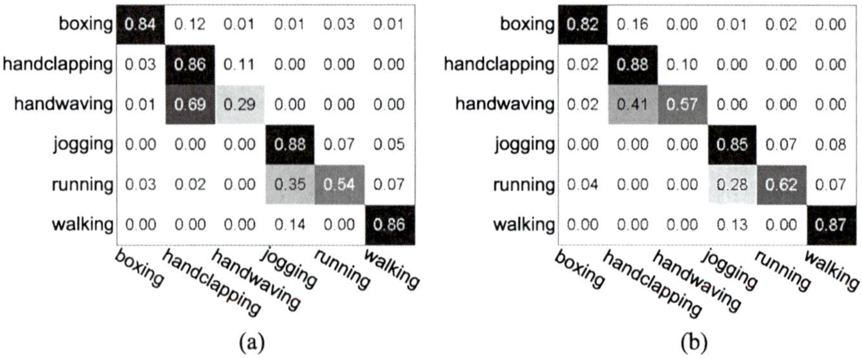

Fig. 5 Confusion matrices on KTH of experiment round 2. *Horizontal rows* are ground truths, and *vertical columns* are predictions; (**a**) direct comparison; (**b**) training on Weizmann and testing on KTH

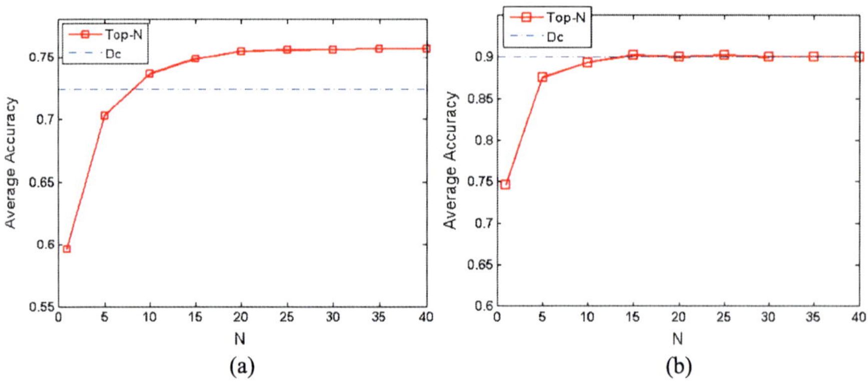

Fig. 6 (**a**) The average accuracy of five rounds of experiments on KTH using only top N patches of each frame. (**b**) The average accuracy of five rounds of experiments on cluttered action dataset using only top N patches on the frame. The *dash–dot line* denotes the average accuracy of the direct comparison using all patches

exist 1,000 patches over the sequence of 25 frames. With learned distance function of the query sequence, we can sort the patches on each frame by their weights. Instead of using all patches for matching, we only choose the top N patches with high weights from each frame. We change N from 1 to 40 and compute the average accuracy over the five rounds of experiments. The results are illustrated in Fig. 6(a). Using only ten patches on each frame, we can achieve a better result than the patch-based direct comparison using all patches on the frame. This would save 3/4 matching time, significantly increases the efficiency of whole recognition process.

Table 3 The accuracy of five rounds of experiments on Weizmann using patch based direct comparison. The top row denotes the round index. Std. denotes the standard deviation

	1	2	3	4	5	Avg.	Std.
Dc	0.928	0.892	0.916	0.819	0.795	0.870	0.060

Table 4 Comparison of the average accuracy on Weizmann using one exemplar per action with [40]

	Dc	1NN [40]	1NN-M [40]
FE-1	0.8699	0.5300	0.7231

Table 5 The accuracy of five rounds of experiments on the cluttered human action dataset. The top row denotes the round index. Std. denotes the standard deviation

	1	2	3	4	5	Avg.	Std.
Dc	0.944	0.900	0.844	0.900	0.911	0.900	0.036
Tr	0.944	0.900	0.856	0.900	0.900	0.900	0.031

5.2.3 Direct Comparison on Weizmann

The setup we use in this experiment is exactly the same as the direct comparison experiment on KTH. In each round of the experiment, we randomly select one actor and use one clip per action with 25 frames from this actor as the template. The sequences of the remaining actors are used as the query set. The results are shown in Table 3. We compare our results with the work of Tran and Sorokin [40], as shown in Table 4. Our result outperforms both "1-Nearest Neighbor + motion context descriptor (1NN)" and "1-Nearest Neighbor with metric learning + motion context descriptor (1NN-M)". Note that we only use a 25 frame clip as the template rather than the whole video as in [40].

Unfortunately, a fair transfer learning experiment training on KTH and testing on Weizmann is not possible. After removing overlapping actions, there are only three actions left in the KTH (boxing, hand-clapping and jogging). The number of actions is too small to contain enough generic knowledge. So we do not run the experiments of training on KTH and testing on Weizmann.

5.2.4 Direct Comparison on Cluttered Action Dataset

The goal of this experiment is to evaluate the robustness of our patch based direct comparison on more challenging datasets with cluttered backgrounds. For each action, we randomly choose one clip with 25 frames as the template and the remaining sequences as the query set. The same patch decomposition scheme is used. Similarly, we perform five rounds of experiments by choosing different templates. The results are shown in the **Dc** row of Table 5. We can see the patch based direct comparison achieves very high accuracy on this dataset.

5.2.5 Training on KTH and Testing on Cluttered Action Dataset

This experiment follows the same protocol as training on Weizmann and testing on KTH. We first remove the two-hand waving action from KTH since it also exists in the cluttered action dataset. KTH contains a large number of sequences, we choose only five actors' sequences to form the source training set. The results are shown in the **Tr** row of the Table 5. As expected, the transferable distance function learning achieves almost identical results as the direct comparison, since direct comparison has achieved very promising results. However, the transferable distance function can be used to sort the patches and choose the patches with top N highest weights, and thus improve the efficiency of the recognition system. As illustrated in Fig. 6(b), we are able to use only top 5 patches on each frame and achieve 86.67% accuracy. The efficiency is boosted significantly (saving 7/8 matching time) with the cost of only 3% accuracy decrease.

6 Conclusion

In this chapter, we have presented an action recognition algorithm based on a patch-based matching scheme. A set of motion patches on input query clips and template clips with known actions is matched. This matching scheme proves to be effective for action recognition in the difficult case of only a single training clip per action. Further, we have demonstrated that learning a transferable weighting on these patches could improve accuracy and computational efficiency. These weights, based on patch hyper-features, are generic, can be directly applied to novel video sequences without further learning, and hold promise for recognition in small training set scenarios such as video retrieval and surveillance.

References

1. Ahmed, A., Yu, K., Xu, W., Gong, Y., Xing, E.P.: Training hierarchical feed-forward visual recognition models using transfer learning from pseudo-tasks. In: European Conference on Computer Vision (2008)
2. Argyriou, A., Evgeniou, T., Pontil, M.: Multi-task feature learning. In: Advances in Neural Information Processing Systems. MIT Press, Cambridge (2006)
3. Argyriou, A., Micchelli, C.A., Pontil, M., Ying, Y.: A spectral regularization framework for multi-task structure learning. In: Advances in Neural Information Processing Systems. MIT Press, Cambridge (2007)
4. Argyriou, A., Evgeniou, T., Pontil, M.: Convex multi-task feature learning. Mach. Learn. **73**(3), 243–272 (2008)
5. Ben-David, S., Schuller, R.: Exploiting task relatedness for multiple task learning. In: Proc. of Computational Learning Theory (2003)
6. Blank, M., Gorelick, L., Shechtman, E., Irani, M., Basri, R.: Actions as space-time shapes. In: IEEE International Conference on Computer Vision (2005)
7. Caruana, R.: Multitask learning. Mach. Learn. **28**, 41–75 (1997)

8. Dai, W., Yang, Q., Xue, G.R., Yu, Y.: Boosting for transfer learning. In: International Conference on Machine Learning (2007)
9. Dalal, N., Triggs, B.: Histogram of oriented gradients for human detection. In: IEEE Computer Society Conference on Computer Vision and Pattern Recognition (2005)
10. Daume, H. III, Marcu, D.: Domain adaptation for statistical classifiers. J. Artif. Intell. Res. **26**, 101–126 (2006)
11. Dollár, P., Rabaud, V., Cottrell, G., Belongie, S.: Behavior recognition via sparse spatio-temporal features. In: ICCV'05 Workshop on Visual Surveillance and Performance Evaluation of Tracking and Surveillance (2005)
12. Efros, A.A., Berg, A.C., Mori, G., Malik, J.: Recognizing action at a distance. In: IEEE International Conference on Computer Vision, pp. 726–733 (2003)
13. Evgeniou, T., Pontil, M.: Regularized multi-task learning. In: ACM SIGKDD. ACM, New York (2004)
14. Farhadi, A., Tabrizi, M.K.: Learning to recognize activities from the wrong view point. In: European Conference on Computer Vision (2008)
15. Farhadi, A., Forsyth, D., White, R.: Transfer learning in sign language. In: IEEE Computer Society Conference on Computer Vision and Pattern Recognition (2007)
16. Fathi, A., Mori, G.: Action recognition by learning mid-level motion features. In: IEEE Computer Society Conference on Computer Vision and Pattern Recognition (2008)
17. Fei-Fei, L., Fergus, R., Perona, P.: One-shot learning of object categories. IEEE Trans. Pattern Anal. Mach. Intell. **28**(4), 594–611 (2006)
18. Felzenszwalb, P., McAllester, D., Ramanan, D.: A discriminatively trained, multiscale, deformable part model. In: IEEE Computer Society Conference on Computer Vision and Pattern Recognition (2008)
19. Ferencz, A., Learned-Miller, E., Malik, J.: Learning to locate informative features for visual identification. Int. J. Comput. Vis. **77**(1–3), 3–24 (2008)
20. Frome, A., Singer, Y., Malik, J.: Image retrieval and classification using local distance functions. In: Advances in Neural Information Processing Systems, vol. 19. MIT Press, Cambridge (2007)
21. Frome, A., Singer, Y., Sha, F., Malik, J.: Learning globally-consistent local distance functions for shape-based image retrieval and classification. In: IEEE International Conference on Computer Vision (2007)
22. Jhuang, H., Serre, T., Wolf, L., Poggio, T.: A biologically inspired system for action recognition. In: IEEE International Conference on Computer Vision (2007)
23. Ke, Y., Sukthankar, R., Hebert, M.: Efficient visual event detection using volumetric features. In: IEEE International Conference on Computer Vision, vol. 1, pp. 166–173 (2005)
24. Ke, Y., Sukthankar, R., Hebert, M.: Event detection in crowded videos. In: IEEE International Conference on Computer Vision (2007)
25. Ke, Y., Sukthankar, R., Hebert, M.: Spatio-temporal shape and flow correlation for action recognition. In: Visual Surveillance Workshop (2007)
26. Lampert, C.H., Nickisch, H., Harmeling, S.: Learning to detect unseen object classes by between-class attribute transfer. In: IEEE Computer Society Conference on Computer Vision and Pattern Recognition (2009)
27. Larochelle, H., Erhan, D., Bengio, Y.: Zero-data learning of new tasks. In: AAAI Conference on Artificial Intelligence (2008)
28. Liu, J., Shah, M.: Learning human actions via information maximization. In: IEEE Computer Society Conference on Computer Vision and Pattern Recognition (2008)
29. Mansour, Y., Mohri, M., Rostamizadeh, A.: Domain adaption with multiple sources. In: Advances in Neural Information Processing Systems. MIT Press, Cambridge (2008)
30. Niebles, J.C., Wang, H., Fei-Fei, L.: Unsupervised learning of human action categories using spatial-temporal words. In: British Machine Vision Conference, vol. 3, pp. 1249–1258 (2006)
31. Nowak, E., Jurie, F.: Learning visual similarity measures for comparing never seen objects. In: IEEE Computer Society Conference on Computer Vision and Pattern Recognition (2007)
32. Nowozin, S., Bakir, G., Tsuda, K.: Discriminative subsequence mining for action classification. In: IEEE International Conference on Computer Vision (2007)

33. Palatucci, M., Pomerleau, D., Hinton, G., Mitchell, T.M.: Zero-shot learning with semantic output codes. In: Advances in Neural Information Processing Systems. MIT Press, Cambridge (2009)
34. Pan, S.J., Yang, Q.: A survey on transfer learning. Technical Report, HKUST-CS08-08, Department of Computer Science and Engineering, Hong Kong University of Science and Technology (2008)
35. Quattoni, A., Collins, M., Darrell, T.: Transfer learning for image classification with sparse prototype representations. In: IEEE Computer Society Conference on Computer Vision and Pattern Recognition (2008)
36. Raina, R., Battle, A., Lee, H., Packer, B., Ng, A.Y.: Self-taught learning: transfer learning from unlabeled data. In: International Conference on Machine Learning (2007)
37. Schindler, K., Van Gool, L.: Action snippets: how many frames does action recognition require? In: IEEE Computer Society Conference on Computer Vision and Pattern Recognition (2008)
38. Schuldt, C., Laptev, I., Caputo, B.: Recognizing human actions: a local SVM approach. In: IEEE International Conference on Pattern Recognition, vol. 3, pp. 32–36 (2004)
39. Shechtman, E., Irani, M.: Space-time behavior based correlation. In: International Conference on Computer Vision and Pattern Recognition (2005)
40. Tran, D., Sorokin, A.: Human activity recognition with metric learning. In: European Conference on Computer Vision (2008)
41. Weinland, D., Boyer, E.: Action recognition using exemplar-based embedding. In: IEEE Computer Society Conference on Computer Vision and Pattern Recognition (2008)
42. Yang, W., Wang, Y., Mori, G.: Efficient human action detection using a transferable distance function. In: Asian Conference on Computer Vision (2009)
43. Yang, W., Wang, Y., Mori, G.: Human action recognition from a single clip per action. In: ICCV Workshop on Machine Learning for Vision-based Motion Analysis (2009)
44. Yu, K., Tresp, V., Schwaighofer, A.: Learning Gaussian processes from multiple tasks. In: International Conference on Machine Learning (2005)
45. Yu, S., Tresp, V., Yu, K.: Robust multi-task learning with t-processes. In: International Conference on Machine Learning (2007)
46. Zhu, X.: Semi-supervised learning literature survey. Technical Report TR1530, University of Wisconsin at Madison (2005)

Index

L. Wang et al. (eds.), *Machine Learning for Vision-Based Motion Analysis*,
Advances in Pattern Recognition,
DOI 10.1007/978-0-85729-057-1,